• • • • • • • • • • • • • • • •

Paleoethnobotany
A Handbook of Procedures

• • • • • • • • • • • • • • • •

Paleoethnobotany
A Handbook of Procedures

Deborah M. Pearsall
Department of Anthropology
American Archaeology Division
University of Missouri–Columbia
Columbia, Missouri

Academic Press, Inc.
Harcourt Brace Jovanovich, Publishers

San Diego New York Berkeley Boston London Sydney Toronto

• • • • • • • • • • • • • • • •

COPYRIGHT © 1989 BY ACADEMIC PRESS, INC.
ALL RIGHTS RESERVED.
NO PART OF THIS PUBLICATION MAY BE REPRODUCED OR
TRANSMITTED IN ANY FORM OR BY ANY MEANS, ELECTRONIC
OR MECHANICAL, INCLUDING PHOTOCOPY, RECORDING, OR
ANY INFORMATION STORAGE AND RETRIEVAL SYSTEM, WITHOUT
PERMISSION IN WRITING FROM THE PUBLISHER.

ACADEMIC PRESS, INC.
San Diego, California 92101

United Kingdom Edition published by
ACADEMIC PRESS LIMITED
24-28 Oval Road, London NW1 7DX

Library of Congress Cataloging-in-Publication Data

Pearsall, Deborah M.
    Paleoethnobotany : a handbook of procedures / Deborah M. Pearsall.
      p. cm.
    Includes bibliographies and index.
    ISBN 0-12-548040-7 (alk. paper)
    1. Plant remains (Archaeology) 2. Ethnobotany. I. Title.
CC79.5.P5P43 1989
930.1—dc19   88-28814
                                                                      CIP

PRINTED IN THE UNITED STATES OF AMERICA
89 90 91 92   9 8 7 6 5 4 3 2 1

• • • • • • • • • • • • • • • •

To Richard Ford, who led me into paleoethnobotany; to Donald Lathrap, who encouraged me in the search for my own niche in archaeology; and to my husband, Mike, for his loving support

Contents

Preface ix

Chapter 1 The Paleoethnobotanical Approach 1

Introduction 1
Historical Overview 3
Nature and Status of Ethnobotany 6
References 9

Chapter 2 Techniques for Recovering Macroremains 15

Introduction 15
In Situ Collection of Material 16
Screening Techniques 17
Water Recovery: Flotation Techniques 19
Building and Operating Flotation Systems: Sample Designs 35
Issues in Recovery of Macroremains 75
Sampling for Macroremains 95
References 102

Chapter 3 Identification and Interpretation of Macroremains 107

Introduction 107
Initial Processing of Samples 108
Building a Comparative Collection 128

viii • Contents

Basic Identification Techniques   144
Specialized Identification Techniques   183
Presenting and Interpreting Results   194
References   231

Chapter 4   Archaeological Palynology   245

Introduction   245
Nature and Production of Pollen   246
History of Pollen Analysis   256
Field Sampling   258
Laboratory Analysis   269
Presenting and Interpreting Results   284
References   304

Chapter 5   Phytolith Analysis   311

Introduction   311
Nature and Occurrence of Phytoliths   312
Phytoliths and Archaeology: A Brief History   328
Status of Phytolith Analysis   340
Field Sampling   345
Laboratory Analysis   356
Presenting and Interpreting Results   404
References   428

Chapter 6   Integrating Paleoethnobotanical Data   439

Introduction   439
Strengths and Weaknesses of Paleoethnobotanical Data   439
Examples of Paleoethnobotanical Interpretation   443
Special Topic: Reconstructing Prehistoric Diet   447
Special Topic: Reconstructing Paleoenvironment   454
References   457

Index   461

Preface

> *Being something of an anthropologist and something of a botanist, one is looked upon as not quite either. One goes through life feeling miscellaneous. Confucius say: "When man desire catch particular mouse, he not seek cat with two heads."* Volney Jones, 1957

In this book I describe the approaches and techniques of paleoethnobotany, the study of the interrelationships between human populations and the plant world through the archaeological record. Paleoethnobotanists are truly grounded in two worlds. If trained as anthropologists, we must struggle to learn techniques for identifying archaeological plant remains and to understand the ecology of human–plant interactions. If trained as botanists, we must endeavor to view the plant world from a cultural perspective and to understand the strengths and weaknesses of the archaeological record. Although this need to master two disciplines may be considered a weakness by some (Jones's lament rings true today), in truth the diversity of training and experience of its practitioners is at the heart of the exciting contributions made by paleoethnobotany to archaeology and botany. One goal of this book is to make the approaches and techniques of this field more accessible to the general anthropological and botanical audience. A greater understanding of the field, its contributions, and its potential, should result.

Another goal is to provide an overview of paleoethnobotany for those wishing to learn some or all of its approaches. Whether one's interest is in the study of macroremains, pollen, or phytoliths, it is important to begin with a basic understanding of each data base. Each complements and strengthens the others.

Finally, archaeologists will find here a handbook of field sampling and flotation techniques as well as an introduction to methods of analysis in paleoethnobotany that will guide critical evaluation of research in this field.

I begin in Chapter 1 with a brief overview of the field of paleoethnobotany and the history of its development. Chapters 2 and 3 are dedicated to recovery and analysis of macroremains—the charred, waterlogged, and dried botanical remains recovered from sites by flotation or sieving. In Chapter 4, I turn to an overview of archaeological palynology. This chapter presents the basic techniques of analyzing pollen from archaeological sites and offers guidelines for understanding the work of stratigraphic palynology. Chapter 5 presents the newest area of paleoethnobotany, phytolith analysis—the recovery and identification of plant silica bodies. Whereas in the early 1970s we could only discuss the "potential" of this technique, today phytolith analysis is a proven contributor to archaeology and paleoecology. In the final chapter, I discuss how the results of analyzing diverse types of botanical data can be integrated to address questions of interest in archaeology.

Acknowledgments

I would like to thank the many people who have contributed to the development of this book. Laboratory facilities for my macroremain and phytolith research are provided by the American Archaeology Division of the Department of Anthropology, University of Missouri–Columbia. I thank our director, Michael J. O'Brien, and office staff, Peggy Loy and Judy Atteberry, for their support and assistance. I also thank all the archaeologists who entrusted botanical remains to me; their research contributed greatly to this book.

Elizabeth Dinan provided invaluable help in coordinating production of figures for the manuscript. She organized photo sessions of laboratory activities, selected photos and designed layouts, and mocked up figures for the artists, contributing in some way to the development of every figure. In addition, she created Figures 2.34, 2.35, 2.36, and Table 2.2, worked with Dolores Piperno to produce the original drawing for Figure 5.16, and adapted Table 2.1. She and fellow laboratory assistant Renee Roberts also took the phytolith photographs that appear in Chapter 5.

Janet Miller photographed the scenes that appear in Figures 2.11, 2.14, 2.17, 2.18, 2.19, 3.4, 3.5, 3.6, 3.7, 3.13, 3.14, 3.15, 3.28, 5.18, 5.19, 5.20, and 5.21. In addition, Janet developed and printed all black-and-white photographs and reproduced Figures 4.9 and 4.10.

Lisa Harrison created the original art in Figures 2.3, 2.15, 2.16, 3.18, 3.19, 3.20, 3.23, 3.24, 3.25, 3.26, 3.27, 3.29, 3.32, 3.39, and 5.14. Her illustrations greatly enhance the text. Lisa also reproduced Figures 2.5, 2.20, 2.21, 2.22, 2.31, 2.33, 3.10, 3.16, 3.21, 3.22, 4.1, 4.2, 4.3, 4.4, 5.6, 5.9, 5.10, 5.16, and 5.28.

Tom Holland drew Figure 3.17 and the *Achras* berry in Figure 3.20. Brian Deevy took the photos in Figure 3.33. Eric Voigt drew the graph in Figure 3.42 and re-

Acknowledgments • xi

produced Figures 3.45, 5.34, 5.35, 5.36, 5.37, and 5.38. Renee Roberts created Figure 5.24. Andrea Hunter put together Table 5.12.

I thank paleoethnobotany lab assistants Marcelle Umlauf, Andrea Hunter, Eric Hollinger, and Elizabeth Dinan for "reenacting" various laboratory activities and some manual flotation sequences depicted in the text. I also thank Andrea Hunter for her assistance in compiling and entering references.

Finally, special thanks are due to Naomi Miller, Christine Hastorf, Glenna Dean, Vaughn Bryant, Jr., and Dolores Piperno, who read chapters of the book and offered many useful suggestions and clarifications, and to Richard Ford, who reviewed and commented on the entire manuscript. Many improvements on the original text are due to the efforts of these colleagues and to those of the staff of Academic Press. I also thank all the other friends and colleagues who have shared procedures and ideas with me over the years. I alone bear responsibility for errors or omissions.

The following tables and figures are included in the text with permission of the authors and publishers:

Table 2.1, from Thomas 1969, Society for American Archaeology
Table 3.3, from Asch and Asch 1985, University of Michigan, Museum of Anthropology
Table 3.4, from Pearsall 1980, Academic Press
Table 3.5, from Johannessen 1984, University of Illinois press
Tables 3.8, 3.9, 3.10, 3.11, from Flannery 1986, Academic Press
Table 3.12, from Pozorski 1983, Society of Ethnobiology
Table 3.13, from Pearsall 1988a, University of Calgary, Archaeological Association
Table 3.14, from Pearsall 1988b, University of Chicago Press
Table 4.1, from Schoenwetter 1974, Academic Press
Table 5.1, from Pearsall 1978, American Association for the Advancement of Science
Table 5.2, from Pearsall 1982, American Anthropological Association
Table 5.4, from Pearsall and Trimble 1984, Academic Press
Tables 5.5, 5.6, from Smith and Atkinson 1975, Paul Elek
Table 5.10, from Pearsall 1985a, US Army Corps of Engineers
Table 5.11, from Piperno 1985c, Academic Press
Table 5.12, from Dinan 1988, with permission of the author
Table 5.13, from Piperno 1984, Society for American Archaeology
Figure 2.5, from French 1971, The British Institute of Archaeology at Ankara
Figure 2.6, from Diamant 1979, Boston University Scholarly Publications, and the Franchthi Cave Archives, courtesy of T. Jacobsen
Figure 2.7, from Davis and Wesolowsky 1975, reproduced with permission of the *Journal of Field Archaeology* and the Trustees of Boston University
Figure 2.20, from Watson 1976, *Midcontinental Journal of Archaeology,* with permission of the Kent State University Press

Figure 2.21, from Williams 1973, *Antiquity*, Girton College, England
Figure 2.26, from Gumerman and Umemoto 1987, Society for American Archaeology
Figure 2.27, from Crawford 1983, University of Michigan, Museum of Anthropology
Figure 2.30, from Ramenofsky et al. 1986, Society for American Archaeology
Figure 2.31, from Kenward et al. 1980, Science and Archaeology, RCCA—North Straffordshire Polytechnic
Figures 3.2, 3.3, from Bohrer and Adams, 1977, Eastern New Mexico University
Figures 3.30, 3.31, 3.37, 3.38, from Esau 1977, John Wiley and Sons
Figure 3.36, from Catling and Grayson 1982, Chapman and Hall
Figures 3.44, 3.47, from Johannessen 1984, University of Illinois Press
Figures 3.41, 3.48, 3.49, from Pearsall 1988b, University of Chicago Press
Figures 3.43, 3.46, from Pearsall 1983b, Society of Ethnobiology
Figure 3.45, from MacNeish 1967, University of Texas Press and the Robert S. Peabody Foundation
Figure 3.50, from Hillman 1984, A. A. Balkema
Figures 4.1, 4.2, 4.3, from Kapp 1969, by permission of the author
Figure 4.4, from Faegri and Iversen 1975, Munksgaard
Figures 4.6, 4.8, from Dimbleby 1985, Academic Press
Figure 4.7, from Hansen et al. 1984, Geological Society of America
Figures 4.9, 4.10, from Schoenwetter and Smith 1986, Academic Press
Figure 4.11, from Birks and Gordon 1985, Academic Press
Figure 5.6, from Twiss et al. 1969, Soil Science Society of America
Figure 5.7, from Lewis 1981, Society of Ethnobiology
Figures 5.9, 5.10, from Brown 1984, Academic Press
Figures 5.17, 5.33, from Pearsall and Trimble 1984, Academic Press
Figure 5.23, from Pearsall and Trimble 1983, Bernice P. Bishop Museum
Figures 5.27, 5.30, from Terry and Chilingar 1955, Society of Economic Paleontologists and Minerologists
Figure 5.31, from Piperno 1988a, Academic Press
Figure 5.34, from Chiswell 1984, with permission of the author

The photographs used in Figures 2.8 and 2.28 are courtesy of G. Crawford. Figure 2.9a was provided by P. Watson. The sketch on which Figure 2.22 is based is courtesy of G. Hillman. Figure 2.33 was redrawn from lab documents provided by S. Johannessen. Photographs used in Figure 3.8 are courtesy of M. Cornman. Figure 3.40a is courtesy of B. Cumbie; Figure 3.40b is courtesy of G. Brown. E. Wohlgemuth kindly gave me permission to discuss results of his unpublished research.

Chapter 1 **The Paleoethnobotanical Approach**

Introduction

In 1941, Volney H. Jones published a short article, "The Nature and Status of Ethnobotany," in which he formalized a field of inquiry into mankind's knowledge and use of plants: ethnobotany, "the study of the interrelations of primitive man and plants" (1941:220). Although the term "ethnobotany" was first used by J. W. Harshberger in 1895 to refer to use of plants by aborigines, the focus on ecological interactions of human populations and the plant world which characterizes modern ethnobotany may be traced to the influence of Jones and the Ethnobotanical Laboratory of the University of Michigan Museum of Anthropology. Jones was also among the first to call for an interdisciplinary approach to the field: "Ethnobotanical studies can be most successfully made when ethnobotanical problems are paramount in the investigation and when the worker or workers are familiar with the techniques, methods and approach of both anthropology and the plant sciences" (1941:220).

Jones's concept of ethnobotany was soon expanded to include ancient man and contemporary cultures at whatever level of complexity. Margaret Towle's definition is typical: "This all-pervading association [between humans and plants] has come to be known as ethnobotany, a term applied to the study of the relationship between man and the plant world, without limits to time or to the degree of his cultural development" (Towle 1961:1).

Paleoethnobotany (the term was introduced by Helbaek in 1959) is part of the field of ethnobotany—specifically, that aspect concerned with elucidating human–

plant relations in the past through study of archaeological plant remains such as pollen grains, phytoliths, charred wood, seeds, and the like. In Richard Ford's words, "Paleoethnobotany . . . is the analysis and interpretation of the direct interrelationships between humans and plants for whatever purpose as manifested in the archaeological record" (1979:286).

The subject matter of this book, paleoethnobotany, has two distinctive components inherent in this definition. First, it is an archaeological approach. Research materials of paleoethnobotany, archaeological plant remains (also referred to as archaeobotanical remains), must be recovered from sites and identified. Much of a paleoethnobotanist's energy and time may be devoted to discussing sampling strategy, floating or sieving soil to recover charred seeds and wood, collecting pollen or phytolith samples, compiling comparative collections, and processing and identifying materials in the lab. If the paleoethnobotanist is not trained as an archaeologist, then he or she must learn to think like one, or at least to communicate with archaeological field personnel and project directors.

The nature of these archaeological research materials also demands expertise in botany. Much as an archaeological ceramic specialist or lithic specialist must learn about parent materials, manufacturing technology, and the like, so the archaeological botanical specialist must learn plant taxonomy, anatomy, and laboratory skills necessary to recover and identify plant remains. Even paleoethnobotanists whose primary training is in botany must adapt their skills to deal with fragmentary materials and the incomplete archaeological record.

Second, paleoethnobotany uses an ecological approach. Once fieldwork and identifications are done, data are interpreted to elucidate the nature of human–plant relationships. These relationships may take many forms: how plants are used as fuels, foods, medicines, or in ritual; how seasonality of plant availability affects settlement systems; the extent and nature of human–plant interdependency, and the impact of humans on vegetation. Problems addressed using paleoethnobotanical data depend, not only on the nature and quality of remains, but on overall objectives of research. This point brings us back to the importance of interaction between botanical specialist and archaeologist: data, including plant remains, are only as good as the archaeology; interpretations are constrained by sampling strategy.

Much as Jones recognized the importance for ethnobotany of cooperation between anthropologists and plant scientists, so recently has the importance of cooperation and communication among all specialists who study relationships between humans and living organisms been recognized with formalization of the field of ethnobiology. Weber defines ethnobiology as "work that draws on both biology and anthropology to make statements about the interrelationship between living organisms and human culture, whether prehistoric, historic, or contemporary" (1986:iii). The first conference of the Society of Ethnobiology in 1978 and the subsequent appearance of its journal, *Journal of Ethnobiology*, mark recognition of the diversity of approaches to the study of human relations with the biotic environment. Diversity of approach has led inevitably to specialization; in the area of prehistoric approaches alone there are pollen analysts, phytolith analysts, specialists in analysis of botanical macroremains (seeds, wood), vertebrate faunal analysts, invertebrate specialists, and so on. It is difficult to master more than one area in the plant or animal kingdom, impossible to have expertise in all areas of study of mankind's interaction with the living world.

In spite of this diversity, however, there are similarities in method and approach and common problems which unify the field of ethnobiology. Chief among these are the shared ecological approach, and in the case of prehistoric applications, limitations imposed by the nature of the archaeological record. For example, many sampling, identification, and quantification problems that I discuss for botanical macroremain analysis apply also to analysis of faunal materials. By staying in communication with fellow ethnobiologists and noting methodological developments in related fields, we can help lessen the isolation that led Dimbleby to write in the introduction to *Plants and Archaeology*, "In principle, I am opposed to the writing of this book. Being trained as an ecologist makes me constantly aware that an artificial distinction is being made by dealing only with man's relationships with plants and omitting the animal kingdom, geology, soils and other components of the environment" (1978:11).

In the remainder of this chapter, I first look briefly at the field of ethnobotany from a historical perspective, then summarize current thinking on the nature and status of the field.

Historical Overview

There are a number of journal articles, reviews, and book chapters that include reviews of the development of paleoethnobotany (e.g., Bohrer 1986; Ford 1979, 1981, 1985a, 1985b; Helbaek 1970; M. Jones 1985; V. Jones 1957; Nabhan 1986; Renfrew 1973; Towle 1961; Yarnell 1970). The following overview relies on several of these sources.

The history of development of paleoethnobotany is actually the history of two paleoethnobotanical traditions, one European, one American. Today these two may be distinguished by the focus of many Old World ethnobotanists on precise botanical description and taxonomic treatment of remains, especially of cultivated materials, and by the emphasis of many Americanists, especially those with anthropological training, on cultural aspects such as use or presence of plants at a site.

The European paleoethnobotanical tradition is the older. Interest in analysis of archaeological plant remains was sparked by Kunth's (1826) study of desiccated material from Egyptian tombs and Heer's (1866) analysis of waterlogged material from lakeside Swiss villages. Among materials identified by Heer (1878) were a number of varieties of barley, wheat, and millet, numerous common field weeds, vegetables such as pea and lentil, fruits and berries such as apple, pear, plum, grape, and cherry, nuts, including walnut and water chestnut, and a variety of fibers, woods, mosses, and aquatic taxa. Heer used these data to discuss cultural connections, seasonality of site occupation, and the differences between ancient plants and modern types.

Study of macroremains continued in the late nineteenth and first half of the twentieth centuries in Europe on sites in Switzerland, central Europe, Germany, Italy, Greece, Anatolia, and Egypt, as well as in coastal Peru (see Renfrew 1973:1–6, and Towle 1961:1–13, for more detail on these early studies). It was also during this period that pollen analysis began to develop as a discipline in northern Europe, and that its application in archaeology was realized (see Chapter 4 and Bryant and Holloway 1983).

Beginning in the 1950s, the geographic scope of paleoethnobotany in Europe expanded to include the Near East. Much work in the 1950s and 1960s was carried out by Helbaek, working on materials excavated by Braidwood in Iraq (Helbaek 1959, 1960a, 1960b), and Hole, Flannery, and Neely in Iran (Helbaek 1969), among other projects. This research and work of other ethnobotanists such as van Zeist (1975; van Zeist and Casparie 1968), Hopf (1969), and the Cambridge economic prehistory group (Higgs 1972) provided the basic botanical evidence for domestication of many food crops of the Near East.

Recent years have seen a dramatic increase in paleoethnobotanical research by European scholars, a trend paralleled by expansion of the field in the New World. This activity was related in part to development of the flotation technique for recovering macroremains (see Chapter 2) and to greater application of pollen analysis in archaeology. Both developments were in turn related to increased interest in archaeology in agricultural origins and dispersals and human interaction with the environment. In 1968, the International Work Group for Paleoethnobotany was set up as a forum to bring together researchers who were working in comparative isolation. Symposia are held every three years. I discuss a number of papers from the proceedings of the sixth symposium (van Zeist and Casparie 1984) in Chapter 3.

The first paleoethnobotanical study of New World materials was conducted by Saffray (1876), who examined contents of a Peruvian mummy bundle. This was followed in 1879 by a report by Rochebrune on plant specimens from Ancon, Peru, and work by Harms (1922) and Yacovleff and Herrera (1934–1935), among others. Towle (1961) conducted the first analyses of stratigraphically excavated Peruvian materials.

In spite of the early studies of Peruvian materials by European scholars, American archaeologists and botanists did not show much interest in archaeological plant remains until after 1930. A very similar situation existed for pollen analysis. Although a few macroremain studies were carried out prior to 1930, such as Young's (1910) identification of plant material in coprolites (human feces) from Salts Cave, Kentucky, it took Guthe's (1930) invitation for archaeologists to send botanical material for identification to the Museum of Anthropology of the University of Michigan to spark interest in the field. Identifications at the museum were made by Melvin Gilmore and Volney Jones.

Analyses by Gilmore (1931) and Jones (1936) of desiccated plant remains from rockshelter sites in the United States helped fuel interest in paleoethnobotany in North America. In particular, Jones's report of plant remains from Newt Kash Hollow, Kentucky, demonstrated the great potential of plant remains for paleoenvironmental and archaeological interpretation. In this study, which set the standard for subsequent ethnobotanical analyses, Jones identified a variety of wild and cultivated plants, including seeds extracted from coprolites. He discussed size differences in chenopod seeds recovered from the site and suggested that larger seeds were the product of cultivation. Differences in size between archaeological and wild modern *Iva* and *Helianthus* seeds were also discussed from the viewpoint of their potential cultivation. Jones drew inferences about gathering practices from the nature of maygrass remains and suggested a possible extension of prairie based on presence of grassland indicators.

Although research of the Michigan lab during the 1930s and 1940s led more archaeologists to save botanical materials, it was publication of *Excavations at Star Carr* by British archaeologist J. G. D. Clark (1954) that convinced many of the importance of biological remains for archaeological interpretation. With increased interest in American archaeology on reconstructing subsistence and paleoenvironment, greater emphasis was put during the late 1950s and 1960s on recovering and analyzing macroremains and pollen. I have already mentioned the work by American archaeologists in the Near East; research by MacNeish and colleagues in Mexico is another example. A number of American botanists became involved with the study of crop plants recovered from New World sites, especially those in the American Southwest and Mexico where dry conditions gave excellent preservation. Among these were, for corn, Mangelsdorf (1974), Cutler (1952, 1956; Cutler and Blake 1976), and Galinat (Galinat and Gunnerson 1963); for cucurbits, Whitaker and Cutler (1965; Whitaker *et al.* 1957) and Heiser (1973); for beans, Kaplan (1956, 1963); and for cotton, Stephens (1970, 1975). Additional references to publications by these and other researchers studying archaeological remains of crop plants may be found in Chapter 3.

After Struever (1968) described flotation, archaeologists systematically began to look for botanical macroremains from a great diversity of sites, not just those where dry or waterlogged conditions preserved quantities of material. Where flotation quickly became a routine part of excavation, for example in the Midwest and Southwest of the United States, paleoethnobotany labs such as those at the University of Michigan (R. Ford), Texas A. and M. University (V. Bryant, Jr.), and Eastern New Mexico University (V. Bohrer) were almost overwhelmed with data. This situation became even more acute in the late 1970s and 1980s as a result of data generated by cultural resource management (CRM) projects. A number of large projects, such as Dolores in Colorado (Bohrer 1986) and FAI-270 in Illinois (Bareis and Porter 1984), employed their own professional paleoethnobotanists. So many important data have come from recent CRM projects that it is difficult to keep abreast of developments in North American paleoethnobotany. As Nabhan (1986) points out in his review of Ford's (1985c) *Prehistoric Food Production in North America*, a number of papers in this synthesis of horticultural evolution were outdated by new CRM data even before it was published. A similar situation has occurred in England as a result of rescue archaeology.

Nature and Status of Ethnobotany

It should be clear from the brief historical overview presented here that paleoethnobotany on both sides of the Atlantic is in a period of rapid development. Many changes in the field may be viewed as positive developments—for example, an increasing focus on quantification in macroremain and pollen analysis and advances in a new technique, phytolith analysis. Other developments are disturbing in their implications—for example, limited availability of many CRM botanical reports and increasing difficulties in training and employing students. In this final section of Chapter 1 I address some of these issues.

To start with the positive, American paleoethnobotany has maintained and strengthened the ecological focus pioneered by Jones. Bohrer (1986), for example, reviews recent efforts at understanding the relations between people and maize, one of the most important New World cultivated plants. The story is far from completely understood, and debate continues on maize evolution (most recently centering around the controversial, catastrophic sexual transmutation hypothesis, see Iltis 1983). In her overview, Bohrer also discusses recent work on the role of biotic factors, including human intervention, in vegetation change.

Ford (1979, 1981, 1985a) reviews the many ways in which archaeological plant remains can elucidate the nature of human–plant relations. Wood charcoal, for example, is used to investigate patterns of firewood selection, to indicate environmental disturbance, and as one source of data for vegetation reconstruction. The geographic distribution of plant remains in archaeological deposits is a potential source of information on cultural contacts. Numerous other examples could be given. Interpreting data in terms of what can be learned about the impact on humans of using certain plants, and on plants of being used by humans, remains the focus of American paleoethnobotany.

European paleoethnobotany seems to be moving away from its traditional focus on morphology and taxonomy of remains toward emphasis on cultural interpretation (Ford 1985a). This is a positive development from the perspective of American researchers. For example, Hillman's (1984) and G. Jones's (1984) applications of ethnographic models to interpretation of macroremain assemblages are exciting and innovative. Not only are these researchers' quantitative approaches rarely used by American paleoethnobotanists, but the detail of ethnographic observation and comparative sampling used to formulate their models is not available in any New World ethnobotany. This research is a model for others to emulate.

Another positive area in the discipline as a whole is development of new field and laboratory techniques. In the domain of macroremain analysis, for example, innovations and refinements in recovery techniques continue to be made. Flotation technology now includes machine-assisted systems such as the SMAP (Watson 1976) and froth (Jarman et al. 1972) rigs. Few manual systems look much like the system described by Struever (1968); current systems, like the Dayton Museum/IDOT system (Wagner 1977), use smaller mesh to enhance seed recovery. Siphons have been used to recover semibuoyant material (Gumerman and Umento 1987); fine sorting using liquids of varying densities (Bodner and Rowlett 1980) or air pressure (Ramenofsky et al. 1986) has been proposed. Although most identification of seeds and charcoal is still done by low magnification, with visual comparison to securely identified modern voucher specimens, techniques that can expand what is "identifiable" are available. Scanning electron microscopy, for example, allows identification of very fragmentary remains by minute anatomical structure. Trace element analysis of charred cooking residues can indicate food source, at least to a broad group such as legume or C_4 grass (DeNiro and Hastorf 1985; Hastorf and DeNiro 1985). Analysis of organic residues such as lipids represents another innovative approach (Rottlaender and Schlichterle 1980).

In pollen analysis, perhaps the most important development in technique is the increasing use of quantitative approaches to aid interpretation of pollen assemblages. Although much of this work has been done in Quaternary palynology (e.g., Birks and Gordon 1985), applications in archaeological palynology are clear. For example, Schoenwetter and Smith's (1986) analysis of pollen data from Archaic sites in Oaxaca, Mexico, utilizes discriminant function analysis to distinguish among pollen assemblages representing distinctive ecological conditions.

Development of phytolith analysis, the identification and interpretation of opaline plant silica, represents the major addition to paleoethnobotanical technique in recent years. If I were writing this book ten years ago, Chapter 5 would not exist.

Phytolith analysis has progressed to the point that it has a body of technique and research results which must be considered seriously by archaeologists and paleoethnobotanists. Its major contributions lie ahead.

A positive development for paleoethnobotany has been a trend in archaeology toward interdisciplinary research. It is certainly true that recovering botanical remains is still an afterthought in some projects, with the analyst being contacted for the first time after budgeting and fieldwork are done. But that scenario is becoming much less frequent. Paleoethnobotanist, zooarchaeologist, and soil expert may still be referred to as "ancillary specialists," but they are often called in at the planning stage of projects, participating with the "excavation specialists" in this critical phase of research. Similarly, more effort is made to incorporate results of all data analyses into final interpretations and conclusions. It is becoming common to see results of macroremain or pollen analyses presented as chapters in publications, rather than relegated to the appendix, or not presented under the analyst's name at all.

Although I agree with Bohrer that "ethnobotany is flying high in the 1980's" (1986:27), I also feel, as she does, that there are problems. Concerns about quality of training and research led me to write this book.

The structure of the remainder of this book mirrors the structure of a course I teach in paleoethnobotany. That course is a broad overview of the field, one in which I give equal time to the three primary types of archaeological botanical data—macroremains, pollen, and phytoliths—and discuss recovery, identification, data presentation, and interpretation. Whatever aspect of the field may attract future ethnobotanists in the class, I feel strongly that they should be familiar with all aspects of the field and able to read and understand studies outside their own area. This addresses one problem in contemporary paleoethnobotany—that in which specialization, either in technique or geographic focus, leads to narrowness of view and intolerance of innovation.

I make no claim to be a fount of wisdom in paleoethnobotany, but I have had the benefit of supervised training as an undergraduate at the University of Michigan Ethnobotany Laboratory. In Chapters 2 and 3, on the recovery and analysis of macroremains, I try to impart some of this training and the benefit of experience (my own and others') to the novice who is faced with handling his or her first flotation analysis without help from Richard Ford. Given the great expansion of flotation-generated botanical data in recent years, especially as a result of CRM projects, many analyses are handled "in house" rather than sent to already overloaded ethnobotany laboratories. This creates a problem in maintaining products of quality. A guide to technique and interpretation for novice ethnobotanists and supervising archaeologists may help in this regard.

Let me make it clear, however, that I am not advocating "cookbook" paleoethnobotany; working with an experienced researcher is the best way to learn the techniques and approaches of paleoethnobotany. Rather, I am responding to a problem: how are we in the field to maintain consistency of technique and quality of interpretation among relatively isolated researchers? Although this problem is not limited to macroremain analysis, it is more acute in this area than among pollen analysts or phytolith researchers. Guides such as this and increased participation in professional societies such as the Society for Ethnobiology and the International Work Group for Palaeoethnobotany may be part of the solution.

Finally, there is a problem of provincialism in paleoethnobotany—specifically, a lack of communication between American and European researchers. Ford (1985a) gives as an example the fact that few American researchers cite the systematic experiments carried out by Europeans on the effects of charring botanical materials. Better communication through thorough literature review and attendence at international meetings would enhance the research of all.

Given that there are problem areas in paleoethnobotany, some of which are shared by the disciplines we draw upon, the status of the field is healthy. Both the New and Old World traditions are undergoing a period of rapid growth and development, with exciting new areas of research and increasing quality of work in traditional areas of interest. Those of us who work in Latin America are seeing increasing interest in the field among local students and archaeologists and new efforts to establish laboratories. Paleoethnobotany is growing in many parts of the world—for example, in India, where a well-established ethnobotanical tradition already exists (Maheshwari 1983). In the chapters that follow I try to bring together the many techniques of paleoethnobotany, to illustrate how charred seeds, pollen grains, and phytoliths can shed light on human–plant relationships.

References

Bareis, Charles J., and James W. Porter (editors)
    1984   *American Bottom archaeology*. Urbana: University of Illinois Press.
Birks, H. J. B., and A. D. Gordon
    1985   *Numerical methods in Quaternary pollen analysis*. London: Academic Press.
Bodner, Connie C., and Ralph M. Rowlett
    1980   Separation of bone, charcoal, and seeds by chemical flotation. *American Antiquity* 45(1):110–116.
Bohrer, Vorsila L.
    1986   Guideposts in ethnobotany. *Journal of Ethnobiology* 6(1):27–43.
Bryant, Vaughn M., Jr., and Richard G. Holloway
    1983   The role of palynology in archaeology. In *Advances in archaeological method and theory* (Vol. 6), edited by M. Schiffer, pp. 191–223. New York: Academic Press.

Clark, J. G. D.
    1954   *Excavations at Star Carr: An early Mesolithic site at Seamer, near Scarborough, Yorkshire.* London: Cambridge University Press.

Cutler, Hugh C.
    1952   A preliminary survey of plant remains in Tularosa Cave. In *Mogollon cultural continuity and change: The stratigraphic analysis of Tularosa and Cordova caves,* edited by Paul S. Martin et al. Fieldiana: Anthropology 40:461–479.
    1956   Plant remains. In *Higgins Flat Pueblo, western New Mexico,* by Paul S. Martin et al. Fieldiana: Anthropology 45:174–183.

Cutler, Hugh C., and Leonard W. Blake
    1976   Plants from archaeological sites east of the Rockies. *American Archaeology Reports* 1. Columbia: University of Missouri–Columbia.

DeNiro, Michael J., and Christine A. Hastorf
    1985   Alteration of $^{15}N/^{14}N$ and $^{13}C/^{12}C$ ratios of plant matter during the initial stages of diagenesis: Studies utilizing archaeological specimens from Peru. *Geochimica et Cosmochimica Acta* 49:97–115.

Dimbleby, G. W.
    1978   *Plants and archaeology.* Second edition (1967). Atlantic Highlands, New Jersey: Humanities Press.

Ford, Richard I.
    1979   Paleoethnbotany in American archaeology. In *Advances in archaeological method and theory* (Vol. 2), edited by M. Schiffer, pp. 286–336. New York: Academic Press.
    1981   Ethnobotany in North America: An historical phytogeographic perspective. *Canadian Journal of Botany* 59:2178–2188.
    1985a   Paleoethnobotany: European style. *Quarterly Review of Archaeology* 6(1):1.
    1985b   Preface. In *Prehistoric food production in North America,* edited by R. I. Ford, pp. xi–xv. Anthropological Papers of the Museum of Anthropology, University of Michigan 75.

Richard I. Ford (editor)
    1985c   *Prehistoric food production in North America.* Anthropological Papers of the Museum of Anthropology, University of Michigan 75.

Galinat, W. C., and J. H. Gunnerson
    1963   Spread of 8-rowed maize from the prehistoric Southwest. *Botanical Museum Leaflets* 20:117–160, Harvard University.

Gilmore, Melvin R.
    1931   Vegetal remains of the Ozark Bluff-Dweller culture. *Papers of the Michigan Academy of Science, Arts, and Letters* 14:83–102.

Gumerman, George, IV, and Bruce S. Umento
    1987   The siphon technique: An addition to the flotation process. *American Antiquity* 52(2):330–336.

Guthe, C. E.
    1930   Identification of botanical material from excavations. *National Research Council, Division of Anthropology and Psychology Circular* 6 (cited in Ford 1979).

Harms, H. von
    1922   Übersicht der bisher in altperuanishen Graben gefundenen Pflanzenreste. In *Festschrift Eduard Seler,* pp. 157–186. Stuttgart (cited in Renfrew 1973).

Hastorf, C. A., and M. J. DeNiro
   1985   Reconstruction of prehistoric plant production and cooking practices by a new isotopic method. *Nature (London)* 315:489–491.

Heer, O.
   1866   Treatise on the plants of the lake dwellings. In *The lake dwellings of Switzerland and other parts of Europe*, edited by Ferdinand Keller, translated by John E. Lee. London: Longmans, Green. (cited in Renfrew 1973).
   1878   Abstract of the treatise on the "Plants of the Lake Dwellings." In *The lake dwellings of Switzerland and other parts of Europe* (Vol. 1), edited by Ferdinand Keller, translated by John E. Lee, pp. 518–544. London: Longmans, Green.

Heiser, C. B., Jr.
   1973   Variation in the bottle gourd. In *Tropical forest ecosystems in Africa and South America: A comparative review*, edited by B. J. Meggers, E. S. Ayensu, and W. D. Duckworth, pp. 121–128. Washington, D.C.: Smithsonian Institution Press.

Helbaek, H.
   1959   The domestication of food plants in the Old World. *Science* 130:365–372.
   1960a   The palaeoethnobotany of the Near East and Europe. In *Prehistoric investigations in Iraqi Kurdistan*, edited by R. J. Braidwood and B. Howe, pp. 99–118. *Studies in Ancient Oriental Civilization* 31. Chicago: University of Chicago Press.
   1960b   Cereals and weed grasses in phase A. In *Excavations in the plain of Antioch I*, edited by R. J. Braidwood and L. J. Braidwood, pp. 540–543. *Oriental Institute Publications* 61. Chicago: University of Chicago Press.
   1969   Plant collecting, dry-farming, and irrigation agriculture in prehistoric Deh Luran. In *Prehistory and human ecology of the Deh Luran Plain*, edited by Frank Hole, Kent Flannery, and James Neely, pp. 383–426. *Memoirs of the Museum of Anthropology, University of Michigan* 1.
   1970   Palaeo-ethnobotany. In *Science in archaeology: A survey of progress and research* (2nd Ed.), edited by Don Brothwell and Eric Higgs, pp. 206–214. New York: Praeger.

Higgs, E. S. (editor)
   1972   *Papers in economic prehistory*. London: Cambridge University Press.

Hillman, Gordon
   1984   Interpretation of archaeological plant remains: The application of ethnographic models from Turkey. In *Plants and ancient man*, edited by W. van Zeist and W. A. Casparie, pp. 1–41. Rotterdam, Holland: A. A. Balkema.

Hopf, M.
   1969   Plant remains and early farming in Jericho. In *The domestication and exploitation of plants and animals*, edited by P. J. Ucko and G. W. Dimbleby, pp. 355–359. London: Duckworth.

Iltis, Hugh H.
   1983   From teosinte to maize: The catastrophic sexual transmutation. *Science* 222:886–894.

Jarman, H. N., A. J. Legge, and J. A. Charles
   1972   Retrieval of plant remains from archaeological sites by froth flotation. In *Papers in economic prehistory*, edited by E. S. Higgs. London: Cambridge University Press.

Jones, G. E. M.
   1984   Interpretation of archaeological plant remains: Ethnographic models from Greece. In *Plants and ancient man,* edited by W. van Zeist and W. A. Casparie, pp. 43–61. Rotterdam, Holland: A. A. Balkema.

Jones, Martin
   1985   Archaeobotany beyond subsistence reconstruction. In *Beyond domestication in prehistoric Europe,* edited by Graeme Barker and Clive Gamble, pp. 107–128. London: Academic Press.

Jones, Volney H.
   1936   The vegetal remains of Newt Kash Hollow shelter. In *Rock shelters in Menifee County, Kentucky,* edited by W. S. Webb and W. D. Funkhouser. *University of Kentucky Reports in Archaeology and Anthropology* 3(4):147–167.
   1941   The nature and status of ethnobotany. *Chronica Botanica* 6(10):219–221.
   1957   Botany. In *The identification of non-artifactual archaeological materials,* edited by Walter W. Taylor. *National Academy of Sciences, National Research Council Publication* 565:35–38.

Kaplan, Lawrence
   1956   Cultivated beans of the prehistoric Southwest. *Annals of the Missouri Botanical Gardens* 43(2):189–251.
   1963   Archeoethnobotany of Cordova Cave, New Mexico. *Economic Botany* 17(4): 350–359.

Kunth, C.
   1826   Examen botanique. In *Catalogue raisonné et historique de antiquités découvertes en Egypte,* edited by J. Passalacqua. Paris (cited in Renfrew 1973).

Maheshwari, J. K.
   1983   Developments in ethnobotany. *Journal of Economic and Taxonomic Botany (India)* 4(1):i–v.

Mangelsdorf, Paul C.
   1974   *Corn; Its origin, evolution, and improvements.* Cambridge, Massachusetts: Harvard University Press.

Nabhan, Gary
   1986   Book review of *Prehistoric food production in North America,* edited by R. I. Ford. *Journal of Ethnobiology* 6(1):44–45.

Ramenofsky, Ann F., Leon C. Standifer, Ann M. Whitmer, and Marie S. Standifer
   1986   A new technique for separating flotation samples. *American Antiquity* 51(1):66–72.

Renfrew, Jane M.
   1973   *Palaeoethnobotany.* New York: Columbia University Press.

Rochebrune, A. T., de
   1879   Recherches d'ethnographie botanique sur la flore des sepultures péruviennes d'Ancón. *Actes de la Société Linnéenne de Bordeaux* 3:343–358 (cited in Ford 1979).

Rottlaender, R. C. H., and H. Schlichterle
   1980   Gefässinhalte—eine kurz Kommentierte Bibliographie [contents of vessels—a briefly annotated bibliography]. *Archaeo-Physika* 7:61–70.

Saffray, Dr.
   1876   Les antiquités Péruviennes à l'exposition de Philadelphia. *La Nature (Paris)* 4:401–407 (cited in Renfrew 1973).

Schoenwetter, James, and Landon Douglas Smith
    1986   Pollen analysis of the Oaxaca Archaic. In *Guilá Naquitz: Archaic foraging and early agriculture in Oaxaca, Mexico*, edited by K. V. Flannery, pp. 179–237. Orlando, Florida: Academic Press.

Stephens, S. G.
    1970   The botanical identification of archaeological cotton. *American Antiquity* 35(3):367–373.
    1975   A reexamination of the cotton remains from Huaca Prieta, north coastal Peru. *American Antiquity* 40(4):406–419.

Struever, Stuart
    1968   Flotation techniques for the recovery of small-scale archaeological remains. *American Antiquity* 33:353–362.

Towle, Margaret
    1961   *The ethnobotany of pre-Columbian Peru.* Chicago: Aldine.

van Zeist, W.
    1975   Preliminary report on the botany of Gomolava. *Journal of Archaeological Science* 2:315–325.

van Zeist, W., and W. A. Casparie
    1968   Wild einkorn wheat and barley from Tell Mureybit in northern Syria. *Acta Botanica Neerlandica* 17(1):44–53.

van Zeist, W., and W. A. Casparie (editors)
    1984   *Plants and ancient man: Studies in palaeoethnobotany.* Rotterdam, Holland: A. A. Balkema.

Wagner, Gail E.
    1977   The Dayton Museum of Natural History flotation procedure manual. Ms. on file with author, Center for American Archaeology, Kampsville, Illinois.

Watson, Patty Jo
    1976   In pursuit of prehistoric subsistence: A comparative account of some contemporary flotation techniques. *Mid-Continental Journal of Archaeology* 1(1):77–100.

Weber, Steven A.
    1986   The development of a society: An introduction to the special issue. *Journal of Ethnobiology* 6(1):iii–vi.

Whitaker, Thomas W., and Hugh C. Cutler
    1965   Cucurbits and cultures in the Americas. *Economic Botany* 19(4):344–349.

Whitaker, Thomas W., Hugh C. Cutler, and R. S. MacNeish
    1957   Cucurbit materials from three caves near Ocampo, Tamaulipas. *American Antiquity* 22(4):352–358.

Yacovleff, Eugenio, and Fortunato L. Herrera
    1934–1935   El mundo vegetal de los antiguos Peruanos. *Revista del Museo Nacional, Lima* 3(3):241–322; 4(1):29–102.

Yarnell, Richard A.
    1970   Palaeo-ethnobotany in America. In *Science in archaeology: A survey of progress and research* (2nd Ed.), edited by Don Brothwell and Eric Higgs, pp. 215–228. New York: Praeger.

Young, B. H.
    1910   The prehistoric men of Kentucky. *Filson Club Publication* 25 (cited in Ford 1979).

Chapter 2   **Techniques for Recovering Macroremains**

Introduction

The most widely applied paleoethnobotanical approach is analysis of macroremains, botanical materials visible to the naked eye and large enough to be identified at low-power magnification. Identification of charred, desiccated, or waterlogged wood, wild plant seeds, fruit pips, nut shells, and cultivated plants has contributed substantially to our knowledge of human diet, subsistence strategies, and the history of plant domestication.

Analysis of macroremains (as well as of pollen and phytoliths) can be thought of as a three-part process: recovery, identification, and interpretation. I discuss recovery of macroremains in some detail, devoting this entire chapter to the topic, because this step is usually carried out by the field archaeologist rather than a paleoethnobotanical specialist. There are a variety of techniques to choose from, and choice of technique, as well as correct execution, greatly affects the quality of data recovered and the end results of analysis.

Botanical macroremains can be recovered from archaeological sites in three ways: (1) by collection of material in situ during excavation, (2) through screening, and (3) by using water recovery techniques, or "flotation." Each of these approaches varies in the type of data obtained, in recovery biases, and in analytical or physical problems.

In Situ Collection of Material

Archaeologists have long realized that perishable biological materials may occur in archaeological sediments. During late nineteenth-century excavations at waterlogged lake dwellings in Switzerland and other parts of Europe, for example, quantities of burned and nonburned botanical materials were recovered and studied. Heer (1878) describes plant remains recovered from the "relic bed" of one such lake site, identifying cereal grains, common crop weeds, vegetables, fruits, berries, nuts, oil plants, and a variety of other perishable materials preserved by the wet environment. On the other end of the spectrum, early accounts of opening of tombs on the desert coast of Peru include descriptions of the quantities of textiles and other organic remains preserved by dry conditions there (e.g., Towle 1961; Yacovleff and Herrera 1934–35).

Even at sites without such remarkable preservation it is usually possible to spot concentrations of charcoal flecks or isolated pieces of charred wood during excavation. Such finds can help identify the function of a feature (e.g., helping to identify hearth areas) and are often important for dating. It is usually preferable to date charcoal with known point provenience than to rely on samples taken from the screen or concentrated by flotation. In situ recovery of botanical material can also give interesting data on the association of plant remains and nonbotanical artifacts.

Although one can argue that in situ recovery of botanical remains has a number of advantages, relying solely on the naked eye of even the most experienced excavator for recovery of biological materials introduces a number of biases into data. A major bias relates to size of materials recovered. Trowel recovery of plant remains, bones, or artifacts is generally limited to what can be easily seen in the soil. Corn cobs and projectile points are often detected visually, but *Chenopodium* seeds and small beads are another matter. Small seeds, fragments of larger fruits and nuts, small wood charcoal flecks, and cupules detached from cobs are difficult or impossible to see while troweling, especially when charred material blends into dark soil. Visual recovery is heavily biased toward larger remains or material present in caches; naked-eye plant assemblages lack many taxa entirely, and others are underrepresented.

Recovery by naked eye may also be spatially uneven. Fewer remains are recovered from darker, heavier soils, which obscure visibility of material, from lower excavation levels, where dim lighting becomes a problem, or from units excavated quickly (overburden, fill, lower priority areas, etc.).

Another source of bias is the skill or interest of the excavator. Inexperienced or careless workers recover fewer small materials of all sorts; pick and shovel or poor trowel technique can reduce dramatically recovery of remains.

Screening Techniques

The obvious problems that arise by relying solely on in situ recovery of botanical material, bone, and small artifacts have lead archaeologists to screen soil for systematic recovery of all classes of artifacts. Although screening is now a normal part of excavation procedures in many regions of the world, there has been considerable debate about whether, or under what circumstances, screening should be used. I describe one such debate, among British archaeologists in the early 1970s, in connection with the development of flotation systems in the Old World. A series of sieving experiments carried out by Payne (1972) were instrumental in convincing archaeologists that screening not only recovered less biased samples of all artifact types but also recovered whole classes of material previously missed.

Screening systems set up to process bulk soil from excavation allow more systematic recovery of macroremains than does the naked eye. The primary limiting factors for recovery are screen mesh size and force used to process soil through the screen. If one were to pass the same bucket of soil through a $\frac{1}{2}''$, $\frac{1}{4}''$, $\frac{1}{8}''$, and $\frac{1}{16}''$ screen, the quantity and type of material recovered in each instance would be quite different. The bias introduced by choice of mesh size can be easily illustrated with an example from faunal analysis. Thomas (1969) presents data from three Great Basin hunting sites to illustrate how bones of different size classes are lost through screen meshes often used by archaeologists and to test whether correction factors could be developed for each common screen size. Table 2.1 illustrates the results of tests at one site, Smoky Creek Cave. Note how recovery of animal class I, the smallest bone class (mammals weighing less than 100 g, such as field mouse), varies among the three screen sizes. While all bones (0% loss) are recovered in the $\frac{1}{16}''$ screen in each level, 3–54% are lost using a $\frac{1}{8}''$ screen, and 74–93% are lost using a $\frac{1}{4}''$ screen. Although Thomas demonstrates that correction factors can be developed by experiment for use at a specific site, or for one cultural component at a multiple-component site, he concludes that there is no universal correction factor for under-representation by size.

Recovery of floral remains can be enhanced by bulk screening of soil, especially if $\frac{1}{8}''$ or smaller mesh is used, but only if remains survive the screening process intact. The practice of mashing dirt clods through the screen is ruinous to charcoal. It is usually easy to distinguish in situ collected material from remains recovered in the screen; the latter, if intact, have a smoothed, worn look. Water screening is often the answer to better recovery of all artifacts from difficult soils, but charcoal can be pulverized if water screening is done too forcefully.

The real importance of screening as a macroremain recovery technique is in the form of fine sieving carried out when water flotation is inappropriate, such as when

Table 2.1 • Numbers of Bones of Different Size Classes Recovered by Screens of Different Mesh Sizes, Smoky Creek Cave Site.

Faunal size class[a]		1			2			3			4			5			6	
	$\frac{1}{4}$	$\frac{1}{8}$	$\frac{1}{16}$	$\frac{1}{4}$	$\frac{1}{8}$	$\frac{1}{16}$	$\frac{1}{4}$	$\frac{1}{8}$	$\frac{1}{16}$	$\frac{1}{4}$	$\frac{1}{8}$	$\frac{1}{16}$	$\frac{1}{4}$	$\frac{1}{8}$	$\frac{1}{16}$	$\frac{1}{4}$	$\frac{1}{8}$	$\frac{1}{16}$
I																		
no.	31	86	31	17	75	3	36	94	8	15	81	24	6	44	4	13	69	97
% loss	79	21	—	82	3	—	74	6	—	88	21	—	89	7	—	93	54	—
II																		
no.	61	84	40	44	21	7	77	61	16	56	71	46	33	51	9	62	107	73
% loss	67	22	—	39	10	—	50	10	—	68	27	—	64	10	—	74	30	—
III																		
no.	97	79	16	86	56	—	114	98	—	65	88	6	88	53	1	62	45	2
% loss	49	8	—	40	0	—	46	0	—	59	4	—	38	1	—	47	2	—
IV																		
no.	21	—	—	6	—	—	8	—	—	8	—	—	4	—	—	6	1	—
% loss	0	—	—	0	—	—	0	—	—	0	—	—	0	—	—	14	0	—
V																		
no.	130	—	—	84	—	—	117	—	—	65	—	—	39	—	—	75	1	—
% loss	0	—	—	0	—	—	0	—	—	0	—	—	0	—	—	0	0	—

[a] I–V, smallest to largest.
Source: data from Thomas (1969:394).

materials of interest are waterlogged or desiccated. I describe specific techniques for fine sieving in these situations below (see "Issues in Recovery of Macroremains"). Flotation, as discussed below, is a technique that allows recovery of all size classes of botanical macroremains. To achieve this degree of recovery by screening, very small screen meshes, generally down to 0.5 mm, must be used. Both fine-screen dry screening and water screening face the same basic dilemma: how to eliminate as much soil as possible from samples without losing or damaging botanical materials. Samples that retain soil take much longer to sort, increasing the time and cost of analysis. The solution most often used is to pass samples through a series of increasingly fine sieves, so that only the smallest mesh sizes contain soil. Standard geological sieves, available in a wide variety of mesh openings, can be used, and wire cloth can be built into larger screens. Small geological sieves can be used with a mechanical sieve shaker, which allows fairly rapid sieving of small samples. In general, fine sieving is not used as a bulk soil processing procedure; rather, it is used to process systematically collected subsamples (analogous to flotation samples) for recovery of all size classes of botanical remains.

Water Recovery: Flotation Techniques

Water flotation, recovery techniques utilizing differences in density of organic and inorganic material to achieve separation of organic remains from the soil matrix, greatly enhances both the quantity and range of botanical materials that can be recovered archaeologically. When properly practiced, flotation allows recovery of all size classes of botanical material preserved in a sediment sample, making quantitative analysis possible. Ease of processing samples and simplicity of equipment make flotation easy to apply in the field. Today it is quite common to read in site reports that flotation was used for recovering botanical remains. It may come as a surprise to the student that the key publications for the development of water recovery techniques (flotation, water separation, water sieving) are just twenty years old. The impact on archaeology of the development of means of recovering small botanical and faunal materials in quantity can hardly be overstated. As Watson succinctly put it, "It is no great exaggeration to say that the widespread and large scale use of . . . [flotation] has amounted to a revolution in recovery of data relevant to prehistoric subsistence" (1976:79).

Before describing in detail the major water recovery techniques and their operation, a look back at the development of these systems will give the reader some insight into why such variation exists, which systems developed independently to meet similar needs, which developed from one another, and which have been found to be best suited to certain field situations. A review of terminology will help clarify this discussion.

Terminology

A number of terms describe what I am considering here under the generic category of water recovery techniques. At one time, Struever's (1968) use of the term "flotation" to describe the Apple Creek or "tub" system was widely accepted. Now it is usually applied to all systems by which manual agitation of a soil sample immersed in water results in botanical material being released from the soil and floating to the surface, where it is skimmed off (Fig. 2.1). The body of water into which the sample is immersed can be small and enclosed (a basin of water, garbage can, oil drum) or large and moving (an irrigation ditch, river, ocean). Flotation systems of this simple type are similar in (1) relying on human labor to agitate the flotation bucket or tub holding the soil sample, (2) requiring hand removal of botanical remains floating on the surface of the water (light fraction) (but see the siphon system described below), (3) creating a small heavy fraction, caught on the flotation bucket screen, which may contain nonbuoyant macroremains, (4) being inexpensive to construct and easily adaptable to field or laboratory conditions, but (5) being labor intensive and usually fatiguing. Most manual flotation systems are less consistent and effective in recovering small seed remains than mechanized or machine-assisted systems (Wagner 1982). Flotation may also proceed more slowly using a manual system. Manual agitation may not be vigorous enough to float some dense material (nut shells, dense wood charcoal, and the like), resulting in incomplete recovery. Balancing these disadvantages are ease of construction and transport and low water requirements, advantages contributing to continued use of manual flotation systems.

Watson (1976) uses "flotation" to refer also to machine-assisted or mechanically aided systems such as the SMAP machine and the Cambridge froth flotation cell (Fig. 2.2). In this sense, systems that *float* botanical material out of a soil matrix, either by manual agitation, the pressure of water coming from below, or bubbling of air and lift of frothing agents, are flotation systems. Mechanized flotation systems differ from manual systems in that they (1) replace manual agitation by pressure of water or air passing through the sample from below to achieve flotation, (2) recover floating material by washing material out of the unit into a second flot box or screen, (3) create a heavy fraction that may serve as a bulk water screen for the excavation, (4) usually require the services of a welding shop and purchase of pumps, bubblers, and storage tanks, but (5) are more efficient in terms of volume or weight of soil processed per operator hour. Tests of seed recovery indicate that machine-assisted systems can achieve very effective recovery rates (Wagner 1982), but some are also prone to operating problems (e.g., see Williams 1976, on the froth flotation cell).

The use of the term "water sieve" by French (1971) to describe the Ankara flotation system, which greatly influenced development of botanical recovery systems in the Old World, can be confusing to American archaeologists to whom water

Figure 2.1 Manual water flotation: (a, b) small bucket and irrigation canal, Panaulauca Cave, Peru; (c, d) wood flotation device in Lindberg Bay, U.S. Virgin Islands; (e, f) 55-gallon drums at the El Bronce project, Puerto Rico. In all cases, floating material is removed by hand.

 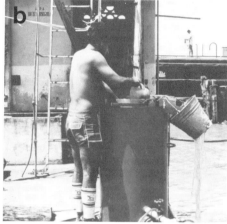

Figure 2.2 Machine-assisted flotation: (a) in froth flotation an air bubbler lifts plant materials from matrix, Salango project, Ecuador; (b) a SMAP-style flotation system uses water pressure to separate materials, Escuela Superior Politécnica del Litoral, Guayaquil, Ecuador. In each system, botanical remains are floated out of the system by water overflow.

sieving or screening means hosing down wet excavation soil in a screen to wash silt through and reveal clean artifacts. This latter is not a procedure usually carried out using screen sizes for recovering small botanical remains; rather, it is a way to screen soil that cannot pass through a dry screen. The water sieve, or Ankara machine, as published by French (1971) and widely adopted (Diamont 1979; Limp 1974; Williams 1973, 1976), uses a continuous water flow from *beneath* to float botanical remains from bulk excavation soil while at the same time cleaning heavier artifacts. It is both a flotation device and a water screen. This system was devised to secure unbiased samples of *all* size classes of *all* artifacts, not just botanical material. The development of this system is linked to progress in object recovery in general. Payne (1972), who was concerned with biases in bone samples secured by excavation without screening, is credited as being a driving force behind the acceptance of systematic sieving. Although the Ankara machine and related systems do separate botanical material by flotation, the fact that they may function as the major bulk sieving system for an excavation, rather than as a device to recover botanical materials, justifies a distinctive terminology.

This discussion of terminology has an important point, that of recognizing potential bias in recovery of botanical remains (Fig. 2.3). A technique in which soil is washed on a screen by water spray from above or shaken on a screen under water, with material remaining on the screen then removed for study, is not flotation. Such recovery of botanical remains, as well as all other material, is directly dependent on

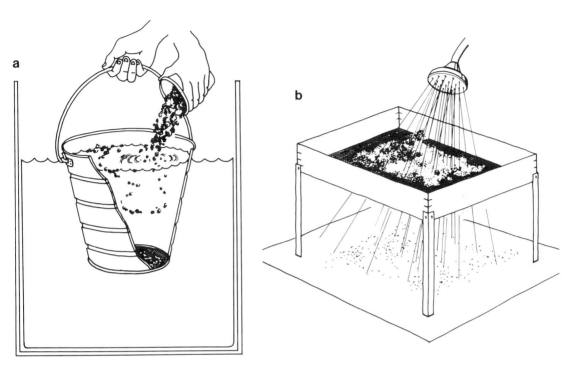

Figure 2.3 Water flotation contrasted to water screening: (a) In a flotation system, soil is poured into a body of water. Agitation breaks up soil, allowing light materials (seeds, charcoal) to float and heavier materials (pebbles, potsherds) to sink. Although some light materials smaller than the bucket mesh may be carried through the bottom, many float. (b) In water screening, soil is placed on a screen and washed with a water jet. All materials smaller than the screen mesh are washed through the screen; only large light and heavy materials are recovered.

the size of the mesh of the screen; most material smaller than the diagonal measure of the screen mesh is lost. A technique in which soil is immersed in water and in some way agitated so that light material is buoyed to the surface and skimmed off or washed out and collected is flotation. Such recovery of botanical remains is dependent on a variety of factors, including screen mesh size in the flotation bucket, mesh size in the hand sieve or catch sieve, the extent and consistency of agitation, and the range of densities of charred material. Whatever the technique is called, this distinction affects recovery and must be understood.

Development of Flotation in the New World

Begun in the 1960s, initially parallel developments of water recovery techniques in the New World and the Old World have converged in recent years, leading to establishment of three basic ways of doing flotation: by hand in a body of water, by machine-assisted water pressure, and by compressed air with frothing agents.

Most American archaeologists are familiar with Struever's 1968 article, "Flotation Techniques for the Recovery of Small-Scale Archaeological Remains." This seminal article brought flotation to the attention of archaeologists. Prior to this publication, flotation had been a laboratory procedure for cleaning excavated botanical samples sent still mixed with soil to a botanist. Hugh Cutler, who suggested the field application of flotation to Struever in 1960, used this type of flotation on material from Tularosa Cave, Higgins Flat Pueblo, and Point of Pines Ruin during the 1950s (Watson 1976:79).

At about the same time that Struever and his co-workers were adapting the lab technique of stirring small amounts of soil in a bucket of water to float off seeds and charcoal to a field procedure for processing larger soil samples, Hole, Flannery, and Neely (Hole *et al.* 1969:23–27) tried flotation on samples from Ali Kosh, in Iran. During their 1963 field season, armed with the basic Struever technique, which was already being circulated in the United States, they tried flotation but were stymied by water shortages. On the suggestion of botanist Hans Helbaek, who like Cutler had used flotation for years as a lab procedure, they switched back to smaller-scale flotation, using a basin of water in the field lab (Helbaek 1969:385). Helbaek was accustomed to using carbon tetrachloride (CCl_4), at 1.8 specific gravity, as a flotation medium, but this proved impractical in the hot Iranian field situation.

The development of flotation in the New World thus began with the adaptation of a small-scale laboratory procedure for cleaning botanical specimens for use on a larger scale in the field to recover small botanical and faunal remains. Flotation is based on the principle that materials of differing densities segregate if added to a medium whose density is *between* that of the materials. The Apple Creek or tub flotation procedure devised by Struever is a two-phase flotation process that combines water separation and chemical flotation. Water is used to separate charred botanical material and bone from gravel, sherds, lithic debris, and soil matrix. Zinc chloride ($ZnCl_2$) of 1.62 specific gravity is then used to separate the charcoal and bone mix.

Flotation as practiced in the early 1960s by American archaeologists was from the beginning conceived as a way to recover a sample of small biological materials when these were not present in concentrations visible to the eye during excavation. As Hole *et al.* put it, "Our preliminary report on the 1961 season states confidently that 'plant remains were scarce at Ali Kosh'. Nothing could be further from the truth. The mound is filled with seeds from top to bottom. All that was 'scarce' in 1961 was our ability to find them" (1969:24). Archaeologists were accustomed to excavating caches or concentrations of such material and had been instructed by botanists in the handling of fragile materials (e.g., Barghoorn 1944). Concern about proper identification of faunal and floral materials existed before the beginning of specialized recovery techniques, as indicated by the National Academy of Science conference on identification of nonartifactual archaeological materials (Taylor 1957). But although American archaeologists came to realize that neither the typical ½" or ¼" excavation screen nor the naked eye allows recovery of the full size range of archaeological materials, they emphasized sampling part of excavated soil for missing size fractions rather than devising new bulk screening methods to enhance recovery sitewide. This was the direction flotation took in Great Britain.

Before turning to the Old World, however, a brief summary of post-Struever developments in manual flotation in the New World is in order. There are probably as many adaptations of manual flotation as there are archaeological projects where flotation has been used. The importance of the earliest adaptations, which were published and inspired subsequent generations of flotation personnel, should be acknowledged, however. Two important changes were for low-water field situations and for indoor use.

Robert Stewart, in work done in 1970 at Tell el-Hesi, Israel (Stewart and Robertson 1973), and at Çayönü in Turkey (Stewart 1976), described one of the early adaptations of outdoor flotation to limited water availability. This is a tank flotation system, using 55-gallon oil drums (Çayönü) or a 1- × 1- × 2-m galvanized metal flot tank (Tell el-Hesi) in place of running water. A flotation bucket with screen in the bottom, similar to that devised by Struever, was used. Stewart and Robertson noted that bone did not float in their samples, so a $ZnCl_2$ chemical flotation to separate charcoal from bone in the light fractions proved unnecessary. This has become a common observation, leading to a shift in emphasis in chemical flotation to concern with removing stray charcoal from heavy fractions or separating bone from the gravel and sherd debris in the heavy fractions rather than with separating bone and charcoal in light fractions.

Minnis and LeBlanc (1976) published a similar system developed over several years on projects in the American Southwest. This is again a tank system, using 55-gallon drums as the water source. Their final version was successfully used at the Mimbres, New Mexico, and Black Mesa, Arizona, projects, and an earlier model served as the flotation apparatus for the Cibola Archaeological Project, New Mexico, during the summers of 1972 and 1973. The easy availability of 55-gallon drums, galvanized buckets or laundry tubs, $\frac{1}{16}$" window screen, and fine-mesh sieving material (whether baby diapers, nylon stockings, or brass carburator mesh), as well as the low water requirements, explain the enduring popularity of the "barrel" system. Water use can be further minimized by using two barrels; water displaced by soil in one barrel can be transferred to a second. As the first barrel fills with sediment during flotation, the second is filled by displaced water. When the first barrel is too full of sediment to use, the remaining water is transferred, the barrel is dumped, and the process starts again. This method has also been adapted to salt water (Stemper 1981) and scaled down for smaller water volumes (Fig. 2.4). Materials from Guangala and other Ecuadorian sites (Pearsall 1984b) were floated with a small plastic garbage can, a plastic bucket with mesh sewn into the bottom, and a nylon-stocking-covered kitchen sieve. This variation, reinvented numerous times and requiring only a pocket knife (to cut the bucket) and pliers (to pull the sewing needle) or a thimble (to push it) to construct, is outdoor manual flotation at its simplest.

At times, flotation using the "tub" or Apple Creek system (Kaplan and Maina 1977) is difficult, not because of arid conditions or water shortage, but because of site location. If a site is too far from moving water so that transport of soil, equipment, and personnel becomes difficult, or if it is a situation where use of a stream, even if present, is impractical, moving the water to the samples may be the best solution. A system developed by Wagner for the Illinois Department of Transportation (Wagner 1976) and Dayton Museum of Natural History (Wagner 1977) and later used to float thousands of samples for the FAI-270 highway mitigation project is of this type. Known as the IDOT, Dayton Museum, or Incinerator system in its various manifestations (Wagner 1982), it uses a flotation tank made of a horse trough instead of a series of barrels. An innovation of this system is the use of a wooden flotation box, with side as well as bottom screens, which floats and allows flotation to be carried out by one person. Most other manual systems require one person to agitate the bucket or box and one to scoop material out (but see the discussion below of the Minnis and LeBlanc procedure for another one-person method). In Wagner's system, two flotation stations are manned at the trough, permitting rapid processing.

Figure 2.4 Small-scale flotation systems using enclosed water sources: (a) small garbage can system, Guangala project, Ecuador; (b) saltwater flotation in a cement tank, Salango project, Ecuador (photograph (b) by D. Stemper).

Adaptation of manual flotation to indoor use takes the process a full circle, back to its origin in the laboratory. A major change occurs, however; indoor systems, such as one described by Schock (1971), are set up to handle much higher soil volumes than the lab procedures used by botanists. Schock describes an indoor system devised with the help of two students (Wolynec and Mertz 1968) which allows flotation to continue when cold weather prohibits use of a tub or barrel system outside. A hose connected to a sink faucet serves as the water source, a metal tub with a drain in the bottom as the barrel, and a standard bucket with a window screen bottom as the flotation bucket. By filling the tub with water and then adjusting drain and faucet to keep a steady flow of water and silt through the system at a constant depth, flotation is easily carried out. A heavy-duty silt drain solves the sludge problem. "What to do with the sludge" is a problem facing all closed flotation systems, making the barrel or tank system unpopular with university groundskeepers and lab directors (when a drain turns out not to be a silt drain). At the Missouri lab, we use an indoor version of the IDOT system. Barrels (garbage cans sealed with silicon caulk) are filled from the tap. Dirty water is siphoned into a floor drain; sludge is shoveled into a wheelbarrel and dumped outside.

Development of Flotation in the Old World

If Struever's 1968 article was the spark that popularized flotation in the United States, David French's "An Experiment in Water-Sieving" (1971) and Jarman, Legge, and Charles's "Retrieval of Plant Remains from Archaeological Sites by Froth Flotation" (1972) did the same for archaeologists working in the Old World, especially the British. These two distinctive machine-assisted flotation systems, which are still in use today, can be thought of as two solutions to the same problem: how can bulk soil from a site be processed to recover all size classes of all artifacts?

Water Sieves and Water Separators

The development of the water separator or water sieve flotation apparatus was linked to efforts to improve recovery of all types of artifacts and was part of a shift to incorporation of screening as a routine part of excavation. Payne's (1972) sieving experiments during the Franchthi Cave, Greece, and Can Hasan III, Turkey, projects were instrumental in this effort.

Payne's experiments, which simply involved water screening soil that had already been visually examined for artifacts through $\frac{1}{8}$" mesh, showed that recovery of pottery, chipped stone, and bones during excavation was only partial. Not only were whole classes of objects missed, but material recovered represented biased samples of each artifact type. This experiment, illustrating that even among large artifact categories much more material was being missed during excavation than most archaeologists realized, generated considerable debate. Barker (1975) presented data from excavations in northern Italy which suggested that careful troweling by trained personnel gave recovery rates equal to those of water sieving (using 3.0-mm mesh). He did not observe the dramatic differences in recovery rates for bone noted by Payne. Barker emphasized, however, both in this article and in a reply to Cherry (1975), that slow, careful troweling may give adequate recovery at a small site, but that "if economic evidence is considered important and is a goal of excavation, certain requirements must be met. The most reliable data will only come from units adequately sieved" (1975:63). This is particularly true if minimally trained workers or hand picks are used for excavation. Commenting on Barker's article, Cherry (1975) raised an issue central to data retrieval: how can sufficient data be collected to answer the questions posed by the archaeologist without exceeding the time and funding available for the project? Cherry described an experiment conducted at the site of Phylakopi on Melos which showed that troweling provided as representative a sample of pottery, as measured by recovery of diagnostic pieces, as water or dry sieving. All these researchers agreed, however, that on-site experimentation should be used to determine which recovery techniques would give data necessary to achieve the goals of excavation, and that all procedures relating to recovery and the biases these introduced should be clearly reported (see Aschenbrenner and Cooke 1978; Diamant 1979).

The debate of "to sieve or not to sieve," to borrow Barker's (1975) paraphrase, has been briefly outlined here because it was part of the force behind development of water recovery systems that not only increased the recovery of larger artifacts but also allowed recovery of seeds and charcoal lost even when $\frac{1}{8}''$ (3.0-mm) screen was used. The first system described in print is the Ankara machine (named for its development at the British Institute of Archaeology at Ankara; Fig. 2.5). French (1971) developed the water sieve for use at Can Hasan III, Turkey, (1) to recover macroscopic materials of all kinds in a practical, efficient, unbiased manner; (2) to use standardized, mechanical means, so that the method is repeatable; (3) to combine in one process the separation and recovery of floatable and nonfloatable materials; and (4) to allow bulk sieving of all excavated earth if desired (French 1971:59). The Ankara machine utilizes constant water flow through two open units, the main box and the flot box, to break down soil, float off buoyant materials, and clean nonbuoyant remains. Floating material is washed from the main box into the flot box, where it is collected in a 1-mm screen. French reported some problem with silt clogging the flot box screen. An apparatus called an "elutriator" was developed by Weaver (1971) to clean the flot fraction further. The elutriator, a type of ore dresser, is used to separate particles of different densities by adjusting the rate at which water rises in a column. Field use of this device proved difficult, but a version was used successfully at Nichoria to wash silt from samples (Aschenbrenner and Cooke

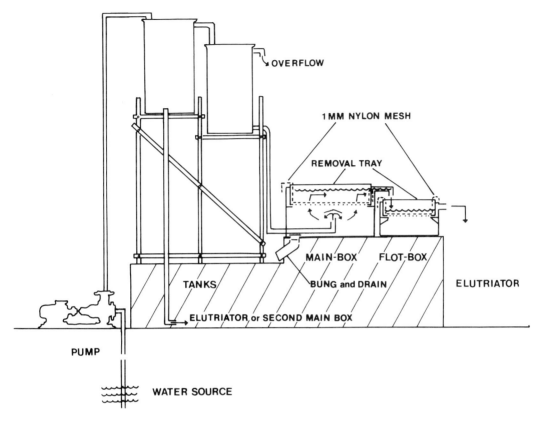

Figure 2.5 Schematic of the Ankara machine (from French 1971).

1978). Most versions of the water sieve omit the elutriator, relying instead on hand rinsing or adjustment of water flow to minimize silt in the light fractions.

The water separator (Fig 2.6) used at Jacobsen's Franchthi Cave excavations (1967–1976) and other Mediterranean sites (Diamant 1979; Limp 1974) is very similar to the Ankara machine. The version used at Franchthi Cave, called the Indiana Machine by Watson (1976), did not include an elutriator. This unit used slightly saline water, since its water source was a spring located near sea level. Limp (1974) suggested adding a sloping floor to the main box to aid in sludge removal; the idea was adopted in a number of later versions of the machine.

The final two systems to be mentioned here are variants of the Ankara machine which were more transportable and cheaper and required less-specialized construction. The Siraf machine, described by Williams (1973), was developed during the

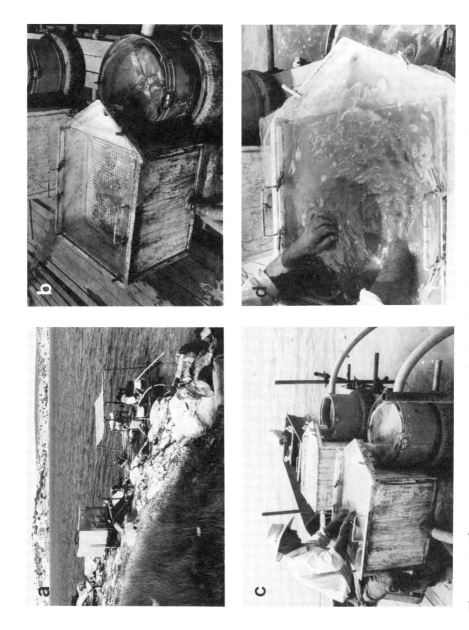

Figure 2.6 The water separator system used at Franchthi Cave: (a) water separators and reservoir; (b) main box (left) and flot box (right); (c, d) in operation, agitation and overflow are assisted by the operator (from Diamant 1979, and project archives; photographs provided by T. Jacobsen).

Water Recovery: Flotation Techniques • 31

1972–73 field season at the Siraf site in southern Iran. The Siraf machine used a 55-gallon drum as the main box. Water flowed from beneath, removable 1.0-mm nylon mesh was used in the main box and flot box screens; the latter used water to cushion floating materials. Modifying a 55-gallon drum to hold the necessary plumbing fixtures proved easier than welding a main box.

The Izum, developed at the Stobi site in Yugoslavia (Davis and Wesolowsky 1975), is probably the most primitive of the water separators. It is almost identical to the Siraf machine; the water outflow is not welded onto the 55-gallon drum but is instead held manually beneath the main box screen by means of an L handle (Fig. 2.7). This system, like all the water separation systems described here, utilizes pressure generated by the fall of water from an elevated reservoir to break up soil samples. Reservoirs are filled by a gasoline-driven pump, which is run only when the water supply must be replenished. Since the Izum is not welded, it is easily constructed and the key parts are transportable. Although recovery rate tests have not been reported for the Izum, I suspect that the use of a hand-held water outflow

Figure 2.7 The Izum flotation system, developed for use at the Stobi site, Yugoslavia (from Davis and Wesolowsky 1975; Stobi Archives Negative 75-146-6A, photograph provided by A. Wesolowsky).

may introduce enough of a human error or fatigue factor to make this system somewhat less efficient than a comparable completely mechanized system.

Froth Flotation

The late 1960s also saw development of a technique to recover charred botanical remains which did not rely solely on the principle of density separation in water. It had been observed that some charred plant remains, being denser than water, did not float even when water pressure from below was used to break up soil. Refloating heavy fractions in heavy liquids such as $ZnCl_2$ or CCl_4 can recover those remains that sink and do not pass through the main box–flotation bucket screen, but what of small, nonbuoyant materials lost through the $\frac{1}{16}''$ (1.0-mm) screen, the time and expense of refloating samples if much large dense material is present, and the obnoxious nature of many heavy fractionating liquids?

The froth flotation cell, or Cambridge machine, was designed to overcome the limitations of water and chemical flotation (Jarman *et al.* 1972). Although the cell creates some new problems (outlined by Williams 1976; see below), it affords excellent recovery of small botanical material and impressive quantities of soil can be processed.

The notion of using froth flotation to separate charred botanical materials from soil came from use of this technique in industry to remove coal from associated shale or other oxides. In froth flotation, charred remains are floated out of the soil matrix after being coated with a "collector," often kerosene, which is added to the water before flotation begins. The collector increases the contact angle between remains and rising air bubbles, which causes remains to become attached to air bubbles and to be lifted to the surface. Addition of a frothing agent to the water lowers air–water surface tension so that the air bubbles can cluster on the surface without coalescing.

The froth flotation cell as originally published has been used successfully at a number of sites. Some modifications of the original design have been made, however. Crawford (1983), for example, modified the flotation cell chamber and settling tank (Fig. 2.8; see Fig. 2.27). These changes are discussed in more detail below.

Machine-Assisted Flotation in North America

I have outlined the parallel developments in flotation techniques which occurred in North America and in the Old World, primarily in England, during the 1960s and early 1970s. These two processes were brought together in the mid-1970s when American researchers led by Watson introduced the use of machine-assisted systems into the United States. Variations on the SMAP machine (Shell Mound Archaeological Project), designed by Robertson and described by Watson (1976), are now commonly used at sites in the New World.

Water Recovery: Flotation Techniques • 33

Figure 2.8 The froth flotation devise used by Crawford to recover botanical materials from Jomon period sites in Japan (photograph by G. Crawford).

The SMAP machine (Fig. 2.9) is a version of French's Ankara water sieve, similar in design to the Siraf machine developed by Williams. Watson credits the European machines with the inspiration for design of the SMAP machine and the clayey sediments of the Green River sites for the impetus behind its development (1976:82). Finding that the "garbage can" system used successfully at Salts Cave could not handle the sediments of the Green River sites, Watson and her co-workers experimented with the machine-assisted system during 1974 and 1975.

Several hardware differences exist between the Ankara and Siraf systems and the SMAP machine and its variants. The SMAP machine utilizes a rigid barrel insert in place of the frame and removable soft mesh common to all the European systems (Fig. 2.9b). The SMAP main box insert is designed like the flotation bucket or tub of the Struever manual flotation system. The heavy fraction accumulates on the screen of the insert, which is removed as a unit, and the material is dumped out. The SMAP system does not have a flot box, as this term was originally used. The floating material carried over the sluiceway is caught directly in geological sieves held in a bucket clamped to the sluiceway or positioned beneath it. The inflow of water to the SMAP system also differs from that in the European systems. Rather than using

Figure 2.9 The SMAP flotation system: (a) original machine in use by Patty Jo Watson and Gail Wagner on soils from Green River sites; (b) a rigid barrel insert, with screen bottom, replaces removable mesh used in European systems; (c, d) the flot box is replaced in a SMAP system by geological sieves held in a bucket. [Photograph (a) provided by P. Watson.]

water flow generated by the fall of water from storage tanks elevated above the system to clean the soil samples, the SMAP machine uses pumped water under pressure, with rate of flow regulated by valve. A gasoline pump works throughout the flotation process. Finally, the SMAP flotation barrel (main box) has a baffle or sloping base and a large-diameter sludge drain to facilitate desludging.

A number of changes in the original SMAP design have been suggested to improve its operation. These are discussed in the next section.

Building and Operating Flotation Systems: Sample Designs

In this section we get down to the basics: how does one actually build and operate a flotation system of the Apple Creek, SMAP, or Cambridge type? In presenting the diagrams, procedures, and recommendations associated with these systems, I relate my personal experiences but also rely extensively on the experiences of those who designed and built the basic systems and those who participated directly in the innovations of the second- and third-generation systems. The treatment of each system is organized as follows: (1) equipment description, (2) processing procedures, and (3) hints for good recovery.

Manual Flotation

For this most basic of all flotation techniques I focus on outdoor systems designed to process soil in quantity. The reader can refer to Schock (1971) for one way to adapt flotation to the lab, but most of the equipment and procedures described here can be adapted to inside use if the primary constraints of sludge removal and water source can be dealt with.

Equipment

Flotation tub or bucket. The function of the flotation bucket is to create an enclosed area of water into which soil can be introduced and agitated, light fractions contained, heavy fractions caught, and soil washed away. The bucket should be light and small enough for one person to manipulate. It must be constructed of material that can tolerate repeated immersion in water and that is strong enough to withstand the weight of heavy materials on the screen, thumping to remove material, and the general wear and tear of a field season.

The choice of size, shape, and material for the frame of the flotation bucket depends on availability of materials, planned duration of use, and the particular technique and water source to be used. The Apple Creek flotation system (Struever 1968) utilized galvanized steel washtubs, square in shape, of unstated dimensions. Tub bottoms were cut out and replaced with heavy-duty $\frac{1}{16}$" window screen supported by two crossed steel rods welded to the bottom. Circular washtubs of similar size, with welded screen bottoms and side handles, are commonly used in barrel flotation, where a circular shape is more practical. Experience with both square and circular flotation buckets has convinced me that a square design is not as easy to manipulate as a circular one. There is more water resistance on the outside of a square tub, and a rotating agitation to wash soil creates more turbulence inside the tub, which increases the danger of washing floating materials out.

Washtub-size flotation buckets are useful for processing samples in the larger volume range (e.g., ½ bushel as at Apple Creek; more than 10 l, in general), since there is more water volume to break up the soil and more room on the screen to hold the heavy fraction. However, standard volume galvanized buckets (Fig. 2.10) make adequate flotation devices for processing small samples (1–10 l) or samples from which larger artifacts and rocks have been removed by prescreening. Construction is the same as for washtub buckets. Plastic buckets are often useful for short field seasons or in areas where welding facilities are not available. One removes the bottom of a thick-gauge plastic bucket and wires a screen to it. Such buckets do not have a tight seal at the bottom, and small bits of stone must be washed out of the folds of the screen between samples.

One can also make flotation buckets rather than modify a manufactured bucket or tub. For flotation at the Krum Bay site, U.S. Virgin Islands, we fabricated a bucket from wood sealed with fiberglass. A $\frac{1}{16}''$ mesh screen was attached to the bottom. Although wood was used so that the bucket would float, ocean floating proved to be

Figure 2.10 Manual flotation bucket: small-mesh wire cloth (0.5 mm) soldered to the bottom of an aluminum bucket.

quite fatiguing. As mentioned above, the square design had other drawbacks as well (Pearsall 1983).

Wagner (1976, 1977) designed wooden flotation buckets for the IDOT and Dayton Museum flotation systems. An IDOT bucket is made from 1" white pine, joined by wood screws. In the original design, it is a four-sided box, 10" × 10" on a side, with two sides of solid wood and two with screen inserts (approximately 5" × 8"). The bottom was covered by fine wire cloth supported by ¼" screen. More recent modifications include use of screen inserts on all sides of the box or use of a box that is virtually all screening, such as that in Figure 2.11. A variety of mesh sizes have been used in IDOT flotation boxes. As reported by Wagner (1982), the size of mesh in the sides and bottom of the box affects rate of recovery of remains. A mesh appropriate to the nature of the soil matrix should be chosen. Fine, silty, or clayey soils allow use of finer mesh (0.42 mm, #40 mesh) on both sides and bottom; sandy soils limit bottom mesh size to 0.59 mm (#30 mesh) or window screen (1.0 mm). Our University of Missouri IDOT system utilizes 0.5-mm wire cloth for both sides and bottom, so that the system is compatible with our SMAP machine. The IDOT/Dayton Museum system requires vigorous shaking of soil in the flotation bucket rather than a circular or side-to-side motion to wash soil. This difference of motion makes screen size in the bottom of the box a particularly important consideration, since seeds that can pass through the mesh are more likely to do so when soil is forced down onto the screen as the bucket is shaken.

Figure 2.11 IDOT-style flotation device with 0.5-mm wire cloth on all but two sides; the wire cloth is supported on a wooden frame.

Hand sieve. The hand sieve is used to scoop floating material from the flotation bucket. In laboratory flotation systems such as that described by Helbaek (1969), a fine-mesh sieve was used to catch the floating material as the flotation water or chemical was slowly poured out of the small basin in which flotation occurred. Most manual systems now use a sieve with a handle to scoop material.

Various materials can be used to make flotation scoop sieves. Small tea strainers were used for the Apple Creek system. Tea strainers come in many diameters, so sieve diameter can be fitted to the size of the flotation bucket. I recommend using flotation scoops of smaller mesh size than that available in tea strainers, however. Brass or copper wire cloth, available in a variety of small mesh sizes, is an excellent, durable material for constructing a flotation sieve. One can make a shallow sieve by attaching a circle of wire cloth to a metal hoop (bent wire, such as a coat hanger) whose ends are formed into a handle or inserted into a handle (Fig. 2.12). Wire cloth is easily cut with scissors and can be stitched with fine wire or sturdy thread onto the hoop. A tea strainer can also be used as a base, with wire cloth fitted into it and attached at the edges. Wagner (1977) has used #40 mesh (0.42 mm) in hand sieves; I usually use #60 mesh (0.25 mm). Because wire cloth for hand sieves and flotation buckets can be difficult to find (it is sometimes carried under the name "carburetor mesh"), order ahead and take several pieces into the field. (Wire cloth is available in the U.S. from Tyler, the manufacturer of geological sieves.)

Wire cloth is not the only satisfactory material for making a hand sieve. A quick, short-term solution is to tack a piece of nylon stocking onto a standard tea strainer. A length of stocking leg, long enough that the sieve of the strainer can be inserted into the stocking, is tacked to the sieve on the inside with thread. The edges are rolled over and sewn down, with any excess nylon gathered and secured. The edge of the sieve used for tapping wears rather quickly, so it should be rein-

Figure 2.12 Hand sieve of 0.25-mm wire cloth on a wire frame is used to scoop the light fraction.

forced with stronger material or mended periodically. The nylon must be replaced fairly frequently (we replaced the nylon two or three times for 100 samples at Krum Bay).

Minnis and LeBlanc (1976) report successful use of wet cloth diapers in the hand sieve. In this system, the sieve serves the dual purpose of scooping and containing the botanical materials; they are not tapped out onto cloth or paper as they are with the sieves described above. When all material is scooped off, the diaper is removed (the cut and shaped cloth is held by a rubber band onto a strainer) and the sample dried in it. Minnis and LeBlanc report no difficulties scooping surface material using the cloths; it may be difficult to scoop beneath the surface for semibuoyant materials. For cloth or wire mesh to work in a hand sieve, the mesh must be large enough to allow water to pass through easily but small enough to keep seeds from being lost. I advise a mesh 0.5 mm or smaller.

Water container. If an open body of water is not to be used for flotation, a third piece of equipment, a barrel or tank to serve as water reservoir, is needed. There is little to add by way of specification. Specially constructed tanks, watering troughs, 55-gallon drums, garbage cans, and basins all work. Selection depends on such things as local availability, whether transporting the system limits its size and weight, the size and number of samples to be floated, and whether water is scarce. Smaller containers can be dumped for sludge removal; larger ones can be shoveled out after water is siphoned or drained; both are messy jobs. For a few small samples, a small container, dumped after every sample or two, is an easy solution. For larger numbers and volumes of samples, two or three 55-gallon drums used in rotation allow flotation to proceed all day without delays while silt settles and require sludge dumping at less frequent intervals. Minnis and LeBlanc report floating 60–75 l of soil per drum (using two drums alternately) before dumping. The watering trough used in the Dayton Museum system can hold 480–640 l of soil before being drained and cleaned, with two flotation stations in use at once. Larger volumes of water tend to spread silt out, allowing flotation of more soil before tanks must be allowed to settle. The formation of dirty froth on the surface and increasing quantities of silt in the scoop indicate that settling is needed. Sludge drains are a useful addition to a barrel- or trough-style reservoir. Metal garbage cans may have to be sealed with silicon caulk to keep seams from leaking.

Processing Samples

I begin with a generalized, two-person manual flotation procedure, one that assumes that a flotation bucket with 0.5-mm mesh is being used in a drum.

Preflotation preparations

1. Clean all equipment; be sure that the flotation tank has settled and that its surface is free of debris or material from the previous day's work. Correct the water level.

2. Assemble all materials for processing on a work surface (Fig. 2.13a) near the flotation station: indelible pen for labeling heavy fraction newspaper and light fraction papers or tags; supply of newspaper, two pages closed and folded in half, with a clean surface on the inside, or a supply of clean cloth squares (15" × 15", muslin or similar tightly woven but water-permeable cloth); notebook or clipboard of forms for recording flotation data; pens or pencils; rocks or other weights to hold flotation cloths or papers; measuring scoop or bucket marked in liters; plastic sheet for holding measured soil; and tags with string. If cloth squares are used to hold light or heavy fractions, a clothesline for hanging samples should be put up in the shade. If newspapers are used, beer or cola flats, cafeteria trays, or the like are helpful to hold the wet papers, which should be weighed down while drying.

3. Assemble and order soil samples to be processed during the flotation session (Fig 2.13b): locate all bags belonging to the same sample and place them together; determine flotation order (if ordering by date, flotation number, provenience, etc. has been decided upon); and open bags to check that soil is dry and reasonably friable. Remove wet samples or samples hardened into clay peds which require preflotation soaking.

Flotation

1. Select sample and enter provenience data, flotation number, and any other required information (determined by consultation with the project archaeologists) into the flotation notebook or form. Note flotation personnel for the day.

2. If a standard soil volume is not being used, or if a sample deviates from the standard, measure soil volume (Fig. 2.14). Measure out soil using the liter measure and enter the total number of liters present in the notebook. If a standard volume (or weight) is being used, this figure should be recorded.

3. Spread out the light fraction cloth square or newspaper labeled with the sample's provenience or flotation number on the flotation work surface and anchor it securely. Prepare the newspaper for the heavy fraction.

4. The person handling the flotation bucket (the "agitator") immerses the bucket about half its depth in the water and begins agitation. For a round bucket, this agitation should be a circular motion, with the bucket held level, turned clockwise 90° or so, then counterclockwise 90° (Fig. 2.15). For an IDOT-style device (screen on sides and bottom), a back-and-forth motion is used (Fig. 2.16).

5. Once agitation begins, the second person (the "pourer") slowly pours soil into the flotation bucket (Fig. 2.17). After several liters have been

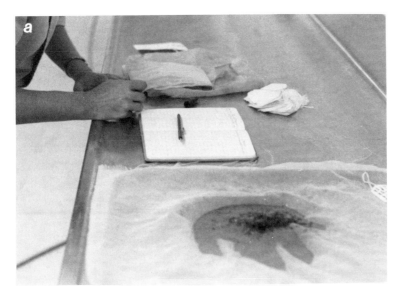

Figure 2.13 Preparation for flotation: (a) necessary materials for recording, labeling, and holding samples are assembled; here, flotation samples are entered into a notebook, tags are prepared with full sample provenience information, and a cloth square is positioned to receive light fraction; (b) before starting flotation, bags are checked to be sure soil is dry, samples are properly ordered, and cloth squares for light and heavy fractions are securely anchored.

Figure 2.14 Measuring soil volume prior to flotation.

poured, the pourer waits while the agitator works some of the soil through the system. After a few minutes, bucket weight will decrease and sounds of rocks or sherds on the screen will usually be heard, signifying that more soil can be poured in. This procedure continues until all soil is poured, or, for large samples, until the bucket sieve fills. I usually pour a 10-l sample in two or three parts; the nature of the soil matrix determines how rapidly it may be introduced into the water. Agitation must continue throughout the pouring stage, especially if bucket mesh is larger than 0.5 mm.

6. When the agitator feels that most of the soil has worked through the bucket sieve, he or she stops bucket motion and drops the bucket down so that water is within a few inches of the top. This helps float botanical material. The pourer now scoops the botanical material floating on the surface. Scooping is done in **S** curves over the water surface, with the scoop held somewhat upright, pushing as well as scooping the remains. At the end of the last **S**, the remains are lifted out. The scoop is emptied by rapping its upper or side edge sharply on the flotation cloth or paper while holding the scoop face downward. One or two raps should knock off all charcoal. The scooping is repeated several times, until most floating mate-

Building and Operating Flotation Systems • 43

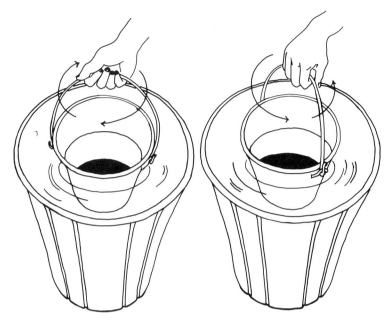

Figure 2.15 Rotation used to agitate a soil sample in a screen-bottomed bucket flotation device. The bucket may be held by its handle or its sides.

Figure 2.16 A back-and-forth motion is used to agitate a soil sample in an IDOT-style flotation device. An up-and-down motion can also be used, but this creates more turbulance in a small water barrel.

Figure 2.17 Manual flotation: (a) the flotation bucket, here an IDOT-style device, is placed in the water; if window screen mesh is used in the bucket, it is important to begin rotating it while the sample is poured in; a bucket with 0.5-mm mesh may be held stationary; (b) the soil sample is poured slowly into the bucket; (c) released floating botanical material is skimmed off with a hand sieve; (d) the light fraction is tapped off onto a cloth square; (e) a subsurface scoop recovers material floating near the bottom screen.

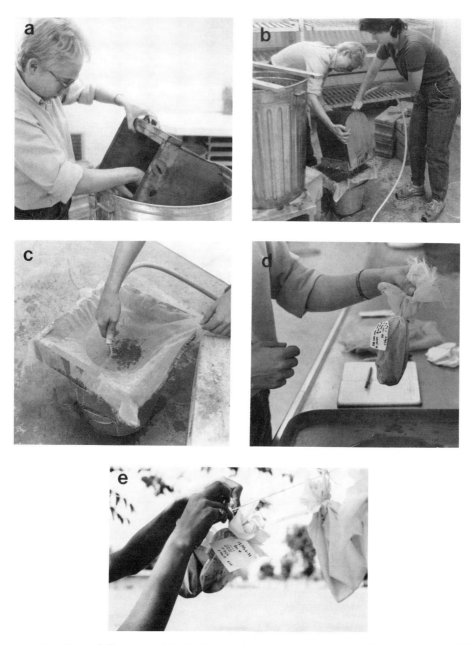

Figure 2.18 Manual flotation: (a) the heavy fraction is concentrated on one side of the bucket; (b) heavy fraction material is dumped onto a cloth square held in a screen box; all residue is washed into the square from the bucket; (c) heavy fraction material is washed with a gentle stream of water into the center of the cloth square; (d) heavy and light fraction squares are tied closed and provenience tags are attached; (e) cloth bags are hung in the shade to dry.

rial is removed. The pourer then signals the agitator to resume bucket rotation. Agitation–scooping–agitation–scooping continues until negligible material rises to the surface. If mesh larger than 0.5 mm has been used in the flotation bucket, it should not be raised out of the water until floatation is completed. If scooping is interrupted or delayed, the bucket should be agitated to keep small seeds from being carried out through the bottom screen.

7. A final subsurface scoop is done when the sample is almost complete. The agitator raises the bucket almost out of the water, then drops it very rapidly, forcing semibuoyant material to rise off the screen. The pourer then scoops these materials up. The procedure is repeated until no charcoal remains. A siphon may also be used to remove semibuoyant material from the bucket screen (see "Machine-Assisted Flotation," below).

8. After the sample is finished, the agitator consolidates the heavy fraction. Dip the bucket in and out of the water at a slight angle, concentrating material on the screen at one end. Upend the bucket and rap the edge on the heavy fraction sheet (Fig. 2.18); most material will slide out. Remove stray pieces by hand or by gentle washing. Take the heavy fraction to a shaded drying area with good air circulation. Thoroughly wash out the flotation bucket.

9. The pourer consolidates the light fraction. If cloth squares are being used, carefully gather up the edges of the square so that charcoal is not caught in the folds (tapping out should be done in the center of the sheet). Tie a provenience tag on to close the cloth, then hang the bundle on the drying line. If newspaper, paper towels, or the like have been used, fold the paper to secure the sample, take it to the drying area, and weigh it down securely. I prefer cloth squares, since they can be hung to dry (avoiding problems of wind, animals, or small children) and permit rinsing of residue on the hand sieve.

10. The final step in flotation is to note in the flotation notebook or sample form any problems or observations on processing. Such notes might include an estimate of charcoal and seed abundance, any unusual or noteworthy items in the heavy fraction, any accidents during flotation which influenced recovery, and whether chemical flotation of the heavy fraction (charcoal in the heavy fraction) or soaking and reflotation of the heavy fraction (clay peds) is necessary.

Postflotation cleanup. After the day's flotation is done, all heavy and light fractions should be taken to a secure location for final drying. Dry samples can be put into permanent storage containers: vials, baby food jars, plastic or paper bags, or boxes, with proveniences checked and transferred. Samples must be completely dry, or residual moisture will support growth of bacteria or fungi within the airtight container. Papers or cloths are brushed clean for reuse or new ones prepared. Heavy fractions that require soaking or chemical flotation are set aside (see "Problem Soils" and "Chemical Flotation," below).

Notes on the day's flotation are made. These might include names of flotation personnel, number of samples processed, soil volume floated, time required, problems encountered, changes in procedure to be implemented, sample numbers of recovery efficiency tests, if carried out, and so on.

Samples for which the field crew needs immediate feedback are examined, either cursorily for an idea of recovery efficiency, or in detail, depending on what information is needed.

Finally, all equipment is checked and any repairs to the flotation bucket or hand sieve carried out. If a barrel or trough system is used, the need for sludge removal is evaluated and cleaning and refilling scheduled.

Variations on the generalized procedure. Although the Struever (1968) Apple Creek system is cited as the prototype manual flotation system, and rightly so, that procedure differs somewhat from the generalized procedure presented above. The major differences in the Apple Creek procedure are the following:

1. The flotation tub is held so that the screen is only 1" or so below the water while soil is poured in. Agitation proceeds as described above.
2. After soil is washed away, the tub is raised out of the water, then dipped in (Struever 1968: Fig. 2, suggests that one corner of the tub is dipped in and out). A second individual scoops off material, both that on the surface and that suspended temporarily beneath the surface. This process is continued until examination of the tub screen shows that all bone and charcoal have been removed.
3. Chemical flotation is used to remove charcoal from bone in the light fractions.
4. Tea strainers are used as the flotation scoop.

This procedure has three drawbacks. Use of a tea strainer does not allow capture of seeds in the smaller size ranges. The use of a very shallow water level in the tub and the repeated dipping of the tub in and out of the water washes all material smaller than the $\frac{1}{16}$" mesh openings through the bottom screen. The inclusion of bone in the light fraction means that all charcoal (if all light fractions are chemically floated) is subjected to $ZnCl_2$ immersion, rewashing, and redrying. By contrast, if bone is left in the heavy fraction, then only the nonbuoyant charcoal in this fraction is subjected to chemical flotation and potential damage.

Minnis and LeBlanc (1976) describe a procedure that differs in the following ways from the generalized procedure described above:

1. A cloth attached to the hand scoop is used to both lift out and retain the light fraction.
2. The flotation tub is hooked onto the top of the barrel, so that it is mostly submerged, while soil is introduced. The water is agitated during soil pouring, presumably with the arm or a stirring stick of some kind. The tub is unhooked and rotated after all the soil has been poured.
3. The tub is rehooked to the barrel while the light fraction is scooped.

This procedure is a good example of flotation carried out by one person. Hooking the tub to the barrel allows an operator to pour in soil and stir the water at the same time. Figure 2.19 illustrates a similar system. Likewise, after rotation brings floating material up, rehooking the tub to the barrel allows the operator to scoop out material. Although the procedure does not mention reagitation during the scooping

Figure 2.19 A manual flotation system can be operated by one person. In this example, the IDOT-style flotation device is held in the barrel on two boards.

phase, this could be easily done by unhooking the tub again. As mentioned in the discussion of equipment, the use of a cloth attached to the scoop may inhibit scooping of semibuoyant material or repeated scooping of particularly abundant samples, leaving more charcoal in the heavy fraction and perhaps leading to some loss of small seeds. No recovery tests have been reported for this method, so efficiency of recovery cannot be compared directly to the other variants.

The procedure described by Stewart and Robertson (1973) is quite similar to the generalized procedure presented here. A large flotation tank divided by a baffle into two compartments allows two flotation stations to be operated at once. Carburetor mesh (fine wire cloth) sieves and circular flotation buckets with $\frac{1}{16}''$ mesh in the bottoms are used. A back-and-forth motion of the bucket, rather than circular rotation, is used for agitation. Both surface and subsurface scooping are carried out. These authors report that kerosene added to the water of the tank reduces foaming caused by silt suspension in water. Rather than drying the light fractions in shade, they cover them with wet newspaper to slow drying and place samples in the sun.

The final manual flotation system I discuss is the system developed by Wagner (1976, 1977, 1979), called variously Dayton Museum, IDOT, or Incinerator and used in a number of variations in the Midwest (Figs. 2.17, 2.18, and 2.19). Wagner's original procedure is a one-operator flotation system. A floating wooden flotation device, or box, allows the operator to agitate the sample during the scooping phase. The box is kept still as soil is added. This system uses up-and-down or side-to-side agitation, which is recommended only when the screen in the sides and bottom of the box is smaller than archaeological seeds or other floating materials of interest. Wagner states this dependence on screen size clearly in her discussion of testing recovery rates: "The range of recovery rates within the Incinerator Site System (Wagner 1979) depends upon the use of different screen mesh sizes in the float boxes" (1982: 129). As I discuss below, this applies to all manual flotation systems.

The IDOT system has been shown to be efficient and effective in several large projects, including the FAI-270 project (Bareis and Porter 1984), where thousands of flotation samples were processed. Because many samples had to be soaked before flotation to disperse heavy clays, which resulted in waterlogging of seeds and charcoal, the fine mesh of the flotation box (0.42-mm mesh in this case) assured consistent recovery of all size fractions of remains, regardless of their buoyancy. We now use an IDOT system at the Missouri lab for all indoor flotation. Our system uses 0.5-mm wire cloth in the flotation bucket; agitation is a combination of up-and-down and side-to-side motions. Although using standard $\frac{1}{16}''$ window screen in flotation buckets gives reasonably effective recovery of remains, especially if material is abundant in deposits (and window screen is the only thing available), I recommend using smaller mesh when possible. Wire cloth can be attached easily to a standard bucket and used in place of window screen, or an IDOT-style box can be constructed. Rate of recovery of small seeds can rise dramatically when mesh smaller than $\frac{1}{16}''$ is used.

Hints for Good Recovery

In this section I summarize some points that can make or break application of a manual flotation system.

Monitor the condition and nature of the soil. It is a widely recognized fact that dry, friable soil of sandy-loamy texture is the ideal matrix for manual flotation. Soil should always be completely dry prior to flotation. Any large lumps should be gently broken up by hand before soil is poured into the flotation bucket. Damp excavation soil can by dried by using cloth flotation bags or by leaving plastic bags open. In either case, soil should not be tightly packed. Wet soil may have to be spread on plastic sheets or newspapers to dry.

Soil with high clay content is the nemesis of manual flotation. Although the finer particle size of clays and silts allows the use of smaller bucket meshes, dried soil may consist of hard peds, which sink and require long periods of agitation to break up; even so, complete washing is rarely achieved. Clay content as low as 3% of the total can cause these problems (C. Bodner, personal communication, 1984), and soils with 40–50% clay, such as those processed from the El Bronce site, can only be described as a nightmare to float (Pearsall 1985). If drying soil results in numerous large peds that cannot be broken up by hand, then preflotation soaking of soil in a defloculant is necessary to disperse clay. Although Williams (1976) reports that breaking up clay peds with a hammer did not affect quantity of charcoal recovered by flotation, one must question the quality and identifiability of charcoal freed in this manner. Procedures for preflotation soaking are discussed below. Soils that have been screened prior to flotation and have dried into small clay granules may break up when processed with longer agitation time and slower pouring of soil. It is best to carry out flotation tests on typical site soils to determine whether special treatment is required. Preflotation soaking waterlogs charcoal and seeds and decreases flotation efficiency. It should not be done unless necessary, since an extra procedure, chemical flotation of heavy fractions, may then be required. Using fine wire cloth mesh in the flotation bucket minimizes loss of small waterlogged materials.

Set procedure standards and monitor performance. All manual flotation systems depend primarily on the skill and consistency of the operators for their success. Equipment breakdown is a rare source of problems. In a project of long duration or high volume, flotation personnel should be trained to follow a set procedure, supervised throughout by a crew chief, and rotated periodically to reduce error due to the fatiguing and repetitive nature of processing. Crew manuals such as Wagner

(1977) and Bohrer and Adams (1977) are excellent. Even a short direction sheet or checklist kept at the flotation station can help improve consistency of processing. Some points in the procedure where human error can reduce recovery include too rapid pouring of soil, insufficient agitation during pouring, raising the bucket screen out of the water before flotation has ended, sloppy agitation resulting in the loss of floating material into the water, breakage or loss of charcoal by inaccurate scoop tapping, premature termination of flotation without deep scooping or sufficient cleaning of floating material, and contamination between samples by not washing equipment and cleaning enclosed water surfaces.

Here is an example from my own experience. When archaeologist Emily Lundberg and I knew that a recovery rate test was under way during the Krum Bay project flotation, we could achieve 67–88% (average 79%, 5 tests) seed recovery. But in blind tests, recovery dropped to 40–82% (average 56%, 6 tests) (Pearsall 1985). We were using the same screen mesh, same type of soil, and identical flotation procedure. Fatigue and variation in length and vigor of agitation, pouring rate, and number of agitation–scooping cycles—all factors dependent on operator performance—accounted for these recovery differences. Using fine wire cloth in the flotation bucket can help minimize seed loss, although material can still end up in the heavy fraction with insufficient agitation and scooping.

Keep accurate and up-to-date records of flotation. I have never heard an archaeologist or ethnobotanist complain about having too many notes from the field phase of a project. Which screen mesh sizes were used, and why, sample size standards, flotation procedure steps, and changes in procedure are among the basic flotation field notes which must be kept to facilitate later analysis. If experiments are run to determine the best soil sample size for adequate charcoal and seed recovery, these should be described and the decision-making process recorded. Deviation from the standard should be noted and explained. If experiments with different bucket or sieve meshes are carried out, these too should be described in detail. Results of recovery rate tests need to be monitored and recorded. Accidents or other factors affecting samples should be noted.

Guard against postflotation sample damage or loss. Too rapid drying of wet charcoal can lead to breakage, decreasing identifiability of samples. Drying in a shaded area—or in sun, under wet paper—is recommended. Rapid drying in hot sunlight causes breakage of seeds and charcoal. Wet charcoal may literally explode. Repeated wetting and drying leads quickly to sample deterioration. Jarman *et al.* (1972:45) describe tests on 500 archaeological charred seeds which showed that, whereas only 4% were broken after a first flotation and drying, 56% more were destroyed by an additional wetting and drying, leaving only 15% of the seeds whole and undamaged. A third wetting and drying destroyed all of these. For a decision

52 • Chapter 2   Techniques for Recovering Macroremains

about use of chemical flotation, which rewets charcoal both in the heavy liquid and in subsequent water rinsing, such warnings must be considered. If samples are soaked before flotation, they should not be allowed to redry for the same reason.

Sample deterioration also occurs in the field lab when material is removed from drying papers or cloths and packaged. Material must be handled gently at all times. If paper or plastic bags are used to hold samples, these should be folded around the samples after being closed to hold material steady, packed loosely in a storage box, and padded carefully for transportation. Jars, vials, or boxes prevent crushing of samples but allow material to bounce around inside the container. Fill the empty part of the container with loosely crumpled tissue. Unsorted flotation samples tend to stick to cotton or other fibrous material. Packaging carbon samples (in situ or screen-collected material) in aluminum foil, although an excellent way of avoiding contamination of radiocarbon samples, often results in breakup of material. This is especially true of seed, fruit, and tuber or root material. Wood charcoal holds up better. I recommend opening field foil packets before shipping and adding a cushioning layer of tissue or cotton around any material not destined for radiocarbon dating.

Machine-Assisted Flotation: Water Separators and SMAP Machines

In this section I describe equipment and procedures for water separator and SMAP-style machine-assisted flotation systems and offer recommendations for their use. Although a number of versions of this system exist, there are two major variants: machines where gravity water flow is used to wash soil with a two-unit setup (main box, flot box), and machines where pressurized water flow (powered by gasoline pump) washes soil, the flot box in its original form is eliminated, and the main box is barrel-shaped.

Equipment

The main box or machine barrel functions as the water reservoir for flotation; soil is washed, nonbuoyant material caught, and light, floating materials released from the soil matrix. The main box has two components: an outer box or barrel that serves as water reservoir, and a screen insert or bucket that is immersed in water and into which soil for flotation is introduced (see Fig. 2.5). The box body is discussed here.

A key feature of the body of the main box is the pattern of water flow which breaks up soil with limited assistance from the human operator. In French's (1971) publication of the Ankara water sieve, a pipe carries water into one end of the box, near the bottom. At the center of the box the pipe turns at right angles upward and water flows out beneath the center of the screen insert and is dispersed toward the sides by a rose. From its center outflow water moves through the screen on which soil lies, breaking it up and floating up buoyant materials. These are carried to one end of the box, where an outflow carries water and floating materials into the flot box. No dimensions or details on construction of this system are given in French (1971) or in Diamant's (1979) description of the Ankara system used at Franchthi Cave.

Additional features of the body of an Ankara machine main box include a corner sludge drain and outflow pipe, flat bottom, and rectangular shape. Diamant (1979) reports that the Franchthi Cave model measured 60 × 60 cm in surface dimensions but gives no height. Medium-gauge galvanized iron, welded together to create a seamless interior, was used at Franchthi Cave to construct the main boxes. This was reported to give better results than an earlier machine made of tin sheets on a steel frame.

Water flow in the barrel variant of the water separator (SMAP or Siraf-style) is similar to that outlined above. Both Watson's (1976) and Williams's (1973) descriptions of this system include blueprint diagrams of the units, reproduced here in Figures 2.20 and 2.21. Water is carried into the barrel via a pipe from one side. In our SMAP machine, pipe height is 30 cm above the barrel bottom; the barrel is 85 cm tall and 60 cm in diameter. Placement of the inflow pipe determines the relative volume of water above and below the pipe. The larger the volume below the pipe, the more soil can be processed before sludge buildup clogs the water outflow. But a sufficient volume of water above the pipe is needed for good washing of soil and clean separation of buoyant material. The water intake pipe has its outflow in the center of the barrel. This is an angled pipe with a rose in the Siraf machine, a showerhead (fixed upright or angled toward the outflow) or similar perforated outlet in the SMAP machine. Water flow is directed at the bottom of the barrel insert— directly in SMAP, more dispersed in the Siraf machine. The distance of the outflow beneath the insert is not specified in either machine. The showerhead in our SMAP machine is immediately below the screen bottom, thus prohibiting the use of an angled showerhead, which could be rotated toward the sluiceway for better water circulation and more efficient outflow of material. After water flows through the screen, it is directed out the barrel outflow carrying any floating material with it. The outflow, or weir, in the Siraf machine is triangular in outline, flat on the bottom, and bolted 5 cm below the rim of the main box (7 cm is suggested as an improved design by Williams). This shape concentrates material as it flows into the flot box in an open stream of water. The same design is utilized in the SMAP machine. A slight downward angle improves water flow.

These barrel-type main boxes also have sludge drains (in the barrel side at the bottom). The larger drain used in the SMAP machine (4" diameter) aids in quick draining and sludge removal. A baffle in the bottom of the barrel, angled down toward the drain outlet, makes sludge removal easier but somewhat decreases the volume of soil which can be processed before the barrel must be cleaned. The SMAP

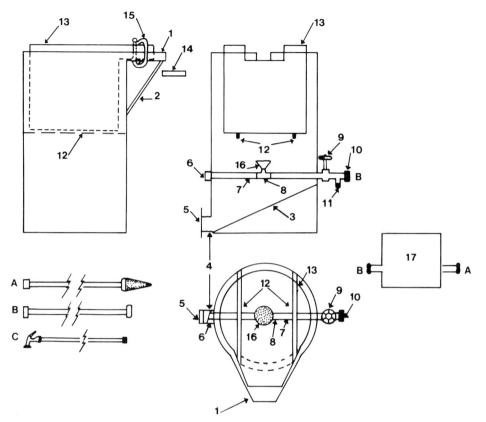

Figure 2.20 Schematic of the SMAP flotation rig (from Watson 1976). A, intake hose; B, discharge hose; C, garden and accessory hose; 1, sluiceway; 2, brace; 3, baffle; 4, threaded drain; 5, drain plug; 6, 7, 8, 16, showerhead apparatus; 9, gate valve, pump to barrel; 10, intake hose attachment; 11, discharge hose attachment; 12, angle iron supports; 13, barrel insert; 14, 15, brass geological screens and C-clamp support; 17, water pump.

machine also includes light-weight angle iron supports to hold the barrel insert. These are welded across the inside of the barrel above the water outflow showerhead. The other water separators use screen boxes suspended from the rim of the box or barrel (see below).

Main box screen insert. Two types of main box screen inserts are used in machine-assisted flotation systems: a sieve tray with removable fine-mesh screen or a rigid insert with attached screen. Both types serve the same functions: to contain the soil being washed and to catch the nonbuoyant heavy fraction.

The original design of the Ankara system (French 1971) utilizes the first type of insert, as do a number of other European machines (e.g., Davis and Wesolowsky

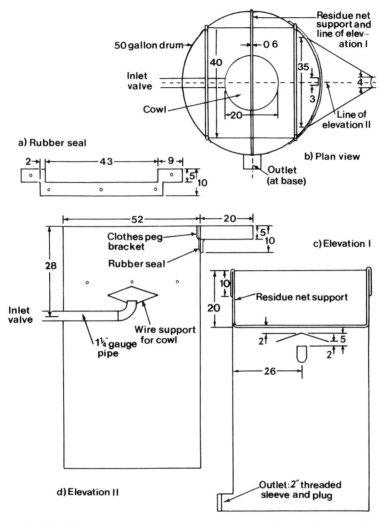

Figure 2.21 The Siraf flotation system; measurements are in centimeters (from Williams 1973).

1975; Williams 1973). A sieve tray or support made of wide metal mesh or other rigid material is constructed to fit the main box. The tray hooks over the lip of the box in such a way that the bottom is immersed into the water to a depth just above the water outflow (see Fig. 2.5). This tray provides support for fine screen or cloth mesh (referred to as the residue mesh by Williams 1973) cut to fit the tray and attached to its top sides. The most commonly used mesh size is 1.0 mm (Davis and Wesolowsky 1975; French 1971; Williams 1973). Mesh must be carefully positioned and secured so that floating material flows easily out of it and over the weir into the flot box or screen without escaping into the main box. The mesh with its heavy fraction is removed from the tray after flotation and a new piece fitted for the next sample.

One of the differences in design between SMAP-style machines and water separators is a rigid, one-piece main box insert in SMAP designs. The barrel insert is designed somewhat like a manual flotation bucket (see Fig. 2.9). Screen, usually $\frac{1}{16}''$ mesh, is attached as the bottom of a metal tub and supported by thin crossbars or heavy large-mesh screen. The original SMAP design called for the screen to be held by a metal band welded around the lower circumference of the barrel insert. Watson now suggests using a metal hoop, tightened with a screw clamp, to hold the screen to the insert. This allows easier replacement of the screen. The screen may also be soldered in place. Unlike a flotation bucket, however, the insert has a sluiceway, constructed to fit snuggly inside the barrel sluiceway, which carries water and floating material out of the main box. The barrel insert is supported from the bottom by angle iron supports welded across the inside of the barrel and positioned directly above the water outflow. The height of the barrel depends on the distance between the water outflow and the top of the unit; it is usually constructed so that the top rises out of the barrel somewhat. Insert diameter should be close to barrel diameter to ensure that most water movement is directed at the sample in the insert. Watson (personal communication, 1982) reports using a partially inflated inner tube, placed around the barrel insert, to create a tighter seal between insert and barrel. As in manual flotation, the heavy fraction is recovered by lifting the insert out of the barrel and upending it onto a heavy fraction paper or cloth.

I know of no discussion of the relative merits of the two types of main box inserts described here. The SMAP-style insert of our machine is awkward and heavy, particularly if the heavy fraction is substantial. On the other hand, the insert sluiceway, which fits snuggly into the barrel weir, ensures that all floating material is carried out of the barrel. Floating material could become trapped in the folds of a nylon mesh insert or escape from the sluiceway back into the main box. In a more recent version of the Ankara machine designed by Hillman, this problem has been lessened by addition of a removable heavy metal weir that fits inside the sluiceway of the tank and over the edge of the residue net (Fig. 2.22). The recycling aspect of this modified Ankara system is discussed below (p. 61). Use of flexible mesh might make removing heavy fractions easier and faster, reducing processing time. Mesh size can be varied in either insert if Watson's suggestion of clamping the screen to the insert is followed. The choice seems a matter of personal preference.

Flot box. As originally designed for the Ankara machine (French 1971), the flot box is a water-filled tank that collects floating materials flowing out of the main box. The Siraf machine flot box (Williams 1973) is similar. Water flows over the

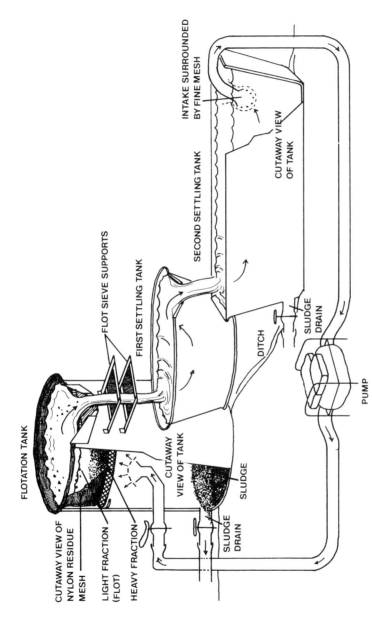

Figure 2.22 A later version of the Ankara flotation system. A metal insert (striped area) holds the residue mesh in the overflow spout. Sieve supports are welded to the barrel to hold flotation sieves. Suspended sediment settles in two settling tanks, permitting reuse of water (after G. Hillman).

sluiceway or out the pipe of the main box and falls into the water of the flot box. Nylon mesh screen attached to a small sieve or support, or a small fine-mesh hand sieve, is positioned so that water falls into it. The bottom of the screen is submerged, but its sides are above water. This feature cushions remains as they fall into the flot box and contains them within the screen. The nylon mesh or sieve is removed and emptied after each sample. Mesh size used in the Ankara flot box is 1.0 mm; in the Siraf system, 1.5 mm. Water flows out of the flot box through a drain in the Ankara machine, over the lip of the tilted flot box (a polyethylene bowl supported on orange boxes) in the Siraf machine.

In later versions of the Ankara machine, and in SMAP-style systems, the water-filled flot box has been eliminated. Water flowing out of the main box passes directly through a screen or series of screens positioned below the sluiceway. In Hillman's modification of the Ankara machine (Fig. 2.22), flot sieve supports are welded onto the flotation tank (now a 40- or 50-gallon barrel) below the sluiceway. The froth flotation rig illustrated in Figure 2.2 also uses this design. These rectangular metal supports must be strong enough to withstand the impact of water flow out of the tank and must extend far enough beyond the end of the sluiceway so that material is caught effectively. If several screens of different mesh size are used, materials can be divided by size fraction at the machine. Screens of 1.0 mm and 0.5 mm, or of 1.0 mm and 0.25 mm, have been used in the newer Ankara machine.

In the SMAP machine (Watson 1976), standard geological sieves are used as flotation sieves. One or several sieves are placed in the bottom of a screen-bottomed bucket. The bucket is suspended by its handle from the barrel sluiceway, using a large C-clamp as a support (see Fig. 2.9). The bucket is positioned so that the water coming over the sluiceway passes through the geological screens. The smallest mesh is usually 0.7 mm or 0.5 mm. In our system, we position the screen bucket on a metal stand beneath the sluiceway and use a stiff foam collar to hold the screen firmly in an open-bottomed bucket. The foam collar makes a tight seal and guides all material into the sieve. Only one screen is used (0.5 mm), since all samples are routinely divided into size fractions in the lab. Hastorf reports using hooks welded to the barrel to hold a bucket with geological sieves below the sluiceway.

I have seen no discussion of why the flot box was eliminated. If there is no difference in damage to seeds between a water-cushioned screen and an uncushioned one, eliminating the flot box reduces the number of pieces of equipment in the system and simplifies it. I know of no tests of seed damage between these systems, however.

Pumps, water tanks, and plumbing. The flotation systems discussed here all use flowing water to break up soil and release plant remains with little or no assistance by the human operator. There are several ways to obtain a sufficient force of water flow. The original Ankara machine, the version used at Franchthi Cave, and the Siraf flotation system all utilized pressure of water flow from reservoirs to wash soil. A sufficient height of water in the reservoir or the placement of the reservoir at an elevation above the main box produces enough pressure to wash soil. At the Franchthi Cave excavation, a Briggs and Stratton 2 HP, four-stroke gasoline-powered pump was used to fill two storage tanks from a spring. Two main boxes were supplied by the storage tanks, which were refilled when necessary. French (1971) reports that a flow of 2.75 m^3/hr was sufficient to wash dry, silty soil. No information on height of the reservoir or hose and pipe diameters are given. Faster flow carried silt or clay into the flot box. Davis and Wesolowsky (1975) obtained a flow rate of 12.5 l/min (0.75 m^3/hr) using a 5-m fall of water from a 400-l reservoir and 15-mm diameter hose. This flow rate was also sufficient for washing soil.

The SMAP machine and later versions of the Ankara machine utilize direct pumping of water to the main box or barrel (Fig. 2.23). This eliminates the need for a reservoir, but it requires a continuously running pump for each machine, whereas more than one machine can be run from a water reservoir. No water flow data are given for these systems, but my own experience indicates that vigorous agitation is easily obtained with a 2 HP pump and 1.5" hoses. The valve regulating flow from the pump to the barrel must in fact be closed much of the way to control agitation. Although Watson (1976) does not state explicitly why direct pumping was chosen over use of a reservoir in the SMAP machine design, it is likely that the clay soil of her research area entered into the decision. This fits with French's (1971) comment that clay soils took longer to wash than silt or loam in the Ankara machine.

Diamant (1979) notes that piped water under pressure can also be used to run a water flow flotation system. Success depends on the degree and constancy of pressure available, however. We ran our SMAP-style flotation machine from the lab water system and discovered that water pressure was insufficient to give vigorous agitation, given the large pipe diameters of the machine. An electric pump could be used to augment water flow in such a case, or smaller pipe diameters could be used.

Detailed description of pipe fittings, valves, hoses, and adapters is available in published form only for the SMAP (Fig. 2.20) and Siraf (Fig. 2.21) flotation systems. In the SMAP machine, pipe and hose diameter are 1.5", chosen to conform to the 1.5" suction and discharge openings of a 2½ HP Briggs and Stratton centrifugal self-priming pump. The SMAP-style machine built by the Escuela Técnica de Arqueología for the Peñón del Rio project (Pearsall 1982) used a 3 HP pump with 2" openings. It is advisable to use the same diameter for all hoses and pipes, not only for easier replacement of parts, but to minimize pump stress caused by changing pipe diameters along the path of water flow. Hillman's version of the Ankara machine (Fig. 2.22) uses 3–4" pipe. Use of a centrifugal pump capable of lifting about 10 m with 2–4" inlet and outlet openings is recommended. Hillman notes that the pump should be able to withstand "back press," that is, the force created when the valve

Figure 2.23 In a SMAP-style flotation system, (a) water is pumped from a water source, such as a stream, directly into the barrel; a gasoline-powered pump is commonly used. A shut-off valve (b) regulates water flow and allows water to be diverted to a small-diameter hose used for washing equipment.

controlling flow from the pump to the machine barrel is closed down. I was advised never to shut the valve to the flotation barrel completely when the pump was running without giving the water another outlet (e.g., by opening the valve to the garden hose used for washing equipment; Fig. 2.23).

In brief, the following pieces of plumbing equipment are necessary to build a water flow flotation system:

1. pump, 2–3 HP: to fill the reservoir or pump water directly to the main box.
2. reservoir tanks: for an original Ankara style system.
3. intake or suction hose: a hard plastic pipe or other noncollapsible hose to carry water to the pump.
4. foot valve: a one-way valve that prevents water from flowing out the suction hose when the pump is shut off.
5. discharge or outlet hoses: flexible hoses to carry water from the pump to the machine barrel or reservoir, or from the reservoir to the main box.
6. galvanized pipe extending into the machine barrel, holding (a) a showerhead or other water outflow inside the barrel, (b) a gate valve on the outside of the barrel to regulate water flow from the pump, and (c) an attachment for a small-diameter cleaning hose located between the pump and the gate valve.
7. small diameter hose: a garden hose with spray nozzle to clean equipment and provide an avenue for water flow.
8. clamps and adapters to attach all pipes and hoses.

Water flow flotation systems may also be set up to recycle water. To accomplish this, water flowing out of the main box or flot box must be contained after botanical materials are removed. Hillman's version of the Ankara machine uses two settling tanks, positioned below the sluiceway, for recycling (Fig. 2.22). A two-stage system allows more silt to settle in a shorter time, resulting in cleaner recycled water. Water is pumped out of the second settling tank, at the same time being filtered through 1.0-mm or finer mesh attached to the end of the suction hose, and returned to the flotation tank. Filtering the water is essential to prevent contamination among samples. Figure 2.24 shows a similar system in operation. A water recycling system is a worthwhile investment of time and money in water-scarce situations or in any setting where water outflow has to be contained (e.g., in an indoor flotation system).

Processing Samples

This discussion of processing is patterned after procedures for the SMAP system. Steps applicable to other systems have been included whenever possible. A

Figure 2.24 Recycling water for flotation using a settling tank system, Maiden Castle, Dorset, England. Two flotation machines are set up to run from a single two-stage settling tank system.

single water flow flotation system can be run by one or two operators. If several machines are run from a single reservoir, one operator is at each main box and an extra person handles paperwork and assists when needed. I have set up flotation with SMAP-style systems using both one and two operators and have gotten satisfactory results both ways. Two people can process more samples in a workday than can one person working alone, but cost in terms of worker-hours/sample rises since a team generally cannot process twice as many samples a day as a single operator.

Preflotation preparations. Steps 2 and 3 of the preflotation preparation for manual flotation reviewed earlier apply also to machine-assisted flotation systems. Additional steps specifically applicable to SMAP- and Ankara-style systems include the following:

1. Clean the main box or barrel and screen insert and check for signs of rust, metal fatigue, and broken screen. Small breaks in screen can be fixed with liquid steel or liquid plastic caulk. The barrel is susceptible to rust at weld joints, especially around the baffle; if water pools under the barrel, rusting and leakage can occur.

2. Position barrel insert or sieve tray with mesh in the machine.

3. Clean and position the flot box and its fine-mesh screen, or the screen bucket with geological screens.

4. Fill the pump with gasoline and check oil level. Change the oil if dirty; Watson (1976) recommends changing the oil daily if the pump is run full-time every day.

5. (*SMAP-style system*) Connect the suction and discharge hoses to the pump and attach the discharge hose to the barrel intake pipe. The garden hose is also attached to the barrel intake pipe. All connections must be watertight and airtight; otherwise the pump will not pull water. Apply plumber's putty to threads before attaching hoses and connectors to help produce a good seal.

(*Reservoir system*) Check the water level in the reservoir; fill tanks by pump if necessary. The reservoir discharge hose should be attached to the main box.

6. The sludge valve in the main box or barrel should be closed.

Flotation

1. Select sample and enter provenience data, flotation number, and other information in the flotation record. Note flotation personnel.

2. If a standard soil volume or weight is not being used, or if a nonstandard sample is being processed, determine and record weight or volume.

3. Label cloth squares or newspapers to hold the light and heavy fractions with sample provenience or flotation number and position them.

4. *(SMAP-style system)* Submerge the pump intake (suction) pipe in water, open the gate valve to the barrel, and start the pump. The pump may be primed by filling the intake pipe with water before starting the pump. Fill the barrel.

(Reservoir system) Open the gate valve to the main box and fill the unit.

5. Once the main box or barrel is filled with water, adjust the gate valve to establish a gentle yet rolling agitation (Fig. 2.25). Check the position of the flot box or screen bucket to make sure that water outflow passes through the light fraction sieves.

6. Pour soil slowly into the center of the barrel insert or sieve tray (e.g., above the showerhead outflow).

7. After the soil has been poured (or during this process, if two operators are used), assist water agitation by slowly stirring water to encourage soil that is sinking and collecting around edges of the screen to move to the center where water flow will break it up. Stir with arm or short blunt pole.

8. As water flow begins to break up soil and carry floating material out the sluiceway, adjust water pressure with the gate valve to maintain even water movement. In a SMAP system, as soil volume decreases less pressure is needed; conversely, in a reservoir system, dropping water levels may require opening the gate valve to maintain even agitation.

9. Once no further botanical material is observed rising to the surface and all floating material has been guided over the sluiceway, shut off water pressure by closing the barrel gate valve. In a SMAP system, the garden hose valve is opened at the same time to reduce back pressure strain on the pump.

10. With a fine wire cloth hand sieve, scoop below the water surface to capture any semibuoyant botanical materials floating beneath the surface in the barrel insert and clean the water surface of any late-floating materials. Deposit these in the light fraction sieve. Alternately, semibuoyant or nonbuoyant botanical materials can be siphoned off the heavy fraction screen using an aquarium siphon (Gumerman and Umemoto 1987). A specially shaped siphon, sold commercially to clean gravel in fish tanks, lifts only lighter material off the bottom sieve, leaving heavier remains behind (Fig. 2.26); this device can also be used during manual-flotation. If fine cloth mesh in a sieve tray was used to hold the heavy fraction, check the folds of this and guide any trapped floating material over the sluiceway.

11. Remove the light fraction caught in the sieves of the screen bucket or in the mesh of the flot box by gently washing it out onto a cloth square. Hang to air dry in the shade. Rinse geological screens or fine cloth mesh which is not used to dry samples.

Figure 2.25 SMAP-style flotation: (a, b) soil is poured slowly into the water; water agitation is assisted by stirring; the light fraction flows through the outflow into a geological sieve; (c) after flotation is completed, the light fraction is carefully rinsed out of the sieve onto a cloth square; (d) the heavy fraction, caught on the barrel insert screen, is placed on a drying tray; (e) the light fraction is hung in the shade to dry; (f) the water surface is cleaned of floating debris between samples.

66 • Chapter 2 Techniques for Recovering Macroremains

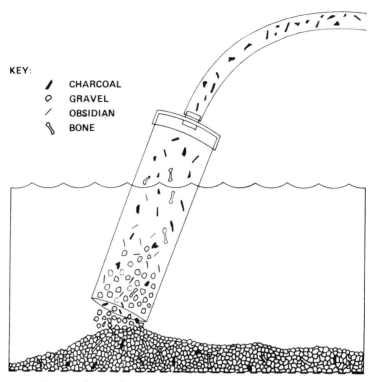

Figure 2.26 A siphon can be used to remove nonbuoyant botanical material from the flotation insert screen (from Gumerman and Umemoto 1987).

12. Remove the heavy fraction by tapping or rinsing it out of the barrel insert or removing it from the cloth mesh of the sieve tray. Tag it with provenience information, and air dry. The barrel insert or cloth mesh, if reused, should be rinsed and checked for damage.

13. If the sludge level in the barrel is high and the water dirty, open the barrel sludge valve and drain sludge and water. Thoroughly clean the unit. Otherwise, reposition the barrel insert or sieve tray and the light fraction screens for processing the next sample.

14. Note any problems or observations on the processing of the sample in the entry for that sample in the flotation record.

Postflotation cleanup. Points discussed for manual flotation cleanup apply also to light and heavy fractions produced by machine-assisted water flotation.

Hints for Good Recovery

This section is a brief summary of important points in the use of machine-assisted flotation systems which can affect the quality and quantity of recovered charred botanical remains.

Monitor condition and nature of soil. Use of pressurized water flow passing through and breaking up soil samples gives good recovery of botanical remains in a variety of soils. As is the case in manual flotation, dry loamy-sandy soils are the fastest and easiest to float. But unlike many manual methods, machine-assisted flotation systems can also effectively process fine-grained, clayey sediments. In general, a longer running time per sample is necessary, soil has to be added more slowly, and agitation must be monitored carefully so that fine sediments are not carried over the sluiceway, dirtying light fractions. Although elutriators (ore separators) can be used to separate fine soil sediments from light fraction remains, they are not easy to use in the field. Refloating dried light fractions in the lab results in seed or charcoal breakage, so it is important that operators be alert to the nature of the soil being processed and adjust processing to produce clean light fractions. If clay soil has dried into large peds, these may not break up entirely, even after a long processing run. It is best to hold such samples for preflotation deflocculation (see below), or, if they have been processed, to refloat the heavy fraction.

Check and test equipment frequently. Unlike manual flotation systems, where equipment breakdown is a rare source of loss of data or processing delays, machine-assisted flotation systems can suffer equipment failure. Because the operator provides little agitation and does not remove light fractions from the barrel, any problems with operation of the machine or damage to screens can result in a dramatic reduction of recovery. Recovery efficiency tests should be run on a frequent basis, and equipment must be inspected regularly for damage and proper functioning. Potential trouble areas include the following:

1. Poor seals at hose and pipe junctures, which cause water leakage and reduced water flow or difficulty in starting pumping action.
2. Placement of suction hose in too shallow water, which results in intake of silt and problems with clogging, dirty samples, and contamination.
3. Rusting or breakage at welds, with subsequent barrel leakage. The area around the baffle is susceptible, as is the weld where the pipe enters the barrel; do not lift the barrel by this pipe. Excessive leakage and the resulting need to increase water flow can lead to dirty light fractions and battered materials.
4. Breakage of the heavy fraction screen (barrel insert type). Cloth mesh screen should also be checked for tears and replaced when worn.

The pump is a vital part of a SMAP-type machine system and should be carefully maintained. If pumping stops during flotation, material can be lost through the heavy fraction screen.

Set standard procedures and monitor performance. Operators of a water separator or SMAP-style flotation system can process samples rapidly and effectively, giving consistently high rates of seed recovery. To ensure comparability of data, however, operating procedures must be held as constant as possible throughout the course of a project. Flotation personnel should be instructed in selection and maintenance of water agitation levels appropriate to sample soil types and monitored to be sure that flotation is continued until no floating material is present and all soil is broken up. If a sieve tray with cloth screen is used to hold soil and heavy fraction, care must be taken in its placement to ensure that no materials are trapped in the folds or lost into the main box. Placement of sieves held beneath the sluiceway to catch light fractions must be adjusted as water flow varies. Finally, cleaning all equipment between samples, and emptying sludge when it nears the level of the water inflow, is vital to guard against contamination among samples. In projects where large soil samples are floated and water is abundant, sludge can be emptied and the barrel hosed out after each sample.

Keep accurate and up-to-date records of flotation and guard against postflotation sample damage or loss. These last points, discussed in detail for manual flotation, also apply to machine-assisted systems.

Machine-Assisted Flotation: Froth Flotation

In this section, equipment, procedures, and hints for good recovery for froth flotation are reviewed. I rely on Jarman *et al.* (1972) and comments by more recent users such as Williams (1976) and Crawford (1983). The froth flotation technique has been criticized, especially by Williams (1976), because of incomplete recovery of charcoal, low sludge capacity, and the necessity of preflotation screening. Good results using froth flotation have, however, been reported.

Equipment

Flotation cell. The flotation cell as designed by Jarman *et al.* (1972) is a two-chamber cylinder constructed of 6-mm polypropylene sheeting. The upper, or flotation, chamber rests on a stand that can be removed for storage. This stand has a side cutout through which the sludge discharge chute extends. Crawford (1983) modified the bottom cylinder into an open workstand for the flotation tank (Fig. 2.27).

The bottom of the flotation chamber tapers into a cone, terminating in a spout to which a butterfly valve is attached. Jarman *et al.* recommend a cone that tapers from 46 to 15 cm in diameter over a distance of 23 cm in order to facilitate sludge removal. Williams reports, however, that sludge removal is still a laborious process,

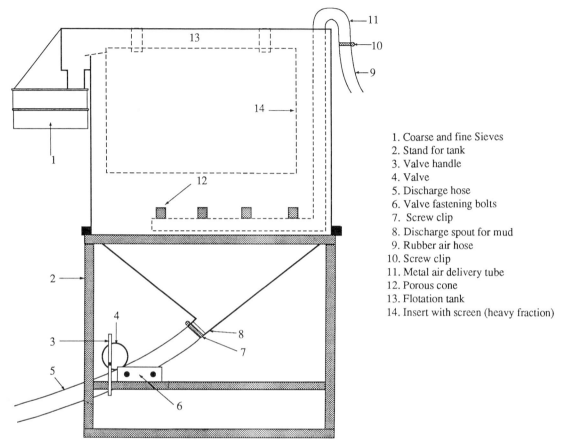

Figure 2.27 Schematic of a froth flotation device. The heavy fraction screen (14) is a recent design improvement (from Crawford 1983:53).

since sludge does not move easily through the discharge pipe. These difficulties may stem from the fact that sludge is a mixture of soil residue and heavy fraction, a bulkier mix than that from a water separator system, and that only part of the water in the upper chamber is released to wash it out. The butterfly valve allows most water to be retained while mud is discharged. Sludge must be removed more frequently in the froth flotation system; Williams reported that the Cambridge machine had only about a tenth the sludge capacity of a Siraf unit.

The top of the flotation chamber in the original Cambridge machine is surrounded by an annular moat that catches charcoal and seeds collected in froth as it overflows the chamber. The moat drains to an outflow spout, below which are positioned geological sieves. The unit used by Crawford (Fig. 2.27) does not have an annular moat; rather, overflow from the unit occurs at one point, the "lip," an outflow pipe similar to that on some water separator units. Catch sieves are positioned beneath the lip.

Bubbler unit. A key element in a froth flotation system is the production of fine bubbles to which charred botanical materials become attached and are carried to the surface. Bubbles are produced by a bubbler unit driven by compressed air. Jarman *et al.* utilized a rotary vane air pump driven by a 2 HP gasoline engine.

The bubbler unit consists of small porous metal bubblers, of sintered bronze, attached to metal tubing. According to Jarman *et al.* bubbler elements can be purchased preformed (sold as flange cones) and are available in various shapes and porosities. These authors obtained good bubble formation using twenty-four elements of 25-micron porosity distributed over a 30- × 30-cm frame of brass tubing. A rigid metal pipe extends upward from the tubing frame and curves over the lip of the flotation chamber. Pipe length is set so that when it is hooked over the tank top the bubbler frame is suspended just above the discharge cone. A flexible hose connects the bubbler pipe to the air pump. The pump should have a pressure relief valve for adjusting air pressure.

Settling Tank. All froth flotation units for which published descriptions are available incorporate a settling tank to catch outflow from the flotation cell sludge discharge. These do not incorporate means of collecting a heavy or residue fraction during flotation. Any nonbuoyant material that is not removed by preflotation screening ends up with soil sludge in the bottom of the unit. If recovery of some or all of this material is desired, sludge must be water screened after flotation. Crawford has recently added a screen insert, suspended above the bubbler unit, to catch the heavy fraction.

A settling tank in the form of a trough 1 m wide, 2 m long, and 0.5 deep was incorporated in the original Cambridge system design. This trough was divided into three compartments. A large 2-mm mesh screen was placed in the first compartment, located under the sludge discharge. Sludge was directed onto this screen and sediment washed by agitating the screen. A three-chamber settling tank was used so that, as water was displaced from the first into the second and third compartments, silts settled out and water could be recycled. Crawford's system was designed so that the flotation cell discharged silt into a garbage can. A screen-bottomed bucket was then placed inside the garbage can to catch nonfloating remains when a sample of this "heavy fraction" was desired. Crawford now recommends a larger settling tank, like the Cambridge system, or inserting a heavy fraction screen.

Processing Samples

The following procedure for the froth flotation system incorporates steps described by Jarman *et al.* (1972) and Crawford (1983). Variation between these published procedures stems mostly from differences in soils and field conditions rather than from any major difference in equipment.

Preflotation preparations. Steps 2 and 3 in the discussion of preflotation preparation for manual flotation also apply to froth flotation systems. Additional preparation steps necessary for using a Cambridge froth machine include the following:

1. Screen samples to be floated to remove heavy objects that might damage bubbler elements. Jarman *et al.* report using either 2.5-cm or 1.0-cm mesh, depending on the nature of the deposits; Crawford used 2.5-mm mesh for preflotation screening.

2. Evaluate sample soil moisture. Crawford reports that soil was floated most successfully when it was processed while still somewhat damp. Dry soils, particularly those with clay content, retained charred material, and damp soil broke apart easily. Wet soil gave the poorest results.

3. After elements of the flotation cell have been checked for cleanliness and damage and assembled, fill the flotation chamber with water. Jarman *et al.* recommend filling to within 3 cm of the lip, Crawford to within 5 cm. Lower water volume is used if the operator wishes to hold overflow until after flotation is completed. Position the bubbler in the cell.

4. Measure out frothing agent (Cyanamid Aerofroth 65 or polypropylene glycol) and a collecting agent (kerosene) and add these to water in the flotation cell. For the original Cambridge machine, 20 cc of frothing agent and 5 cc of collector were used per run.

Flotation

1. Start the air pump and run the bubbler to mix chemicals and build up a good froth bed on the water surface (Figure 2.28).

2. Select a soil sample and record provenience data, flotation number, and other pertinent information into the flotation record. Note flotation personnel.

3. If the soil sample is larger than 15–20 l, divide it into subsamples of that size, each of which are floated as one run of the machine. Enter the total number of liters floated and the number of runs required into the flotation record.

4. Label and set out cloth squares or newspapers for the light fraction.

5. When a good froth bed has built up, pour soil slowly and steadily into the center of the bed; all soil must be thoroughly wetted. Crawford recommends dropping soil from a bucket to the hand, then into the water; Jarman *et al.* suggest pouring soil in a steady stream or using a vibrating feed mechanism. Soil should be broken up by hand if necessary to ensure thorough mixing with the froth.

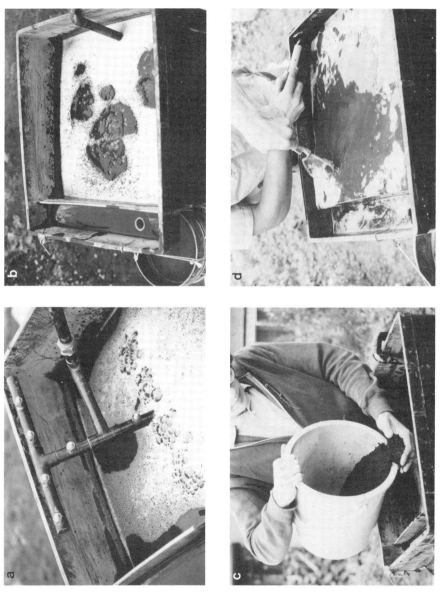

Figure 2.28 Froth flotation: (a, b) fine air bubbles, produced by a compressed air bubbler unit, create a froth bed on the water's surface; the unit is run before soil is added until a good bed builds up; (c) once a froth bed has formed, soil is slowly added; (d) floating froth and botanical remains are washed out of the machine into geological sieves (photographs by G. Crawford).

6. (*Crawford 1983 flotation method*) Allow flotation to proceed for about a minute, with the water level rising to just below the lip of the cell after addition of an entire sample (16-l samples were used). Remove the bubbler and wash froth and botanical remains out of the cell by adding enough water to cause the unit to overflow. Use a hand sieve to help push froth and botanical remains into catch sieves and to clean any floating material off the water surface after all froth is washed out.

(*Jarman et al. 1972 flotation method*) Overflow of the unit begins while soil is being introduced, with froth and remains washing into the moat and then collecting in catch sieves. When flotation is completed, remove the bubbler unit. Williams (1976) notes that complete overflow of froth and remains is still difficult to achieve without adding water to the cell.

7. Allow the bubbler to run in air a few minutes until it is dry; this helps keep pores of the elements from clogging.

8. If the completed run is the only one for a sample, tap out the light fraction caught in the sieves positioned below the cell outflow onto drying paper or cloth. If this material is dirty (e.g., coated with froth and silt, as Williams 1976 found in some instances), spray it gently while in the screen to clean it. Place the light fraction in the shade to dry. If additional runs are necessary to complete the sample, light fractions can be left to accumulate.

9. If nonbuoyant remains (heavy fraction) are to be recovered, position a screen in the settling tank below the flotation cell sludge outflow. Record screen mesh size. Alternatively, use a heavy fraction screen insert.

10. Open the butterfly valve on the basal stand to allow sludge to flow out. Assist the flow by agitating mud from the top with a pole. Practice allows sludge to be removed while conserving water. If only a sample of nonbuoyant material is desired, stop sludge flow after the desired heavy fraction has been caught in the screen in the settling tank, wash this material, and remove the screen. Continue emptying the cell until all mud is removed.

11. When sludge in the cell is agitated during cleaning, additional charred material may rise to the surface (Williams 1976). Remove this material with the hand sieve. Crawford also reports finding charcoal floating in the water of the settling tank after sludge removal. Such late floating charcoal can be added to light fractions if the operator is certain it pertains to the sample being processed.

12. Heavy fraction material, if retained, should be tapped out of the screen onto drying papers, labeled with provenience information, and placed in the drying area. Record estimate of the amount of heavy fraction sample taken (100% of the total run, 50%, etc.).

13. If light fraction sieves have been emptied, clean them thoroughly and reposition them below the cell outflow for the next sample. Clean the heavy fraction screen. The system is now ready for preflotation procedures for the next sample.

14. Note any problems occurring during flotation or procedure changes in the flotation record.

Postflotation cleanup. The points discussed for manual flotation cleanup also apply to froth flotation.

Hints for Good Recovery

The general points raised under this section for manual and water separator or SMAP systems concerning consistency of procedures, cleanliness and maintenance of equipment, accurate record keeping, and careful drying and handling of finished samples apply also to froth flotation. The published procedures and accounts of use of froth flotation cited above (Crawford 1983; Jarman *et al.* 1972) stress the following points:

1. A good froth bed must be present before flotation is started. Botanical material is captured when it comes in contact with the froth, so good recovery is predicated on abundant froth.

2. The way soil is poured into the froth bed greatly affects quality of separation of remains from soil. If soil is poured in too rapidly, there may not be time for wetting to occur before remains sink. In other words, botanical remains will not become coated sufficiently with collector to become attached to rising air bubbles. Since there is little water agitation to assist release of botanical material from soil, lack of adequate coating results in loss of material into sludge. Soil should be added slowly and mixed in manually.

3. There should be two or more operators if possible. With two operators, one person can pour while the other helps soil wetting by mixing the soil and water/froth thoroughly. Because the unit must be desludged after every 15 to 20 l of soil, extra people to take care of light fraction bagging, labeling, and preparations for subsequent samples while desludging is carried out can speed the work.

4. Some botanical remains, particularly wood charcoal, seem not to float readily in froth flotation. Crawford reports that 6–8 times as much wood charcoal was recovered from samples processed by manual flotation as was processed by froth flotation. Some of this difference was due to preflotation screening of froth samples, which removed charcoal, but Crawford reports commonly seeing charcoal lost into the settling tank.

Williams (1976) also reports this problem. However, Crawford reports that seed recovery was comparable for manual and froth flotation, with only one seed recovered from the settling tank. It appears, therefore, that water screening of sludge with a $\frac{1}{16}$″ or 1.0-mm screen is necessary to recover wood charcoal. This class of macroremain would otherwise tend to be underrepresented in samples processed by froth flotation. Williams mentions the possibility of combining froth flotation with water agitation to improve recovery. Keeley (1978) reports that some users of the froth system, such as Lapinskas, have incorporated a sieve inside the drum to improve recovery of nonbuoyant materials. Because of the problem of nonfloating charcoal, it seems advisable to empty sludge after each sample, even if samples are smaller than 15–20 l, to avoid contamination among samples.

5. Machine sludge capacity should not exceed about 20 l; if the bubbler is positioned closer to the lip of the flotation cell, more soil can be floated but desludging becomes difficult. A large-diameter discharge pipe is recommended.

6. The bubbler unit is very susceptible to clogging if it is not allowed to blow dry between samples. If it does become clogged with lime deposits, clear bubbler elements by soaking in 10% acetic acid.

7. All floating material must be carried into the catch screens; overflow often must be assisted by adding water or pushing material out with a hand sieve.

Issues in Recovery of Macroremains

In this section I discuss a number of issues that may arise when one chooses and implements macroremain recovery strategies at an archaeological site. These include how to choose a recovery system that fits both the needs of the project and the nature of deposits, how to deal with problem soils, when to use chemical flotation, how to determine efficiency of flotation recovery, and how to use salt water for flotation.

Choosing a Recovery System

As was discussed earlier in this chapter, botanical macroremains can be recovered from archaeological sites in three ways: (1) by collection of material in situ during excavation, (2) through screening, and (3) by use of water recovery techniques (flotation). For recovering all size grades of macroremains from soil, flotation and fine sieving (dry or wet) are the only reliable alternatives. I begin by discussing how to choose among the options for flotation and then discuss fine-sieving techniques.

Flotation Systems

The techniques of manual flotation, water separator and SMAP systems, and froth flotation each have drawbacks and advantages. Each flotation system was developed to fit the needs of a particular field situation, and attempts to adapt it to different situations meet with varying success. For example, when basic "tub" manual flotation was tried in Iran, lack of running water led to its abandonment and a return to small-scale flotation in the lab (Hole et al. 1969). A flotation system may end up costing a project unnecessary time, money, and headaches because it is not well suited to budget, field conditions, or soils. In deciding what system to use, it is wise to consider the initial cost of equipping the system, the cost of running it, the speed and soil capacity of the system, and its capacity to recover remains adequately from the soils of the site. Considering these factors should at the least minimize unpleasant surprises at the flotation station and at best give good, speedy recovery of remains which keeps pace with excavation.

The cost of equipping a system is the easiest factor to deal with in advance of excavation. Table 2.2 lists estimated costs of building an IDOT-style manual flotation system, a SMAP machine, and a Cambridge froth flotation system in 1988. The higher costs of the SMAP and Cambridge systems would probably lead to a quick decision about what system to use on a low-budget or student project. Closely related to costs of equipping the system is local availability of components—an important consideration for projects operating in remote areas. An additional cost to consider is that of transporting the system to the excavation site.

The costs of actually operating flotation systems are less easy to estimate. Published data on rates of flotation do at least give an idea of this constraint. Watson (1976) and Keeley (1978) have compiled such data, which are summarized in Table 2.3 along with other data available after their compilations. Flotation rates are surprisingly consistent for each type of system, given all the variables of estimating these data. Rates for manual systems, for example, vary between 0.05 and 0.08 m^3/day, and most water separator or SMAP systems cluster at 0.50 to 0.80 m^3/day, with some slower rates (0.20–0.30 m^3/day) and one much higher rate (2.40 m^3/day) reported. A factor of roughly ten times separates machine-assisted and manual water systems. Flotation rates using the froth system are quite variable, from 0.10 to 2.0 m^3/day. It is interesting to note that the slower rates for the froth system are only twice as fast as manual flotation.

The numbers of operators working to give rates cited above are generally one or two. Because it was not always clear how many people were working, I have not converted m^3/day rates into m^3/worker-hour. But the order-of-magnitude difference in rate of flotation between manual and machine-assisted water flotation is such that cost per worker-hour is clearly much higher for a given soil volume in manual flotation. Thus, in projects where large soil volumes are to be processed, the higher

Table 2.2 • Estimated Construction Costs of Flotation Devices

SMAP		Froth		IDOT	
Component	Cost estimate ($)	Component	Cost estimate	Component	Cost estimate
55-gallon drum	17–25	Stainless steel box (flotation tank) with modifications, sluiceway		55-gallon drum	17–25
Steel added to drum design such as sluiceway, baffle				Bucket with screen (ca. 1 ft.² @ $12) or	20
parts	67	parts	820[a]	Plywood/screen (3 ft.²)	49
welding	200	welding	700[a]	Handles	8
Barrel insert		Metal frame (base)			$45–$82
parts (metal and screen), welding	175	parts	28		
		welding	48		
3 HP pump with 2" discharge	200	Air compressor battery powered with regulator (uses car battery),	29–69		
Intake hose	27		24		
Discharge hose	20	or			
Foot valve	28	gasoline powered	169		
Gate valve	18	Discharge hose	20		
50 ft. garden hose with spray nozzle	18	Copper pipe	12		
Garden hose adapter	2	2 screw clips	1		
4" plastic plug	2	Shut-off valve	3		
Galvanized pipe	10		$1685–$1801		
T-fitting for pipe	3				
Cap for pipe end	3				
Shower head	20				
	$810–$818				

[a] Cost at a commercial welding shop; can be reduced by modifying existing barrel.

78 • Chapter 2 Techniques for Recovering Macroremains

Table 2.3 • Comparison of Flotation Rates

Technique	Flotation rate[a]	Notes and references
Manual types		
Apple Creek immersion	0.07	2 people needed (Struever 1968)
Garbage can technique	0.05	(Crawford, 1983)
Ocean immersion	0.06–0.08	Krum Bay (Pearsall 1983)
Barrel method	0.06	2 people needed, El Bronce
Barrel method	0.08	1 person (Minnis and Le Blanc 1976)
Paraffin flotation	0.01–0.02 0.004	two estimates given, 1 person (Keeley 1978)
Water separator types		
Ankara machine	2.40	(French 1971)
Siraf machine	0.80	(Williams 1973)
SMAP machine	0.50–0.60	(Watson 1976)
Indiana machine	0.60	(Limp 1974)
SMAP machine	0.23–0.30	2 people needed, Chapman project (Wiegers 1983)
Izum	0.27	2 people needed (Davis and Wesolowsky 1975)
Froth types		
Cambridge machine	1.00–2.00 (optimal) 0.96 (slow)	2–4 people needed (Jarman et al. 1972)
Cambridge machine	0.13–0.15 0.37 (maximum)	(Crawford 1983)
Cambridge machine	0.10	(Keeley 1978)

[a]Flotation rates in m³ per 8-hour day.

expense of building a SMAP or Ankara system should be more than made up by savings on labor costs, unless a very inexpensive labor pool is available. Some additional operating costs for machine-assisted flotation systems include those for gasoline or diesel fuel, for collector and frothing materials (froth system), and for water supply. This last point may be an expense to consider in deciding whether to use a recycling or low-water-use system.

Speed and soil capacity of flotation systems have been presented as a way of estimating operating costs for these systems. There are other considerations besides cost, however. All archaeologists would agree that, if flotation is set up in the field, processing should attempt to keep pace with excavation. This allows feedback to the field crew on results of processing certain kinds of deposits and means that few bulky soil samples must be transported to the lab at the end of the season. A system that can process higher soil volumes per day may be a better choice if excavations are extensive and rapid feedback to the field crew is desired. Larger samples are more easily processed with a machine-assisted rather than a manual system. If deposits are such that small samples can be used or rate of excavation is low, then manual flotation may be adequate, since less soil must be floated per day to keep up with excavation.

The type of soil to be processed is also a factor for choice of flotation system—one linked closely to efficiency of recovery. All systems can recover a high percentage of seeds if soil is loose and friable, but machine-assisted systems may be better at keeping seed recovery high in heavier, clayey soils. The water separator or SMAP systems seem superior to froth systems in this aspect. The froth system has been criticized as giving incomplete separation of remains in dense soil, although seed loss has not been tested, and proper preflotation treatment might solve this (see "Problem Soils," below). Recovery of charcoal is another matter. Both the manual systems and the froth flotation system tend to lose charcoal, the former because it is difficult to get sufficient agitation to float the denser charcoal, the latter for problems of insufficient wetting.

Fine Sieving

In some situations fine sieving is a better choice for recovering macroremains than flotation; two obvious examples are waterlogged deposits and dry deposits. Flotation is for the most part inappropriate in each of these situations. It is important to remember that, in sieving, recovery efficiency is based solely on mesh size. Even a screening procedure as elaborate as that used at the Nichoria site (Aschenbrenner and Cooke 1978), where a mechanized three-screen system could water screen 0.5 m^3 of deposit each day, is not adequate for recovering macroremains if the mesh used allows escape of small but important materials.

Archaeological sites at which aridity has resulted in the preservation of uncharred, desiccated remains provide opportunities for ethnobotanical analysis not available at sites where uncharred material decays (Fig. 2.29). But desiccated remains may be damaged by wetting, making flotation inappropriate. Although remains may look sturdy and almost modern, tissue breakdown has begun and wetting may accelerate decomposition. During preliminary sorting and identification of dried remains from sites under investigation at the Chincha project in Peru, I compared quantity and quality of recovery of dry-sieved and floated components of one sample. Table 2.4 summarizes the results of this experiment. Overall recovery of remains was similar (46% of material by count in the floated component; 54% by count in the sieved component). Material seemed to be unevenly distributed in a number of instances, however. Loose cupules and seed coat fragments of maize occurred in lower than expected amounts in floated samples, as did peanut pod, lucuma seed, unknown charred seeds, grass florets, and coca seeds. Six types of

80 • Chapter 2 Techniques for Recovering Macroremains

Figure 2.29 The arid coast of Peru is one environment where archaeological middens contain a wealth of dried botanical materials.

remains were more abundant in floated samples. There was no striking pattern of loss of material due to flotation. Maize cob and cupule material did become soft and very weak after wetting; beans became discolored and began to shed seed coats. Loss of dense lucuma seed into the heavy fraction occurred. Mold growth began on some materials. Although flotation was faster than fine sieving (sieving required about 45 minutes per sample using 2.0-, 1.0-, and 0.5-mm sieves) and floated samples required half the time to sort, adding sorting of heavy fractions and preflotation removal of cobs and other larger remains to the process would increase time again. The problem of damage to certain remains outweighed gains in speed of processing, in my view, and flotation was not continued.

In dry deposits, "flotation" samples may be taken as at other sites, but processing should consist of fine screening rather than water immersion. I recommend using regular excavation screens to recover larger botanical materials, while at the same time collecting separate soil samples of a standard volume or weight to be screened for smaller remains. As in water-floated samples, these "flotation" sam-

Table 2.4 • Comparison of Recovery of Dry Sieved and Floated Remains, Chincha Valley

	Floated[a]	Sieved[a]	Total
Zea mays (corn)			
cob fragments	28 (65%)	15 (35%)	43
loose cupules	55 (37%)	95 (63%)	150
kernels	9 (47%)	10 (53%)	19
seed coat fragments	84 (39%)	129 (61%)	213
Capsicum (pepper)			
seeds	116 (44%)	147 (56%)	263
peduncles	13 (65%)	7 (35%)	20
Phaseolus vulgaris (bean)			
pod fragment	6 (67%)	3 (33%)	9
seed	2 (67%)	1 (33%	3
Arachis hypogaea (peanut)			
pod	4 (36%)	7 (64%)	11
seed	0 (0%)	13 (100%)	13
Psidium guajava (guava)			
seed	4 (80%)	1 (20%)	5
Small twigs	110 (52%)	103 (48%)	213
Dicot leaves	48 (56%)	37 (44%)	85
Unknown charred seeds	0 (0%)	12 (100%)	12
Unknown dry seeds	19 (83%)	4 (17%)	23
Grass florets	23 (40%)	35 (60%)	58
Erythroxylon sp. (coca)			
seed/seed coat	0 (0%)	5 (100%)	5
Charred material/non-wood	122 (48%)	133 (52%)	255
	643 (46%)	757 (54%)	1400

[a]Volume floated, 2 liters; volume sieved, 2 liters.

ples represent a sample of smaller remains present at the site. Samples may be fine sieved in the field lab by setting up a tiered sieving apparatus and passing each sample little by little through it. Initial soil volumes or weights are recorded and each size fraction carefully bagged and tagged. Choice of mesh sizes depends on the soil and nature of remains. The smallest mesh should be chosen to concentrate most soil in the catch pan while allowing few or no seeds to pass through; other sieves should segregate materials by type as much as possible. If samples are processed in the field to reduce total weight, a sample of soil residue should also be sent to the analyst so that it can be checked for occurrence of very small materials.

It may also be possible to process soil samples from dry deposits effectively with a seed blower (Ramenofsky *et al.* 1986). Light desiccated remains should be sufficiently different in specific gravity from soil to allow them to be separated by air pressure. Figure 2.30 shows the basic design of a seed blower. Ramenofsky *et al.* tested the effectiveness of this device for separating different classes of material

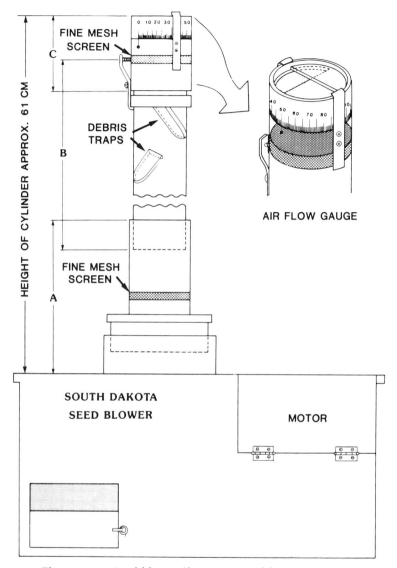

Figure 2.30 Seed blower (from Ramenofsky *et al.* 1986).

(botanical, bone, gravel, lithics) from soil samples that had been deflocculated (clays dispersed) and sieved into graded size classes but not floated. The blower was used to rough sort each size fraction by density. This rough sorting was effective enough to reduce hand-sorting time significantly.

Recovering botanical materials from waterlogged samples can also be difficult using standard flotation techniques. If material is wet because of treatment to dis-

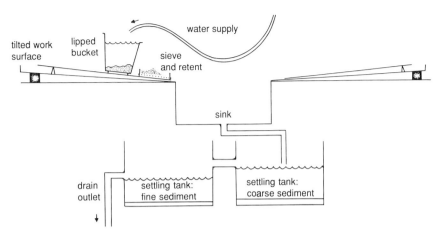

Figure 2.31 Laboratory setup for fine wet sieving of waterlogged botanical samples; scale: sieve, 20 cm dia. (from Kenward et al. 1980).

perse clays (see below) or deposition below the water table, air spaces in plant tissue are filled and material does not float well. Nonbuoyant remains end up in the heavy fraction, necessitating chemical flotation. Rather than going through a two-step flotation procedure for waterlogged samples, fine-sieve water screening may be not only easier but less expensive in chemical and labor costs. Kenward et al. (1980) present an excellent summary of principles and methods of recovering both floral and faunal macroremains from waterlogged sites. Three methods are described in detail: bulk sieving, fine sieving, and paraffin flotation. The bulk-sieving technique recommended is based on use of a water separator similar to that described by Williams (1973). By using 1.0-mm nylon mesh for both the flotation and residue screen, the same size classes of buoyant and nonbuoyant remains are recovered. Bulk-sieved samples are analyzed for shellfish, small bone, and large botanical remains (fruit pits, wood, large seeds).

For recovery of small waterlogged botanical materials (i.e., material lost through 1.0-mm screening), Kenward et al. recommend processing small samples (0.5–1.0 kg) through a graded bank of sieves. Figure 2.31 illustrates the sieving setup used at the University of York laboratory. This setup requires a drain with sump, 10-l lipped bucket, 300-micron sieve, bank of sieves of sizes to fit samples (e.g., 5 mm, 2 mm, 1 mm, 0.5 mm), hot and cold water, vials, labels, and 40% alcohol-glycerin-formalin solution (30:60:2). The following procedure is recommended for fine sieving wet samples (Kenward et al. 1980:8–11):

1. Clean all sieving apparatus thoroughly with hot water, detergent, and scrub brush.

2. Weigh out sediment for processing, and record provenience, sample number, and other pertinent information in the processing log. Label bucket with sample and site number.
3. Place sediment sample in bucket and add hot water (40–50° C) using a hose. Disaggregate sediment by gentle stirring. Once the sample begins to break down, tilt the bucket over the 300-micron sieve and continue to add water so that it flows over the sample and into the sieve. This process is continued until little or no more fine material passes the mesh and the sample is totally disaggregated.
4. Samples may not disaggregate easily in water. Some less friable samples may disaggregate after soaking in water for a day or so (no more than three days, or mold growth will occur). However, it may be necessary to boil sediment for 15 to 30 minutes or even several hours, or to soak it in 5–10% sodium hydroxide or dilute sodium carbonate.
5. If residue in the 300-micron sieve contains a high proportion of inorganic material, this may be removed by "washover." Return residue to the bucket and add cold water. When the bucket is a third or so full, swirl it and pour liquid and suspended organic material onto the 300-micron sieve. Repeat until no further organic material washes out. Set the inorganic component aside.
6. Return residue to the bucket and set up a bank of sieves with the 300-micron sieve at the base. Place the residue in the top sieve and gently wash it through the sieve column. Be careful not to splash material out of the top sieve or to overflow lower sieves. A sprayer nozzle on the hose facilitates sample washing.
7. When separation into size fractions is complete, empty each sieve and store material for sorting. Material should not be allowed to dry out; drying causes materials to shrink and crack, perhaps altering them beyond recognition. For long-term storage, material should be placed in alcohol-glycerin-formalin solution. The entire procedure described here generally takes between 30 minutes and 2 hours for a 1-kg sediment sample.

Kenward *et al.* also discuss paraffin flotation, but as a method of recovering insect remains rather than small botanical materials. These authors feel that paraffin flotation does not recover a representative assemblage of botanical materials, since not all material floats. Also, residual paraffin on remains may limit their use for radiocarbon dating and chemical analysis. Keeley (1978) also reports paraffin flotation as particularly good for recovering insect remains, noting that both floated material and residues must be sorted. Refer to Kenward *et al.* (1980) and Shackley (1981) for detailed procedures for processing small samples in paraffin.

Because it is impossible to collect samples for fine sieving after bulk water screening, one must decide before excavation either to pass all soil through a series of screens, ending with a mesh size small enough to capture seeds, or to bulk screen most soil but retain samples of unscreened soil for fine sieving. If the latter option is chosen, test samples should be processed to choose appropriate soil volumes and sieve sizes.

Problem Soils

The decision whether to float or fine sieve samples may hinge on the nature of soils at the site. When clay content is high enough to impede its easy dispersion in water, flotation can be very difficult and unproductive. If soil does not disperse, botanical remains are not released. Dispersion of soil is dependent on the ion exchange properties of the soil (Brady 1984). A tendency in soil toward granulation (clumping, lack of dispersion) can be combated by changing the ion exchange rate—by adding sodium, for example, which encourages dispersion. Sodium also alters soil pH to a more basic value, which aids dispersion. In practical terms, this means that soils that will not break up completely during routine flotation can be treated before flotation or fine sieving to disperse more readily. This treatment is deflocculation.

During the FAI-270 project, then lab director Connie Bodner conducted tests with flotation crew supervisor Denise Steele to determine what deflocculating agent would work best on the clayey soils of the American Bottom. The results of their tests, cited here from lab documents, guided flotation of many problem samples encountered during this project. Out of many tests, Bodner and Steele found four substances effective in deflocculating clayey soils:

1. sodium hexametaphosphate $(NaPO_3)_n$—available in some dishwasher detergents: 10% solution, 1 cc for each estimated gram of clay in the sample.
2. sodium bicarbonate ($NaHCO_3$, baking soda): 10% solution, approximately 1 cc for each gram of clay in the soil. Hot water aids deflocculation. Sodium bicarbonate is inexpensive when ordered in quantity.
3. mix of ammonia (NH_4OH) and sodium carbonate (Na_2CO_3): a 1:1 ratio of the two chemicals, using household ammonia. A 10% solution of the mix in water is prepared. This defloccculant works as well as those already discussed but is expensive and very smelly.
4. hydrogen peroxide (H_2O_2): a 10–20% solution of 30% H_2O_2 is mixed. This chemical gives the best deflocculation of clays. Extreme caution

must be exercised in mixing the solution, since full-strength (30%) H_2O_2 is a highly caustic oxidizer. Because it is expensive as well as caustic, it should be used only when other methods fail.

Sodium tripolyphosphate ($Na_5P_3O_{10}$) and similar compounds used in detergents and water softeners have effects similar to sodium hexametaphosphate. The FAI-270 project chose hot water and baking soda as the cheapest solution, with H_2O_2 reserved for the worst samples.

After samples have been deflocculated and reduced to a slurry of soil and water, they are still a problem to float because of waterlogging. Using a manual flotation system, I have tested charcoal soaked 24 hours and found it strong enough to go through flotation without undue breakage, but most of it ends up in the heavy, rather than the light, fraction. Similarly, recovery rate tests on deflocculated samples floated wet have shown that small seeds can also become waterlogged and lost through a $\frac{1}{16}''$ heavy fraction screen. If manual flotation is used on wet soils, a system using fine-mesh heavy fraction screens such as the IDOT system should be employed to combat seed and charcoal loss. This was the strategy followed by the FAI-270 project. Chemical flotation or hand sorting of heavy fractions is then necessary to recover nonbuoyant materials. An alternative solution is to set up a wet-sieving apparatus such as that described by Kenward *et al.* (1980) for processing less friable samples.

For soils with moderate clay content, processing with a SMAP machine or water separator may give adequate recovery of remains. Floating wet, deflocculated samples using these systems is problematic, however, since waterlogged material may be lost through the heavy fraction screen. Since Crawford (1983) notes that wet samples floated even more poorly than clayey dry samples using a froth system, there is also a problem using this system on wet, clayey soils.

Chemical Flotation

Struever's (1968) article on the flotation technique includes a description of using zinc chloride ($ZnCl_2$) to float off charred remains from bone and other nonbotanical, nonbuoyant material. At some projects, such as Struever's long-running Koster site excavations, chemical flotation was routinely carried out on most samples. But as costs of chemicals such as $ZnCl_2$ rise (from $1.90/kg in 1979 to $7.50/kg in 1984), it is useful to consider when such treatment is really necessary and how it can be done most efficiently.

One of the reasons cited for development of froth flotation was to minimize the need for heavy liquid flotation to raise materials denser than water. Use of a collecting agent and rising air bubbles in theory allows flotation of materials as dense as coal (Jarman *et al.* 1972), although in practice there are problems with obtaining the necessary wetting to consistently float wood charcoal and other dense material. Chemical flotation of already water-floated materials means that samples are usually rewetted after full or partial drying. As Jarman *et al.* report, rewetting and drying can quickly lead to destruction of seeds and other fragile remains. In dealing with clay soils, which must be deflocculated to release the botanical material, at least some of the charred material is going to be trapped with nonbuoyant remains. How are such materials to be recovered? Hand removal and air pressure separation are the only options to chemical flotation in these cases.

It is always useful to consider hand separating charcoal from heavy fractions or fine-sieved samples instead of using chemical flotation. If nonfloating charcoal is a relatively rare occurrence, then hand sorting may be very economical. If the flotation crew notes which samples have nonfloating charcoal, sorters can easily examine those heavy fractions and remove charcoal. Spot checking heavy fractions can be done to ensure that occurrence of nonfloating material is not widespread. If heavy fractions are to be sorted for bone or small artifacts, removing charcoal can simply be added to that lab step. It is a simple matter to keep track of how much time is needed to check and hand sort heavy fractions, and to compare that to the cost of $ZnCl_2$ flotation.

It may be the case, however, that chemical flotation is the best answer to recovering widespread nonbuoyant botanical remains. If so, how can this be done to minimize sample breakage and to economize? On the first point, if field laboratory facilities permit, it is advantageous to set up chemical flotation so that it follows immediately after water flotation. Figure 2.32 illustrates a simple chemical flotation setup at the field laboratory of the El Bronce project, Puerto Rico (Pearsall 1985). Heavy fractions, placed on screened drying racks to facilitate drainage of excess water, can be chemically floated while charcoal is still damp. As long as little water adheres to material, minimizing dilution of the chemical, flotation can be run routinely on damp samples, removing one cycle of wetting and drying. Samples should be thoroughly rinsed to remove residual chemical.

To economize on the cost of chemical flotation, both in chemicals used and labor expended per sample, an efficient lab setup would have considerable benefits if many samples had to be processed. Figure 2.33 illustrates the chemical flotation lab setup designed by FAI-270 project personnel Lawrence and Hishinuma, under the guidance of project paleoethnobotanist Johannessen. The following procedure is summarized from FAI-270 lab documents:

1. Select a sample and record provenience information and sample number in the flotation record and on the newspaper in the bottom of the flat.
2. Empty the sample into the plastic bottle with window cutout.
3. Lay the filter cloth inside the wire mesh filter support. Hold the plastic

Figure 2.32 Chemical flotation of heavy fractions from the El Bronce site, Puerto Rico: (a) containers for zinc chloride solution and rinse water are arranged in the open air; (b) after a sample is poured into a geological sieve in the zinc chloride solution, botanical material floats to the surface; (c) floating material is scooped out with a hand sieve; (d) the sieve is tapped onto a cloth square to dislodge material; (e) after flotation is completed, the heavy fraction is removed from the geological sieve, placed on a cloth square, and rinsed in water; (f) the light fraction is rinsed.

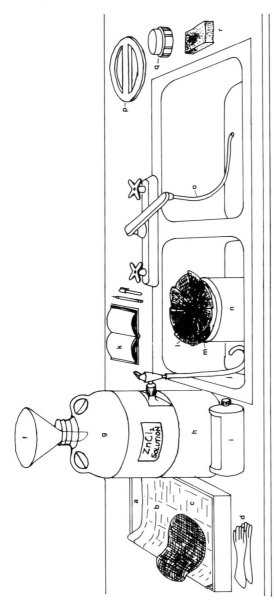

Figure 2.33 Chemical flotation setup used during the FAI-270 project: (a) beer or soda flat; (b) newspaper; (c) filter cloth; (d) rubber gloves; (e) plastic funnel; (g) plastic carboy; (h) stand for carboy; (i) plastic bottle with "window"; (j) rubber tubing from spigot; (k) record book; (l) wire mesh filter support; (m) geological sieve; (n) plastic bucket; (o) rubber tubing from water source; (p) lid for bucket; (q) screw cap for carboy; (r) sponge (redrawn from FAI-270 lab documents, provided by S. Johannessen).

bottle containing the sample so that the window cutout faces up and the bottle mouth is directly over the filter cloth. Open the spigot on the carboy and pour $ZnCl_2$ solution (1.62 specific gravity: 860 g $ZnCl_2$ plus distilled water to 1 l) onto the sample. Shake the bottle gently to float charcoal from the sample matrix. Continue adding solution and tilt the bottle so that solution and floating charcoal flows out onto the filter cloth. Be sure to wash out any charcoal sticking to the inside of the bottle. When all material has floated out, turn off the carboy spigot and pour any remaining solution from the bottle onto the filter cloth.

4. Remove the wire mesh filter support, with filter cloth and "light fraction" inside, and thoroughly rinse the sample (at least one minute) with water. Remove the filter cloth and place it in the flat to dry. Remove the geological sieve from the mouth of the bucket and place it in the sink.

5. Wash the "heavy fraction" residue out of the plastic bottle into the geological sieve and rinse it thoroughly (at least one minute). Empty the residue onto the newspaper in the flat to dry.

6. Pour the $ZnCl_2$ solution from the bucket back into the carboy. Wash sieve, wire mesh filter support, and bottle thoroughly, wiping each dry to minimize dilution of chemical. Reassemble all components for the next sample.

7. When flotation is finished for the day, the $ZnCl_2$ solution should be stored in a closed carboy. When the solution becomes dirty, allow it to settle over night and pour off cleared solution. Wash and dry all metal components of the system to inhibit corrosion.

A heavy liquid solution mixed to a specific gravity between 1.62 and 2.1 (the average specific gravity of gravel) (Bodner and Rowlett 1980) allows flotation of bone from heavy fractions. Bodner and Rowlett recommend ferric sulfate, $Fe_2(SO_4)_3 \cdot H_2O$, for this type of flotation, as it is less expensive than $ZnCl_2$. This chemical, like $ZnCl_2$, is very hygroscopic, and it must be thoroughly rinsed from the material. Wood or seeds coated with these substances never dry and will thus dissolve gelatin capsules and stick to vials.

Workers should always be carefully trained before being allowed to mix heavy liquid solutions or carry out chemical flotation. Mixing $ZnCl_2$ with water produces an exothermic reaction—the mixing container heats up very quickly. Crystalline $ZnCl_2$ burns skin and must be handled very carefully and stored in a tightly sealed container. Flotation should be carried out in a well-ventilated area; rubber gloves should always be worn. In the past, tetrabromoethane and bromoform were used in chemical flotation. *This is not recommended;* these are toxic chemicals and should be used only with fume hood ventilation by an experienced person. I strongly recommend that these substances not be used in chemical flotation of macroremains.

A new technique for separating materials by density, one that uses air pressure rather than chemicals, has been proposed by Ramenofsky *et al.* (1986). Although the seed blower (Fig. 2.30) has not been tested on many flotation samples, it could prove very useful in lifting wood charcoal and nuts out of heavy fractions, and perhaps in separating seeds and charcoal from root debris in light fractions.

Testing Flotation Recovery Rates

As more archaeological projects use flotation to recover plant remains, the body of data on prehistoric plant use and subsistence grows. Although such data are a valuable resource for comparative studies within a region or time period, they are often difficult to use this way because methods used to recover remains vary widely, introducing biases into macroremain assemblages. Testing flotation recovery rates is one means of assessing comparability of data recovered by flotation systems commonly in use.

The poppy seed test (Wagner 1982) is the most common form of testing efficiency of recovery of flotation systems. First suggested as a means of testing recovery by Lawrence Kaplan at the 1976 Society for American Archaeology Ethnobotany Roundtable discussions in St. Louis, the poppy seed test is easy to carry out. Following Wagner, the technique is as follows:

1. Char poppy seeds (*Papaver somniferum*, available in grocery spice departments) by wrapping in aluminum foil and heating in a home oven at 500° F for several hours, or until thoroughly blackened. Seeds can also be heated in sand over a bunsen burner or in a muffle furnace.
2. Working under magnification, count out lots of 50 or 100 seeds, taking care that all seeds are only whole, undamaged, and completely charred. Gelatin capsules make handy containers for seed lots.
3. Add one lot of seeds to each flotation sample to be tested. This can be done immediately after excavation, or while samples are in the lab awaiting flotation. I usually stir seeds into the top few inches of soil. This means that the poppy seeds go into the flotation system early in processing, exposing them to the longest period of potential loss. Record the sample number and seed lot sizes. In general, flotation personnel should not know which samples are being subjected to testing; this ensures that the test samples receive routine treatment.
4. After flotation, examine samples and record the number of poppy seeds present and their condition. Rate of recovery is expressed as a percentage; for example, if 40 of 50 seeds are recovered, a recovery rate of 80% is achieved.

If poppy seed testing is done routinely, problems in recovery can be identified early and corrected. Whatever rate of recovery is achieved, potential loss is at least known for purposes of analysis.

Wagner selected poppy seeds for testing recovery rates because they approximate the size of many small archaeological seeds (0.7–1.4 mm), and they are not native to eastern North America, her primary research area. Pendleton (1979, 1983) has used lettuce, endive, cabbage, and parsley seeds for testing the recovery rates of floating charred materials. He also suggests that nonbuoyant seeds such as catnip (0.7–1.5 mm) and broccoli (1–2 mm) could be used to test how efficiently a flotation system recovers nonfloating botanical materials (Pendleton 1983). Another approach to testing recovery rates is to add charred materials representing the full size range of remains likely to be recovered (e.g., small, medium, and large seeds; materials of differing densities).

Although poppy seed tests do not necessarily give a completely accurate representation of what percentage of archaeological seeds in the soil are recovered by flotation, they are a useful way of determining effectiveness of a particular system and of comparing it to other systems. If poppy seeds are being lost during flotation, then archaeological seeds are surely suffering the same fate. Low poppy seed recovery rates may indicate that rarer archaeological seed types are being lost from the inventory. Table 2.5 presents a summary of available flotation recovery data. Some patterns emerging from these data are the following:

1. Little is known about efficiency of recovery of the froth flotation system. Jarman et al.'s (1972) field test of the Cambridge machine demonstrated recovery of abundant nonbuoyant seeds, but it did not indicate how many seeds had been lost through the system. Pendleton's (1979) tests were not carried out on archaeological soils. Neither of these were blind tests. Little detail is given on tests reported by Kaplan and Maina (1977). The comparability of these to other tests is thus in question.

2. For IDOT-style flotation systems, the shaking motion used for soil washing makes seed recovery highly dependent on mesh size; the smaller the mesh, the better the possible seed recovery. Good recovery is also possible for wet samples.

3. Manual flotation systems that use swirling agitation show quite variable recovery rates. If $\frac{1}{16}''$ or larger mesh is used in the flotation bucket, seeds can be easily lost through the bucket screen. Recovery is highly dependent on soil type, moisture content, and length and vigor of agitation.

4. SMAP-style water flotation systems show the most consistently high recovery of poppy seeds, which are smaller than the mesh size commonly used in SMAP flotation barrel inserts ($\frac{1}{16}''$). Machine-assisted agitation gives consistent flotation of materials. Tests of waterlogged samples show very inconsistent recovery of wet material using this system, however.

Table 2.5 • Comparison of Seed Recovery Efficiency

	No. tests	Recovery (%)	Mean recovery (%)
Wagner (1982)			
SMAP machine	11	84–98	93
Incinerator	1	—	6
1.00 mm mesh (all)			
Incinerator	7	8–78	44
1.00 mm (bottom)			
0.42mm (side)			
Incinerator	3	81–94	87
0.59 mm (all)			
Incinerator	4	84–90	87
0.59 mm (bottom)			
0.42 mm (side)			
IDOT, tank	5	14–72	41
0.42 mm (all)			
IDOT, river	9	45–91	72
0.42 mm (all)			
Kaplan and Maine (1977)			
Modified froth flotation	—	40–90+	—
(air and water)			
Jarman et al. (1972)			
Froth flotation	1	534 seeds recovered: 43% buoyant, 57% nonbuoyant	—
Pendleton (1979)			
Modified froth flotation			
(air and water)			
buoyant seeds	8	62–96	78
nonbuoyant seeds	2	72–80	76
Tub flotation (sieve			
size not specified)			
buoyant	8	24–100	70
nonbuoyant	2	15–34	25
Old Monroe Project			
SMAP machine			
dry floated	35	60–100	89
wet floated, loam	1	—	74
IDOT system			
0.5 mm (all)	2	84–86	85
Krum Bay project			
(Pearsall 1983)			
manual, $\frac{1}{16}$" mesh	3	28–34	30
manual, overlapped $\frac{1}{16}$" mesh	3	37–53	44
manual			
$\frac{1}{16}$" mesh (second test)	11	40–88	69
blind test	5	40–82	56
El Bronce project			
(Pearsall 1985)			
manual, $\frac{1}{16}$" (all)	7	0–88	37
wet floated, clayey	5	0–82	24
dry floated	2	52–88	70
Osage–Missouri project			
(Pearsall et al. 1985)			
SMAP			
wet floated, clayey	2	2–4	3
IDOT			
wet floated, clayey	1	—	84

The overall lower poppy seed recovery rates obtained with a traditional manual tub system using $\frac{1}{16}$" screen should be a factor in the decision of whether to use such a system. If the quantity of charred material per liter of soil is relatively high, such as was the case at Pachamachay and Panaulauca caves (Pearsall 1980, 1984a), then a recovery rate near 50% still gives high seed counts and an adequate sample of rarer types, although there may be differential loss of seeds in smaller size ranges. But if seed occurrence is low, not only should larger soil samples be processed, but a flotation system should be chosen to maximize recovery.

Saltwater Flotation

In dry coastal settings it is often the case that, while salt water is abundant, fresh water may be very limited. Using salt water for flotation may thus be a more attractive option than building a system that recycles fresh water. This is especially true if a simple manual flotation system is planned. The basic issues to be considered in using salt water are whether salt harms botanical materials, both in the short run and during long-term storage, and how using salt water affects recovery rates.

As was mentioned above under discussion of the Franchthi Cave flotation system, tests conducted by Hillman on effects of soaking charred materials in brine solution revealed that prolonged soaking and drying without rinsing had no effect on identification of material. Salt crystals were observed but did not interfere with identification. Grain that was rinsed in fresh water after saltwater soaking rarely had any salt crystals adhering to it. I have found that cloth squares used for light fraction drying can be easily dipped into a bucket of fresh water after being tied shut, making rinsing easy. Saltwater–floated samples treated this way during the Krum Bay project were found to have no adhering salt crystals. On the principle that samples should be left as clean as possible, I recommend rinsing to remove salt, if this can be done before samples dry.

Lange and Carty (1975) report that in side-by-side tests (half the sample floated in fresh water, half in salt water, using a manual tub system), about twice as much charred material was recovered in light fractions processed by saltwater flotation than in those processed in fresh water. Although these data seem to indicate that salt water is a much better flotation medium than fresh water, Lange and Carty's tests also showed more total residue by weight in heavy fractions of four of five test samples floated in salt water than those floated in fresh water, suggesting that, by chance, portions of samples floated in sea water had more material in them. Results of poppy seed tests conducted during saltwater flotation of Krum Bay materials (Table 2.5) show that, while recovery rates are higher for this system than for some manual flotation systems, they are still lower than machine-assisted flotation recovery rates. In addition, salt water did not float some dense wood and large seed fragments (*Sterculia* sp. and *Manilkara* sp.; Pearsall 1983). I disagree with Lange and

Carty's claim that salt water combines the benefits of fresh water and chemical flotation but agree that it is a viable alternative in water-short areas.

Sampling for Macroremains

In most excavation situations it is impractical to recover all size classes of botanical macroremains from every cubic meter of soil removed. Even if all soil is bulk sieved through a water separator system, choice of mesh to allow efficient processing (generally 1.0 mm or $\frac{1}{16}''$) may lead to loss of small remains. If poppy seed testing is used to determine extent of loss, and an acceptable recovery level is achieved, bulk sieving is still time consuming and may produce more material than can ever be analyzed.

Collecting samples of excavated soil for flotation or fine sieving is a practical alternative to bulk sieving for all size grades of macroremains. Sampling produces fewer samples for analysis and speeds processing, since small soil volumes (rather than entire contexts) may be used. In the remainder of this section I discuss sampling strategies for flotation and fine sieving, outline basics of taking samples, and cover special recommendations for successful sampling. Good sampling strategies and processing techniques are the bases of a successful paleoethnobotanical analysis; even the best specialist cannot make up for erratic sampling or sloppy flotation.

Strategies for Sampling

I recommend a "blanket sampling" strategy for flotation: collect soil for flotation from all excavation contexts. There are practical reasons for choosing this strategy, but it also has advantages for later analysis of materials.

During the course of excavation it is usually impossible to predict which contexts (i.e, site areas, feature classes, soil types, cultural components) contain macroremains. This is especially true if only charred remains are preserved, since these are difficult to see during excavation. Because charring occurs through deliberate or accidental burning, collecting flotation samples only from contexts with evidence of burning may seem a viable strategy. In practice, however, sampling only in hearths or obvious ashy deposits does not lead to recovery of a representative sample of macroremains. Hearth samples often contain little more than charred wood, since repeated use of a hearth may result in ashing of fragile remains such as seeds or tubers. Such material may be more abundant on the floor around the hearth or in a pit filled with garbage. Hearths may be periodically cleaned of wood, ash, and spilled foods, further limiting their usefulness. Once charred material is spread around (i.e., moved from primary to secondary depositional contexts), it is difficult to see and therefore difficult to sample. Routinely sampling all contexts avoids the problem of predicting where remains will occur.

Another practical advantage of blanket sampling is that it is an easy strategy to carry out in the field. Excavation crews are simply instructed to take float samples from every level in each unit and from all features. Taking float samples becomes part of the routine of excavation; variation in sample taking among individuals is avoided.

From the perspective of later analysis of materials, blanket sampling gives the analyst maximum flexibility. If, for example, a site turns out to have several discrete temporal components, a subsample of flotation can be chosen for analysis to maximize temporal contrasts. One might compare assemblages from all hearth features or floor samples through time. For single-component sites, samples might be chosen to give greatest information on differential use of space. In such a case, one might group samples by activity area and observe patterning in the data. It is much easier to choose a subsample for analysis from a large population of flotation samples (perhaps analyzing 25% or fewer of total samples) than to predict what sorts of samples will be needed while excavation is proceeding.

In implementing blanket sampling it is important that defined contexts be sampled discretely; hearths must be sampled separately from surrounding house floor, pits from midden, floor from wall trenches, and so on. Although I describe this strategy as sampling of all excavated contexts, there are contexts where sampling is of little value. Areas of clear disturbance, such as rodent burrows, plow zones, or redeposition from looters pits (or backfill of old excavations), need not be sampled. In a multistage project, preliminary analysis from one season may reveal that certain contexts lack useful paleoethnobotanical data. Wall trench samples or post molds, for instance, might prove unproductive. It is valid to reduce sampling in such contexts to a minimum for spot-checking.

Sampling Techniques

There are three commonly used techniques for taking flotation or fine-sieve samples: "pinch" or composite sampling, column sampling, and point sampling.

Composite, or pinch, sampling is appropriate for many common sampling situations. A composite sample is made up of small amounts of soil gathered from all over a context combined in one sample bag. In the excavation illustrated in Figure 2.34, for example, a house floor has been exposed in eight excavation units (a–h). A number of features have been defined (1–5). Each section of floor (unit) and each feature can be considered a separate context for sampling. To collect composite flotation samples for one level of such a house floor, label flotation bags with provenience information for each context. Fill each bag with small scoops of soil from all around the appropriate areas. Soil may be taken toward the bottom of a level, in the upper part of the level, or little by little throughout. The important point is that soil be collected widely over each context so that the sample represents

Sampling for Macroremains • 97

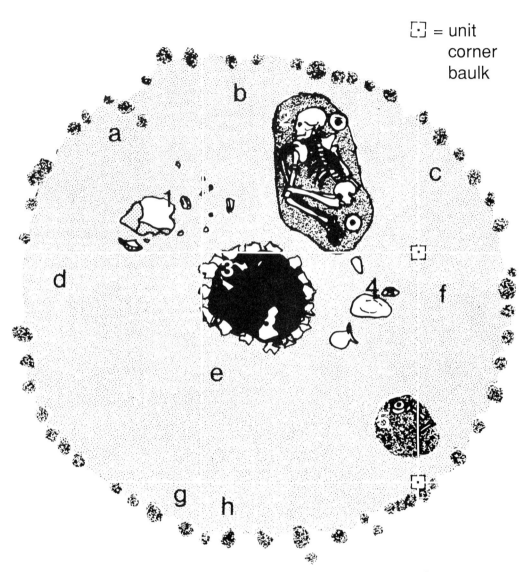

Figure 2.34 During excavation of this hypothetical Neolithic house floor, flotation samples will be taken from each discrete archaeological context. First composite samples will be taken from each 2- × 2-m excavation square (a–h), then each feature will be sampled: (1) stone tool manufacturing area; (2) burial pit; (3) hearth; (4) milling area; (5) trash pit.

the area as a whole. Natural levels in middens, sublevels in large features, and arbitrary fill levels are other appropriate contexts for composite sampling. Composite samples should always be taken from the area around features. A standard soil volume should be collected whenever possible.

Flotation samples for a sequence of fill layers, midden material, or floors in an excavation can also be taken from one area selected at random for sampling purposes. This is column sampling. In Figure 2.35 a sampling column has been left in one unit until excavation is completed. Each natural level of floor (1–5) is to be sampled separately. More than one balk can be left for sampling purposes. The advantages of column sampling are that samples can be left in place until the unit is finished, each level is clearly visible in profile for precise sampling, and all bagging, labeling, and note taking can be done at once. The major disadvantage is that samples represent only remains present in part of the excavation. This is a valid subsample, but of a different sort than a composite sample. When using the column sampling approach, be sure to take separate feature samples and to collect soil from any horizontal strata that do not occur in the column.

There are a number of situations in which sampling small, precisely located areas provides very useful information. This is point sampling. Figure 2.36 illustrates sampling one unit of floor in a 50- × 50-cm grid to obtain detailed information on activity areas. Sampling small features or the soil inside or under ceramic vessels are other examples of point sampling. Soil volume should always be taken.

Hints for Good Sampling

Sampling for flotation or fine sieving can be improved in several ways:

1. Collect a standard-size sample of soil for processing from each sampled context. As is discussed further in Chapter 3, it is important not only that the analyst know how much soil was floated to produce an assemblage of macroremains but also that samples sizes do not fluctuate dramatically among contexts. Sample size fluctuation can affect comparability of rarer remains. Obviously, it may not be possible to take a standard sample from all contexts, especially for point sampling. This is why I recommend measuring sample volume or weight again before flotation. The only way to choose an appropriate sample size is by experimentation and prior experience. Start by floating several 10-l samples collected from different contexts and evaluate the quantity of material recovered. Although there are formulas for calculating optimal sample sizes for assemblages of macroremains (see Chapter 3), a useful rule of thumb is a minimum of twenty pieces of wood. If enough soil is floated to concentrate wood to that extent, chances are that sufficient numbers of seeds are also

Sampling for Macroremains • 99

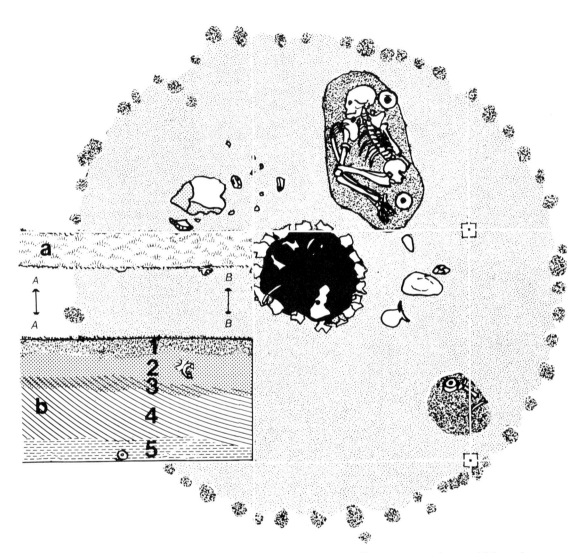

Figure 2.35 During the hypothetical house excavation, flotation samples could be taken from a soil column rather than by composite sampling near the bottom of each level. A narrow soil balk (a) is left in one 2- × 2-m excavation square; (b) shows the balk in profile. Column samples are taken from each stratigraphic level (1–5). Note that a number of important features are not represented in this soil column. These would be sampled separately, as shown in Figure 2.34.

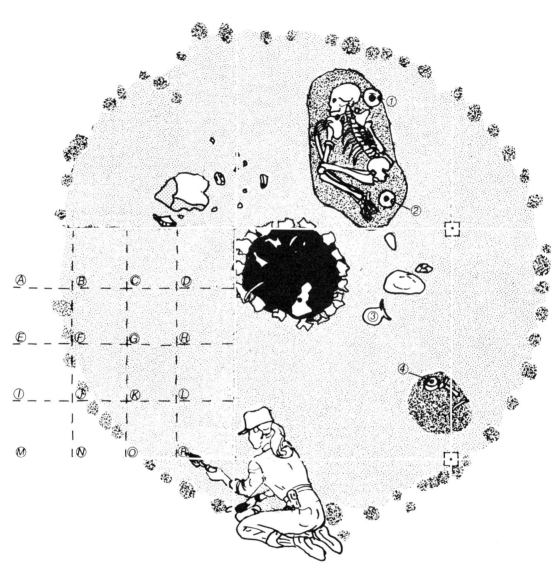

Figure 2.36 Point samples in a 50-cm grid: the archaeologist places a grid in each 2- × 2-m square to facilitate taking samples (A–P). The interiors of four ceramic vessels (1–4) will also be sampled.

present. If test samples consistently contain few seeds, however, sample size should be increased even if wood is abundant. Recovery efficiency of the flotation system should also be tested.

2. Treat soil collected for flotation gently. There is nothing wrong with collecting samples for flotation from soil passed through bulk screens, as long as proveniences are kept straight and soil has not been mashed through the screen. Small seeds and charcoal fragments in friable soil generally pass easily through ¼" screen (hence the reason for flotation in the first place). Do not pack soil tightly into flotation bags, and be careful not to put heavy artifact bags on top of soil bags or stack too many soil bags on top of each other.

3. Double-tag soil bags. Paper tags placed inside soil sample bags disintegrate rapidly in moist soil. Provenience information written with an ink marking pen on the outside of plastic bags fades quickly in sunlight. String tags used to close bags come untied or are pulled off when bags are moved. Double-tagging all bags (one label inside the bag, one outside) maximizes chances that provenience information will stay with the sample until it is processed. For inside labels, use aluminum tags or stiff paper tags wrapped in aluminum foil. Alternatively, double-bag samples, placing the internal tag between inner and outer bags, or write directly on the inner bag with indelible ink. Place the external tag so that it can be easily read. Avoid flimsy paper tags, which tear easily as bags are moved around. I recommend that all provenience data and sample numbers be written in full on both tags; be sure to include the date and excavator's initials; this may help identify a sample when all else fails. This takes a little longer, but it gives the flotation crew an original tag to put with each flotation fraction (light and heavy), two opportunities to decipher poor handwriting or smeared ink, and the chance to catch mislabeling or ambiguities before they are perpetuated.

4. Evaluate the condition of the soil. If flotation samples are moist, plastic bags should be left open for soil to dry while awaiting flotation. It may be necessary to spread out large damp samples for thorough drying. If cloth bags are used, those with moist soil should be placed in open air to dry. If samples are waterlogged and wet sieving is planned, be sure sample bags are tightly closed and check periodically to be sure soil is not drying out. It is also a good idea to close tightly samples containing desiccated remains, since dry soil may begin to pick up moisture from the air, which can lead to mold growth on materials.

5. Process soil samples with the goal of keeping pace with fieldwork. Although flotation invariably runs behind excavation, beginning soil pro-

cessing early in the field season, rather than leaving it all for the end, allows feedback on recovery to guide sampling, both in size and location of samples.

References

Aschenbrenner, S. E., and S. R. B. Cooke
    1978   Screening and gravity concentration: Recovery of small-scale remains. In *Excavations at Nichoria in S.W. Greece* (Vol. 1), edited by G. Rapp and S. E. Aschenbrenner, pp. 156–165. Minneapolis: University of Minnesota Press.

Bareis, Charles J., and James W. Porter (editors)
    1984   *American Bottom archaeology.* Urbana: University of Illinois Press.

Barghoorn, E. S.
    1944   Collecting and preserving botanical materials of archaeological interest. *American Antiquity* 9:289–294.

Barker, Graeme
    1975   To sieve or not to sieve. *Antiquity* 49:61–63.

Bodner, Connie Cox, and Ralph M. Rowlett
    1980   Separation of bone, charcoal, and seeds by chemical flotation. *American Antiquity* 45:110–116.

Bohrer, Vorsila L., and Karen R. Adams
    1977   Ethnobotanical techniques and approaches at Salmon Ruin, New Mexico. *San Juan Valley Archaeological Project Technical Series* 2; *Eastern New Mexico University Contributions in Anthropology* 8(1).

Brady, Nyle C.
    1984   *The nature and properties of soils* (9th Ed.). New York: Macmillan.

Cherry, J. F.
    1975   Efficient soil searching: Some comments. *Antiquity* 49:217–219.

Crawford, G. W.
    1983   Paleoethnobotany of the Kameda Peninsula Jomon. *University of Michigan, Museum of Anthropology, Anthropological Papers* 73.

Davis, E. Mott, and Al B. Wesolowsky
    1975   The Izum: A simple water separation device. *Journal of Field Archaeology* 2(3):271–273.

Diamant, Steven
    1979   A short history of archaeological sieving at Franchthi Cave, Greece. *Journal of Field Archaeology* 6(2):203–217.

Ford, Richard I.
    1979   Paleoethnobotany in American archaeology. In *Advances in archaeological method and theory* (Vol. 2), edited by Michael B. Schiffer, pp. 286–336. New York: Academic Press.

French, D. H.
    1971   An experiment in water-sieving. *Anatolian Studies* 21:59–64.

Gumerman, George, IV, and Bruce S. Umemoto
    1978   The siphon technique: An addition to the flotation process. *American Antiquity* 52:330–336.

Heer, O.
    1878   Abstract of the treatise on 'Plants of the lake dwellings.' In *The lake dwellings of Switzerland and other parts of Europe*, by F. Keller, translated by J. E. Lee. Vol. 1, pp. 518–544. London: Longmans, Green.
Helbaek, Hans
    1969   Plant collecting, dry-farming, and irrigation agriculture in prehistoric Deh Luran. In *Prehistory and human ecology of the Deh Luran Plain*, by Frank Hole, Kent Flannery, and James Neely. *Memoirs of the Museum of Anthropology, University of Michigan* 1:383–426.
Hole, Frank, Kent Flannery, and James Neely
    1969   *Prehistory, and human ecology of the Deh Luran Plain. Memoirs of the Museum of Anthropology, University of Michigan* 1.
Jarman, H. N., A. J. Legge, and J. A. Charles
    1972   Retrieval of plant remains from archaeological sites by froth flotation. In *Papers in economic prehistory*, edited by E. S. Higgs, pp. 39–48. Cambridge, England: Cambridge University Press.
Kaplan, Lawrence, and Shirley L. Maina
    1977   Archaeological botany of the Apple Creek site, Illinois. *Journal of Seed Technology* 2(2):40–53.
Keeley, H. C. M.
    1978   The cost-effectiveness of certain methods of recovering macroscopic organic remains from archaeological deposits. *Journal of Archaeological Science* 5:179–183.
Kenward, H. K., A. R. Hall, and A. K. C. Jones
    1980   A tested set of techniques for the extraction of plant and animal macrofossils from waterlogged archaeological deposits. *Science and Archaeology* 22:3–15.
Lange, Frederick W., and Frederick M. Carty
    1975   Salt water application of the flotation technique. *Journal of Field Archaeology* 2(2):119–123.
Limp, W. F.
    1974   Water separation and flotation processes. *Journal of Field Archaeology* 1:337–342.
Minnis, Paul, and Steven LeBlanc
    1976   An efficient, inexpensive arid lands flotation system. *American Antiquity* 41:491–493.
Payne, S.
    1972   Partial recovery and sample bias: The results of some sieving experiments. In *Papers in economic prehistory*, edited by E. S. Higgs, pp. 49–64. Cambridge, England: Cambridge University Press.
Pearsall, Deborah M.
    1980   Pachamachay ethnobotanical report: Plant utilization at the hunting base camp. In *Prehistoric hunters of the high Andes*, by John W. Rick, pp. 191–231. New York: Academic Press.
    1982   Informe preliminar sobre la primera época del campo del proyecto etnobotánico "Peñón del Rio." Ms. on file, American Archaeology Division, University of Missouri–Columbia.
    1983   Plant utilization at the Krum Bay Site, St. Thomas, U.S.V.I. Ms. on file, American Archaeology Division, University of Missouri–Columbia.

1984a *Prehistoric adaptation to the Junín Puna, Peru: the role of plant resources.* Paper presented at the 49th Annual Meeting of the Society for American Archaeology, Portland, Oregon.

1984b Preliminary report on botanical materials recovered from sites OGSE-Ma-172, OGSE-82c, OGSE-337, and OGSE-406, Ecuador. Ms. on file, American Archaeology Division, University of Missouri–Columbia.

1985 Analysis of soil phytoliths and botanical macroremains from El Bronce archaeological site, Ponce, Puerto Rico. In *Archaeological data recovery at El Bronce, Puerto Rico, final report, phase 2,* by Linda Sickler Robinson, Emily R. Lundberg, and Jeffery B. Walker, pp. B1–B46. Prepared for United States Army Corps of Engineers, Jacksonville District.

Pearsall, Deborah M., Andrea A. Hunter, and Jeffery K. Yelton

1985 Biological data. In *Osage and Missouri Indian life cultural change: 1675–1825,* edited by Carl H. Chapman. pp. 394–692. Final performance report on National Endowment for the Humanities Research Grant RS-20296.

Pendleton, Michael W.

1979 A comparison of two flotation methods. *Bulletin of the Philadelphia Anthropological Society* 31:1–8.

1983 A comment concerning "Testing Flotation Recovery Rates." *American Antiquity* 48:615–616.

Pulliam, Christopher B.

n.d. Flotation recovery tests, Old Monroe Project: Laboratory notes. Ms. on file, American Archaeology Division, University of Missouri–Columbia.

1987 Middle and Late Woodland horticultural practices in the western margin of the Mississippi River Valley. In *Emergent horticultural economies of the eastern woodlands,* edited by William F. Keegan. *Center for Archaeological Investigations, Southern Illinois University, Occasional Paper* 7:183–198.

Ramenofsky, Ann F., Leon C. Standifer, Ann M. Whitmer, and Marie S. Standifer

1986 A new technique for separating flotation samples. *American Antiquity* 51:66–72.

Schock, Jack M.

1971 Indoor water flotation: A technique for the recovery of archaeological material. *Plains Anthropologist* 16:228–231.

Shackley, Myra

1981 *Environmental archaeology.* London: George Allen and Unwin.

Stemper, David M.

1981 *Skimming, tapping, and floating: Anthropological uses of ancient plants from coastal Ecuador.* Paper presented at the 46th Annual Meeting of the Society for American Archaeology, San Diego.

Stewart, Robert B.

1976 Paleoethnobotanical report—Çayönü 1972. *Economic Botany* 30:219–225.

Stewart, Robert B., and William Robertson IV

1973 Application of the flotation technique in arid areas. *Economic Botany* 27:114–116.

Struever, Stuart

1968 Flotation techniques for the recovery of small-scale archaeological remains. *American Antiquity* 33:353–362.

Taylor, W. W. (editor)
: 1957 *The identification of non-artifactual archaeological materials.* National Academy of Science, National Research Council Publication 565.

Thomas, David H.
: 1969 Great Basin hunting patterns: A quantitative method for treating faunal remains. *American Antiquity* 34:392–401.

Towle, Margaret
: 1961 *The ethnobotany of pre-Columbian Peru.* Chicago: Aldine.

Wagner, Gail E.
: 1976 IDOT flotation procedure manual. Ms. on file, Illinois Department of Transportation, District 8.
: 1977 The Dayton Museum of Natural History flotation procedure manual. Ms. on file, Dayton Museum of Natural History, Dayton, Ohio.
: 1979 *The Green River Archaic: A botanical reconstruction.* Paper presented at the 44th Annual Meeting of the Society for American Archaeology, Vancouver, British Columbia, Canada.
: 1982 Testing flotation recovery rates. *American Antiquity* 47:127–132.

Watson, Patty Jo
: 1976 In pursuit of prehistoric subsistence: A comparative account of some contemporary flotation techniques. *Mid-Continental Journal of Archaeology* 1(1):77–100.

Weaver, M. E.
: 1971 A new water separation process for soil from archaeological excavations. *Anatolian Studies* 21:65–68.

Wiegers, Robert P.
: 1983 Flotation processing log: Osage and Missouri culture change project. Ms. on file, American Archaeology Division, University of Missouri–Columbia.

Williams, D.
: 1973 Flotation at Sīrāf. *Antiquity* 47:288–292.
: 1976 Preliminary observations on the use of flotation apparatus in Sussex: Part XII of rescue archaeology in Sussex, 1975, edited by P. Drewett. *Bulletin of the Institute of Archaeology* 13:51–58.

Wolynec, R. B., and Bernie Mertz
: 1968 Recovery of small scale archaeological remains: A workable water-separation technique for the laboratory. Ms. on file, Department of Anthropology, New York State University–Buffalo. Cited in Schock 1971.

Yacovleff, Eugenio, and Fortunato L. Herrera
: 1934 El mundo vegetal de los antiguos Peruanos. *Revista del Museo Nacional, Lima* 3(3):241–322.
: 1935 El mundo vegetal de los antiguos Peruanos. *Revista del Museo Nacional, Lima* 4(1):29–102.

Chapter 3 **Identification and Interpretation of Macroremains**

Introduction

Recovering botanical macroremains from archaeological sites by screening, flotation, or naked eye creates a body of data that must be analyzed to yield useful information on the relationships between human and plant populations in the past. Increasing use of flotation has led to a dramatic increase in quantity of material recovered, resulting in backlogs at laboratories specializing in botanical analyses (Ford 1979). As a consequence, large archaeological projects may employ their own botanical specialists; general archaeologists and students may carry out their own ethnobotanical analyses. Applying a few standard procedures, both in analysis and in final presentation of research results, will increase comparability within the rich body of paleoethnobotanical data being generated today.

In this chapter I describe basic techniques used by paleoethnobotanists to analyze macroremains. This discussion falls into five sections: initial processing of samples, comparative collections, basic identification procedures, specialized techniques, and interpretation and presentation of results. I stress identification and interpretation of macroremains rather than identification of specific taxa. References to identification manuals are included in the discussion. Although I focus on procedures and approaches that I use, credit for development of these is owed to many ethnobotanists.

Initial Processing of Samples

Flotation and fine sieving concentrate botanical materials present in soil. Unfortunately, charred, waterlogged, or desiccated macroremains are usually mixed with rootlets, twigs, and other modern organic materials of similar density or size. In addition, fine sieving adds an inorganic component to each size fraction. Although it is generally not too difficult to determine which materials are ancient, and thus of interest to the archaeologist, separating these from the rest is time consuming and tedious. Hand sorting samples may require much of the time budgeted for macroremain analysis, yet this is the most common method employed. I focus here on sorting flotation samples, since they are the most common sample type. Most procedures apply to wet or dry fine-sieved samples as well; special requirements of these samples are briefly reviewed. I also address alternative separation methods and discuss aspects of sample-size selection. The processing of preserved human feces, or coprolites, is discussed in Chapter 4.

Basic Hand-Sorting Procedures

Although the aim of sorting light flotation fractions is simple—to remove archaeological botanical materials from buoyant rootlets, twigs, and other modern material—several procedures make this process easier and end results more comparable to other analyses.

The first and most important step in sorting is to be sure that complete and correct provenience information is transferred to lab forms and labels. At the University of Missouri lab, for example, sorters fill out the top of a flotation form (Fig. 3.1) before doing anything else. Thus, if there is any problem reading or interpreting provenience information, this is dealt with before errors spread onto sample tags and labels. If the archaeologist has provided an inventory of samples, provenience data on bags are checked against the inventory; otherwise, an inventory is compiled before sorting begins. This may be sent to the archaeologist for provenience checks and additional information on samples.

The form we use at the Missouri lab has spaces for both flotation and carbon (hand-picked or screen material) sample numbers. If flotation samples have not been assigned numbers in the field, numbers are assigned in the lab. This provides each sample with a short, discrete designation that can be used on small gelatin capsule tags. It also provides a means of easily ordering forms and samples. This number or a similar designation should never be separated from any sorted subsample. Similarly, screen samples or charcoal collected in situ during excavation are assigned carbon sample numbers. I refer later to other features of the sorting/identification form shown in Figure 3.1. This form is not perfect, but it does serve its purpose: to

Project _____	Provenience _____
Flotation no. _____	_____
Carbon no. _____	_____
Sample size _____	_____
Wt. of total sample _____	_____
Wt. of sub-sample _____	_____
Percent of sample examined _____	

	<2.0	>2.0	
Seeds	____	____	Sorted by _____
Wood	____	____	Date _____
Nuts	____	____	Identified by _____
Other	____	____	Date _____
None	____	____	

NUTS	COUNT	COMMENT
Unidentifiable		
Total		
WOOD		
Diffuse porous		
Ring porous		
Unidentifiable		
Total		

Figure 3.1 Flotation form used at the University of Missouri, American Archaeology Division, Paleoethnobotany Laboratory. Seeds and other materials are recorded on the back side.

provide a place where *all* information about materials in a sample and the process of sorting and identifying those materials can be kept.

For their research at Salmon Ruin Pueblo, Bohrer and Adams (1977) developed a series of forms for recording paleoethnobotanical data. These forms were set up so that data could be transferred from them to a SELGEM-based computer file of remains. Their Ethnobotanical Cover Sheet (Fig. 3.2) serves a purpose similar to the top portion of the Missouri lab form. Provenience data are transferred from sample tags to the cover sheet, which is then attached to other forms. Their Flotation Data Record (Fig. 3.3) consists of three or more sheets for recording basic data about flotation samples. The first sheet is for recording generalized information about the sample (who sorted it, sorting time, volume of each size fraction). The second sheet is used to record the presence or absence of various nonbotanical items. Identifications are recorded on the third sheet.

After provenience information is transferred, total weight of the light fraction is taken and recorded. The sample is then divided into several size fractions, or splits (Fig. 3.4). I usually divide samples into two splits with a 2.0-mm geological sieve. Additional sieves, creating more size fractions, may be used for flotation sorting and are commonly used in processing dried or waterlogged samples. Dividing flotation samples into size fractions serves two purposes. First, it is easier to sort materials of like size. Most dissecting microscopes have a narrow depth of focus; it is tedious to sort material if one has to continually refocus the microscope on objects of different sizes. Second, distinctive materials may be removed from the different size fractions. For the remainder of this discussion, I describe sorting flotation samples with charred preservation using two size splits—one greater than 2.0 mm, one less than 2.0 mm.

After the sample is divided, each fraction is individually examined. The >2.0-mm split (i.e., material in the sieve) can usually be sorted by naked eye or with an illuminated magnifier. At the University of Missouri lab, sorters remove all charred material from the >2.0-mm fraction (Fig. 3.5). This may include wood charcoal, nut shell fragments, large seeds, corn cupules, tuber fragments, or palm pits, depending on the site. Students quickly learn to recognize wood charcoal and are asked to count and weigh the amount of wood present and to place it in a capsule or vial. As they learn to recognize other remains in the >2.0-mm split, they make finer classifications, until all discrete classes of remains are counted, weighed, and placed in separate containers. The level of classification achieved at this point is fairly broad—wood, nut, corn, amorphous dense tissue, large seed, and so on—but this speeds later final identification of materials. If sorting is done by an experienced ethnobotanist, final identifications may be made during sorting. However, less experienced workers are often assigned sorting duty, making a separate identification process necessary.

Form Key No.

ECS Page 1 of____

ETHNOBOTANICAL COVER SHEET

Item Key No.

001 01

Nature of Material

- [ ] POLLEN
- [ ] BOTANICAL SAMPLE
- [ ] BROADLY PRO-VENIENCED ORGANIC MATERIAL
- [ ] UNDETERMINED DISSEMINULE

005 01 Sample Number 16

010 01 Record Key Number

020 01 Room 16 21 Strat(s) 23 27 – 31 33 37 Quad/T.T. 41

Grid 1 44

Grid 2 54

Exact Provenience E-W N-S 64

020 02 Depth 1 16 Depth 2 21 Date Excavated 26 D D 28 Month 33 Y Y

030 01 Associated Features: 16 21 26 31 36

040 01 Reasons for taking sample & general comments by excavators:

The information below applies to laboratory analysis and varies for Pollen, Botanical Samples, etc.:
Examination Date & Recorder(s); or Date Pollen Extracted & Extractor's Initials

050 01 D D 16 Month 19 Y Y 23 Initials 26 31 36

099 01 General Comments by Laboratory Personnel:

LA 8846
Jan. 10, 1976

Figure 3.2 Ethnobotanical Cover Sheet used during research at Salmon Ruin Pueblo, New Mexico. This form is attached to other forms detailing contents of pollen, botanical, or other organic samples (from Bohrer and Adams 1977:78).

```
┌─────────────────────────────────────────────────────────────────────────────┐
│  ┌─┬─┬─┬─┬─┐                                         FDR  Page 1 of_____   │
│  └─┴─┴─┴─┴─┘                                                                │
│  Form Key No.              FLOTATION  DATA  RECORD                          │
│  ┌─┬─┬─┐                                                                    │
│  │0│0│0│                                                                    │
│  └─┴─┴─┘                                                                    │
│  Item Key No.                                                               │
│       001    FLOTATION                                                      │
│                                                    070      Sorter          │
│       005    Sample Number                                ┌─┬─┬─┬─┐         │
│              ┌─┬─┬─┬─┬─┐                                  └─┴─┴─┴─┘         │
│              └─┴─┴─┴─┴─┘                                  2nd Sorter        │
│       010    Record Key No.                               ┌─┬─┬─┬─┐         │
│              ┌─┬─┬─┬─┬─┐                                  └─┴─┴─┴─┘         │
│              └─┴─┴─┴─┴─┘                                  3rd Sorter        │
│       071    Floater                                      ┌─┬─┬─┬─┐         │
│              ┌─┬─┬─┬─┬─┐                                  └─┴─┴─┴─┘         │
│              └─┴─┴─┴─┴─┘                                                    │
│                                           D D      M M M      Y Y   Date    │
│       072    Hours Minutes               ┌─┬─┐    ┌─┬─┬─┐    ┌─┬─┐   Sorting│
│              ┌─┬─┬─┬─┐                   └─┴─┘    └─┴─┴─┘    └─┴─┘   Completed│
│              └─┴─┴─┴─┘                                                      │
│              Time to Sort                                                   │
│                                                                             │
│              073  ┌─┬─┬─┬─┐ ml. = Total Volume of Sample before floating    │
│                   └─┴─┴─┴─┘                                                 │
│                   ┌─┬─┬─┬─┐ ml. = Total Volume of Sample after floating     │
│                   └─┴─┴─┴─┘                                                 │
│  ─────────────────────────────────────────────────────────────────────────  │
│         Particle Size Volumes              Amount Examined                  │
│       074  ┌─┬─┬─┬─┐                     ┌─┬─┬─┬─┐                          │
│            └─┴─┴─┴─┘ ml.= Particle Size #1  └─┴─┴─┴─┘ ml.= Particle Size #1 │
│  ─────────────────────────────────────────────────────────────────────────  │
│       075  ┌─┬─┬─┬─┐                     ┌─┬─┬─┬─┐                          │
│            └─┴─┴─┴─┘ ml.= Particle Size #2  └─┴─┴─┴─┘ ml.= Particle Size #2 │
│  ─────────────────────────────────────────────────────────────────────────  │
│       076  ┌─┬─┬─┬─┐                     ┌─┬─┬─┬─┐                          │
│            └─┴─┴─┴─┘ ml.= Particle Size #3  └─┴─┴─┴─┘ ml.= Particle Size #3 │
│  ─────────────────────────────────────────────────────────────────────────  │
│       077  ┌─┬─┬─┬─┐                     ┌─┬─┬─┬─┐                          │
│            └─┴─┴─┴─┘ ml.= Particle Size #4  └─┴─┴─┴─┘ ml.= Particle Size #4 │
│  ─────────────────────────────────────────────────────────────────────────  │
│       078  ┌─┬─┬─┬─┐                     ┌─┬─┬─┬─┐                          │
│            └─┴─┴─┴─┘ ml.= Fine Particle     └─┴─┴─┴─┘ ml.= Fine Particle    │
│                          Residue                          Residue           │
│  ─────────────────────────────────────────────────────────────────────────  │
│       099   Comments on this Sample: _____   │
│             _____  │
│  LA  8846   _____  │
│  Jan. 10, 1976                                                              │
└─────────────────────────────────────────────────────────────────────────────┘
```

Figure 3.3 Flotation Data Record used during research at Salmon Ruin Pueblo to record contents of flotation samples (from Bohrer and Adams 1977:71–73). (Figure continues.)

Item Key No.					Location in Screen Size (X one or more boxes)						Approximate No. of item in total Sample			S=separate from residue R= remain in residue
			760	Feces (all types)	762	1	2	3	4	5	1-10	11-25	26+	S
			760	Insect segments; record whole insects under comments	762	1	2	3	4	5	"	"	"	S
			760	Cocoons, larvae, eggs	762	1	2	3	4	5	"	"	"	S
			760	Juniper twigs	762	1	2	3	4	5	"	"	"	R
			760	Juniper scale leaves	762	1	2	3	4	5	"	"	"	R
			760	Bones	762	1	2	3	4	5	"	"	"	S
			760	Secondary retouch flakes	762	1	2	3	4	5	"	"	"	S
			760	Spines or thorns	762	1	2	3	4	5	"	"	"	S
			760	Black spherical items	762	1	2	3	4	5	"	"	"	S
			760	Snail shells	762	1	2	3	4	5	"	"	"	S
			760	Cuticular layer inside seed coats	762	1	2	3	4	5	"	"	"	R
			760	Bud tips	762	1	2	3	4	5	"	"	"	S
			760	Catkins	762	1	2	3	4	5	"	"	"	R
			760	Non-black spherical items	762	1	2	3	4	5	"	"	"	S
			760		762	1	2	3	4	5	"	"	"	
			760		762	1	2	3	4	5	"	"	"	
			760		762	1	2	3	4	5	"	"	"	
			760		762	1	2	3	4	5	"	"	"	
			760		762	1	2	3	4	5	"	"	"	
			760		762	1	2	3	4	5	"	"	"	
			760		762	1	2	3	4	5	"	"	"	
			760		762	1	2	3	4	5	"	"	"	
			760		762	1	2	3	4	5	"	"	"	
			760		762	1	2	3	4	5	"	"	"	
			760		762	1	2	3	4	5	"	"	"	
			760		762	1	2	3	4	5	"	"	"	

FDR Page 2 of ___
Presence–Absence Categories Apply to the Entire Sample

LA 8846
Jan. 10, 1976

Figure 3.3 (*Continued*).

114 • Chapter 3 Identification and Interpretation of Macroremains

FLOTATION DATA RECORD
second sheet

FDR Page____of____

Form Key No.

Sample No. or Record Key No.
not keypunched

Item Key No.

760 01   Name from Plant Taxonomic Dictionary   02 FLOTATION

766 01   Taxonomic Code Number | ID level Taxon | ID No. | Parts Code Number | Quantity of Item | Particle Size

767 01   Descriptors
Burned   Color   Two-Tone   Damaged   Deflated   Fragment   Hairy   Modern
Parched Shiny   Split on Suture   Swollen   Immature   Matrix

775 01   Comments on this entry:_____

Item Key No.

760 01   Name from Plant Taxonomic Dictionary   02 FLOTATION

766 01   Taxonomic Code Number | ID level Taxon | ID No. | Parts Code Number | Quantity of Item | Particle Size

767 01   Descriptors
Burned   Color   Two-Tone   Damaged   Deflated   Fragment   Hairy   Modern
Parched Shiny   Split on Suture   Swollen   Immature   Matrix

775 01   Comments on this entry:_____

Item Key No.

760 01   Name from Plant Taxonomic Dictionary   02 FLOTATION

766 01   Taxonomic Code Number | ID level Taxon | ID No. | Parts Code Number | Quantity of Item | Particle Size

767 01   Descriptors
Burned   Color   Two-Tone   Damaged   Deflated   Fragment   Hairy   Modern
Parched Shiny   Split on Suture   Swollen   Immature   Matrix

775 01   Comments on this entry:_____
LA 8846
Jan. 10, 1976

Figure 3.3 (*Continued*).

Figure 3.4 Samples are split into size fractions before sorting: (a) the sample is poured into the top sieve; (b) the sieve stack is shaken gently so that material falls through; (c) a bag is labeled for each size split; (d) sieve contents are poured carefully into labeled bags.

Once the >2.0-mm split sorting is completed, the <2.0-mm size fraction is examined (Fig. 3.6). The procedure described below is also followed if multiple splits are to be sorted (e.g., 1.0, 0.5 mm). This work is best done at 10–15 × magnification under an illuminated dissecting microscope. Use of a high-intensity light, such as those sold for microscopes, reduces eye fatique and increases sorting reliability. Fiber optic lights are excellent. Fatigue is one of the main factors contributing to poor sorting. Limiting uninterrupted sorting time to no more than two hours also helps combat this problem. At the Missouri lab, sorters are instructed to remove only ancient seeds (whole or broken) and "special" materials from the <2.0-mm fraction—this means only charred seeds in most flotation samples. We generally do

116 • Chapter 3 Identification and Interpretation of Macroremains

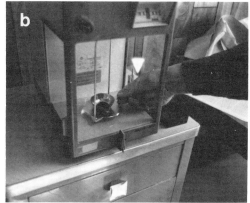

Figure 3.5 Sorting the large (>2 mm) size fraction: (a) all botanical material is removed from the sample with forceps and is placed by type into labeled capsules or vials; (b) after sorting, each type of material is counted, then weighed.

not remove modern seeds from samples. "Special" materials are classes of remains usually found in the >2.0-mm fraction, but which are of particular importance and so removed whenever they are observed. Into this category fall remains of cultivated plants, such as corn kernel and cupule fragments and gourd and squash rind.

We do not usually remove wood from the <2.0-mm split. Nancy and David

Figure 3.6 Smaller size fractions are sorted under low magnification: (a) the sorter brushes material into the illuminated microscope field and examines it; residue is then moved aside and new material is examined; (b) small seeds are removed from the sample with a small paintbrush and placed in gelatin capsules by type.

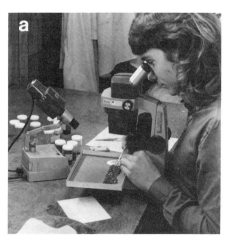

Asch carried out sorting experiments on midwestern samples which showed that removing all charcoal fragments from a sample (including the <2.0-mm split) and using total weight in analysis did not substantually alter ratios such as nut–wood weight based only on >2.0-mm data. Additionally, it often proves impossible to determine species or genus of very small pieces of wood. Unless scanning electron microscopy or high-magnification light microscopy is used to aid identification, sorting out all wood flecks in the small size fractions of flotation is wasted effort. However, certain preservation situations may call for finer sorting of wood. In her sites from Iran, for example, Naomi Miller observed that most wood was present in pieces less than 2.0 mm. She therefore sorted all wood from a >1.0-mm split for identification. One must evaluate preservation conditions and decide what sieve sizes to use at the beginning of a project; it may be necessary to sample or completely sort small size grades for charcoal.

Ethnobotanists use a variety of small hand tools and methods for sorting flotation samples. Small seeds can be picked up with a fine paintbrush; static electricity is usually strong enough to hold the seeds for the journey from sample to container. Fine watchmaker's forceps can also be used to pick up seeds, but these require a gentle touch. Moving the sample through the microscope field for examination can be done with paintbrush, probe, small spatula, or a piece of index card. Sorting trays can be made of paper plates, stationery box lids, or small metal trays. Avoid plastic trays because of the static electricity of plastic surfaces. I prefer a square or rectangular flat-bottomed tray, with a smooth surface and short sides, small enough to rest securely on the viewing platform of a dissecting microscope. I place the sample in the center of the tray and spread it out in a narrow band so that it can be passed from left to right under the viewing field of the microscope (right to left for left-handed sorters). I push examined small split toward the edge of the field, leaving a clear division between examined and unexamined material. I spread materials thinly so that all remains can be easily scanned. A vertical line drawn on the tray makes a useful center point for examining materials.

Bohrer and Adams (1977) recommend a somewhat more systematic method of sorting small splits. They standardize the volume of the sample to be examined at one time as the amount that can be packed contiguously into the field of view used for sorting. This standard volume is then spread over a sorting plate on which a number of lines or squares are drawn. The grid aids in systematic examination of the sample. These researchers feel that examining smaller subsamples increases sorting accuracy, gives the sorter a greater feeling of accomplishment, and creates a standard procedure that increases comparability of results for a long-duration project. Whether or not a standard volume of sample for each sorting session is chosen, large samples should be examined little by little. Thick piles of light fraction cannot be examined accurately.

Large gelatin capsules (size 000) are useful containers for seeds. I ask beginning sorters to place seeds by type into capsules held upright in a holder fashioned from styrofoam, cardboard, or wood. As the student gains expertise at recognizing different kinds of seeds, this type classification becomes finer. Even beginners can sort seeds by size, shape, and surface sculpturing, which helps speed later identification. Although gelatin capsules are handy in the lab, they may not be appropriate for long-term storage of botanical materials. In dry climates capsules become brittle and can shatter; in damp conditions they become sticky and soft. In locations with about 50% relative humidity they are acceptable for long-term storage (R. Ford, personal communication). Small glass or plastic vials and small boxes are alternative containers for long-term storage. If glass vials are used, pad them to avoid breakage.

After the <2.0-mm split has been thoroughly examined and all ancient seeds removed, quantity of each type of seed is entered on the flotation form; descriptions are given if seeds are not yet identified. These counts may not stand as final tallies, since seed types are often mixed by less-experienced sorters. But having the learner briefly describe each seed type and count its occurrence encourages more accurate separation and rapid learning of types.

The final steps in light fraction sample sorting are to label all vials and capsules with sample provenience or flotation number, bag up remainders (residues from <2.0-mm and >2.0-mm splits), and file flotation forms and samples. It may also be necessary to examine heavy fraction portions of samples for nonbuoyant charred materials. As discussed in Chapter 2, certain types of charcoal, notably some nut shell and dense wood, do not readily float in water. If samples were damp when floated, considerable quantities of charred material may be present in heavy fractions. Results of analysis will be skewed by omission or underrepresentation of less buoyant plant material "lost" in heavy fractions. This type of nonrandom data loss can be avoided by spot-checking heavy fractions and removing charred material (by hand or by chemical flotation).

It is fine to be told to remove seeds from flotation samples, but how does one learn to recognize seeds, or to teach someone how to recognize them? During an ongoing project, or at an established lab, new workers can be easily introduced to seeds they will encounter in flotation samples. Photographs, line drawings, and actual charred examples of archaeological seeds can be used to familiarize the beginner. Such visual aids should be accompanied by discussion of diagnostic features of each seed, how the seed appears when broken or without its seed coat, how fresh examples are separated from charred ones, and how and when to try to distinguish among very similar taxa. Once the sorter has seen examples of seeds, recognizing them in samples is just a matter of practice. By checking sample residues, the analyst can determine what types of seeds are being missed and work with sorters to increase recognition of those types. Smaller, less-distinctive seeds are often overlooked by the beginner. Recognition of broken seeds is another skill that comes with experience.

Distinguishing charred seeds from their fresh counterparts is probably the biggest headache for the learner. Seeds that are black in the uncharred state (e.g., *Chenopodium, Amaranthus,* or *Portulaca*) and common in flotation samples from upper strata of sites cause the most problems. If hundreds of such modern seeds are removed inadvertently, this adds hours of sorting labor. Fresh black seeds are in fact quite distinctive from charred seeds. A number of "black" seeds have other hues present in the seed coat, such as red or brown, or have telltale attachments, such as the white tissue near the attachment area in *Portulaca.* The modern seed is often soft or even sticky. One way of determining whether a seed is charred or not is, of course, to break it open. The problem comes when one discovers that the destroyed seed was in fact charred, and rare. I usually instruct beginning sorters to pull out all seeds that they think are charred, including those they are not sure of. I then sort out any modern, black seeds and have the student study them, break them open, and compare them to charred examples. Doing this for a sample or two usually substantially reduces the number of uncharred seeds removed. If sorters encounter what they think is a charred example of a seed usually observed fresh, they take it out for later checking. The lab rule is "If in doubt, pull it out."

Learning to recognize ancient materials in flotation samples is more difficult for the student working in isolation on his or her own project. In this case, published seed identification manuals, such as Corner (1976), Deleroit (1970), Gunn (1972), Martin and Barkley (1961), Montgomery (1977), and Musil (1963), can help the learner get started. Photographs in such manuals help with seed recognition and suggest preliminary identifications. A valuable feature of Martin and Barkley (1961) and Gunn (1972) is discussion of seeds by botanical family. In Martin and Barkley's work, for example, family level seed characteristics are discussed, seeds of common genera described, and drawings of important features such as embryo position and attachment area presented. Published descriptions provide a good lead, but comparison with actual specimens is essential to confirm an identification. Common weed seeds and crop seeds can be obtained from the U.S. Department of Agriculture. These can serve as the core of a study collection for the beginner. Herbaria are also sometimes willing to allow removal of seeds of common plants from voucher specimens. If permission is granted for this, a study collection can be built up of seeds likely to be encountered in samples. By charring some naturally black seeds, one can learn how to distinguish fresh seeds from charred. I return below to the topic of how to build a comparative collection.

Hand sorting flotation samples, especially removing charred seeds from small-sieve fractions, is very time consuming. The experienced ethnobotanist can take advantage of two shortcuts to speed this aspect of sorting. These involve tallying, but not removing, seeds from samples. If time is short to finish a project, I sometimes examine <2.0-mm splits and tally seed occurrences without pulling known seeds from samples. I remove any unknowns or seeds with questionable identifications for further examination. Accurate seed counts can be obtained this way, but counts and identifications cannot be checked except by sorting again. If patterning in seed data is only to be discussed in terms of presence (ubiquity), rather than frequency (see "Presenting and Interpreting Data"), scanning for presence or absence of seeds can further speed analysis. Toll (1988) recommends scanning for presence or absence as a means of quickly evaluating seed occurrence in samples.

Subsampling Large Flotation Samples

Although the volume of soil floated can be adjusted to produce samples of a sufficient, yet manageable size, flotation light fractions sometimes turn out to be larger than necessary. Rather than sorting entire samples, it may be possible to subsample, sorting only 25–50%, and still obtain representative seed, wood, and nut assemblages. Subsampling speeds the sorting and identification process, allowing a greater number of discrete samples to be processed.

Techniques for subsampling are straightforward. Basically, one seeks to produce a random subsample. A riffle box is commonly used (Fig. 3.7). Passing a sample once through the riffle box produces two equal splits, which can then be resplit to produce smaller subsamples. Random samples of flotation light fraction are obtained by pouring material slowly into the riffle box with numerous back-and-forth motions. In this way, all size fractions of material are spread over all riffle box openings.

If a riffle box is not readily available, a series of small boxes, or a large box marked in a grid pattern, can be used. For example, 36 boxes, each the same size and shape, paper-clipped together into a 6 × 6 grid, can be used as a sample splitter. The sample can be poured into the small boxes with a back-and-forth motion, with several passes made over the entire grid. If each box is numbered, a subsample of any size can be easily obtained by randomly selecting the required number of boxes, unclipping those selected, and combining contents into a subsample for sorting.

Van der Veen and Fieller (1982) compared three methods of random sampling seeds: spoon sampling, riffle box sampling, and grid sampling. A bag of charcoal with known numbers of modern seeds was used in their tests. The spoon method (scooping out spoonfuls of material from a thoroughly mixed sample) was found to give unreliable indications of seed composition of the test sample. Both riffle box and grid methods gave good results. These authors noted, however, that the grid method was time consuming and more susceptible to bias (conscious, even spreading of materials) than the riffle box, although both gave equally acceptable results.

Figure 3.7 Sample poured through a riffle box.

As important as how to take a random subsample is how to choose a sufficient sample size. This issue is also relevant to selecting soil volume for flotation, as was discussed briefly in Chapter 2. How many pieces of wood and nut shell, or how many seeds, must be recovered to give an accurate representation of the assemblage of plants preserved? There are two sides to this question. I may wish to know, for example, how many seeds to examine to get reliable ratios of the most common types. If there are ten common seeds, what is the minimum number of seeds per sample for reliable calculation of percentage occurrence of those types? But I will also want to determine the total range of seeds that occur at a site, including types that may be quite rare. What seed count per sample gives a good chance of encountering rarely occurring seeds?

There are several ways to approach these questions. One is to conduct sorting experiments. In this approach, a sample is sorted in its entirety by small subsamples, and materials found in each subsample are identified and tallied. Table 3.1 shows seed data resulting from such an experiment. A light fraction sample from the Old Monroe site weighing 62.7 g was split into four parts using a riffle box. Each split was sorted and the seeds removed and tallied. In this example, the entire sample had to be sorted to recover the full range of seeds present. The final rare seed

Table 3.1 • Distribution of Seed Taxa among Sample Splits

Seed inventory	Count	Sample[a]			
		S_1	S_2	S_3	S_4
Chenopodium	132	36	16	43	37
Hordeum	108	26	28	23	31
Phalaris	43	7	12	8	16
Iva	28	7	9	3	9
Polygonum	24	9	9	0	6
Gramineae	20	11	0	5	4
Compositae	1	0	0	0	1
Portulaca	1	0	0	0	1
Galium	1	0	1	0	0
	358	96	75	82	105
No. missing taxa		3	3	4	1

[a]Light fraction weight, 62.7 g.

types, Compositae and *Portulaca*, were encountered in the last subsample sorted. Note that there were taxa "missing" from each 25% split: Three from S_1 and S_2, four from S_3, and one from S_4. "Missing" taxa included not only rare types (i.e., single occurrence) but also taxa for which 20 and 24 seeds were present. Clearly, sorting only 25% of this sample would have given an insufficient seed sample, since moderately common seeds as well as rare seeds would have been missed. Repeating this experiment several times would give one a fairly accurate idea of minimum sample size for recovery of rare or moderately rare taxa. Further discussion of sorting tests can be found in Green (1979).

We can take analysis of our sorting experiment one step further to see how relative abundances of seeds change as sample size is increased. The last column in Table 3.2 shows ranking by relative abundance of seeds in the sample just illustrated. Comparing this to the other columns shows that sorting 25% gives an inaccurate ranking of four of the six most common taxa, that sorting 50% adds an accurate ranking of *Phalaris* but reverses ranking of the two most common taxa, and that sorting 75% results in accurate ranking of all common taxa. Again, one should repeat the experiment on several more samples to arrive at an acceptable minimum sample size.

Another way to interpret the results of a sorting experiment is to run a Monte Carlo simulation of seed occurrence using data from one sample. In a simulation, a running average is calculated for each seed type using all possible combinations of subsamples. By graphing the averages for a number of taxa (e.g., one common taxon, one moderately occurring taxon, and one rare taxon), one can see when redundancy

Table 3.2 • Percentage Frequency of Seeds within Subsamples

Seed inventory	S_1 (%)	$S_1 + S_2$ (%)	$S_1 + S_2 + S_3$ (%)	Total (%)
Chenopodium	37.5	30.4	37.5	36.9
Hordeum	27.1	31.6	30.4	30.2
Phalaris	7.3	11.1	10.7	12.0
Iva	7.3	9.4	7.5	7.8
Polygonum	9.4	10.5	7.1	6.7
Gramineae	11.5	6.4	6.3	5.6
Compositae	0	0	0	0.3
Portulaca	0	0	0	0.3
Galium	0	0.6	0.4	0.3
Total seeds	96	171	253	358

is reached (i.e., when adding one more subsample does not substantually change the average). An application of simulation is discussed in Chapter 5 (see pp. 396–397).

Van der Veen and Fieller (1982) present formulas for calculating the sample size required to achieve a given confidence level. The variants of the formulas presented below are those used when the number of seeds in the target population (archaeological deposit) can be assumed to be very large. As an example of van der Veen and Fieller's approach, I discuss sampling experiments carried out by Wohlgemuth (1984) using botanical remains recovered from a site in northern California.

The number n of seeds that must be recovered to estimate accurately the proportion of a seed type in an archaeological deposit is given by

$$n = P(1 - P)(Z_\alpha/d)^2 \tag{1}$$

where P is the true proportion of a species in the target population, d is the acceptable error, and Z_α is the confidence interval standard score corresponding to the desired acceptable error. But we cannot know P for any seed type without floating and sorting the whole deposit. Instead, we do one of two things: calculate the upper limit of the necessary sample, or use a subsample p instead of P and calculate its reliability. To determine the upper size limit, we use $P = 50\%$. Thus

$$n = .5(1 - .5)(Z_\alpha/d)^2$$
$$= \frac{(Z_\alpha/d)^2}{4} \tag{2}$$

To use data from a floated subsample, we simply replace P with p:

$$n = p(1 - p)(Z_\alpha/d)^2 \tag{3}$$

To assess the reliability $(1 - \alpha)$ of p as an estimate of P, we use the following formula:

$$1 - \alpha = p \pm Z_\alpha \left[\frac{p(1 - p)}{(n - 1)} \right]^{1/2} \quad (4)$$

Wohlgemuth (1984) sorted several samples from different contexts at a site in Redding, California, and determined the subsample proportion of manzanita seeds from each context. Manzanita seeds were common in the samples. A sample from the House 3 roof, for example, had 45.51% manzanita, and a sample from the House 1 roof had 72.71% manzanita. To calculate the number of seeds that must be identified to get an accurate estimate of the total population, we apply Equation 3. For House 3 roof,

$$n = .4551(1 - .4551)(Z_\alpha/d)^2$$

A common acceptable error is 5%, so $d = .05$ and the corresponding standard score for 95% confidence is $Z_\alpha = 1.96$. Thus

$$n = .4551(1 - .4551)(1.96/.05)^2$$
$$= 381 \text{ seeds}$$

For the House 1 roof sample, for which $p = 72.71\%$,

$$n = .7271(1 - .7271)(1.96/.05)^2$$
$$= 305 \text{ seeds}$$

Because the number of seeds recovered per liter of soil is known, the number of liters needed to produce desired counts can be calculated. Wohlgemuth's initial sorting from the House 3 roof context showed a recovery rate of 47 seeds per liter of soil, so $381/47 = 8.1$ l of soil must be floated to accurately estimate the observed proportion of manzanita seeds. For the House 1 context, 435 seeds per liter of soil were recovered, so only $305/435 = 0.7$ l need be floated.

The reliability of the subsample p of seeds as an estimate of the true proportion P is calculated with Equation 4. For the House 3 roof material,

$$1 - \alpha = .4551 \pm 1.96 \left[\frac{.4551(1 - .4551)}{(383 - 1)} \right]^{1/2}$$
$$= .4551 \pm .0501$$
$$= 40.50\text{–}50.52\%$$

True proportion, P, thus falls between 40 and 50%. Such calculation can be repeated for other seed taxa and other test samples until a seed total figure is determined which ensures adequate sampling in all site contexts and for all seed types.

If P for manzanita had been assumed to be 50%, an upper sample size limit would be calculated, by Equation 2, as

$$n = \frac{(1.96/.05)^2}{4}$$

$$= 384 \text{ seeds}$$

Note that this is greater than that calculated using known proportions of manzanita from House 3 roof samples (381) or House 1 roof samples (305).

This discussion of how to determine an adequate sample size for analysis of seeds and other macroremain data has grown out of discussion of subsampling flotation. In Wohlgemuth's excavation, floating only 0.7 l of soil from the House 1 roof area would have given an adequate sample of seeds, when in fact much more soil was processed. Since it is best to take standard-size flotation samples from all site areas, contexts with exceptional seed preservation should be subsampled in the lab, after soil is floated. But what if sorting tests or calculations show that an insufficient number of seeds for accurate determination of seed proportions were present in flotation samples? This situation can easily arise, especially if examination of flotation samples does not begin until after fieldwork is completed. Sample size chosen to fit deposits tested early in the field season may not be appropriate for contexts encountered later; recovery of wood and nuts may be adequate, while too few seeds are recovered. There are two things to remember if this happens. First, the analyst determines the level of acceptable confidence and error in the study. Although 95% confidence intervals are commonly chosen, 90% or 85% still represent a high degree of statistical accuracy. Second, it is often possible to increase sample size by combining individual flotation samples. In my experience, it is more common to group data by context (e.g., combining several samples from a hearth or house floor) than to base calculations on individual samples. If a blanket flotation sampling strategy is employed, there will be sample redundancy.

Alternatives to Hand Sorting

Removing charred botanical remains from light fraction flotation samples, even small ones, is quite time consuming. Additional time is added if heavy fractions must also be spot-checked or sorted for nonbuoyant remains. As discussed in Chapter 2, the problem of nonbuoyant material can be handled by refloating heavy fractions in a heavy liquid (chemical flotation). Materials recovered by this second flotation can be added to light fractions before sorting or added to appropriate categories (<2.0 mm or >2.0 mm) after sorting.

Bodner and Rowlett (1980) investigated chemical flotation as a method for separating not only dense charcoal (nuts, wood) from inorganic matrix but light seeds from charcoal. Good results were achieved separating charcoal from inorganic matrix using ferric sulfate. Separating seeds from charcoal using ethyl alcohol or acetone gave less consistent results. Some seeds, especially larger types, sank in the flotation medium. It is unclear whether light, uncharred debris, common in many light fractions, floated with seeds. Light seeds can also be separated from wood charcoal by floating the material in a mixture of zinc chloride solutions of different specific gravities. The solution of lower gravity floats on top of the heavier solution; the lighter solution holds seeds, and charcoal is suspended in the heavy layer (R. Ford, personal communication, 1988). Even if total separation of seeds from charcoal cannot be achieved using chemical flotation procedures, it would be interesting to compare sorting times for samples of like size and composition done (1) completely by hand and (2) by chemical separation followed by spot-sorting. For a large-scale project, enough time might be saved to make imperfect chemical flotation of light fractions worthwhile.

Much the same can be said of using a seed blower (see Fig. 2.30) to sort botanical material. This method (see Chapter 2) may not produce completely clean separation of materials of different densities, but it still represents a savings in overall sorting time in comparison to complete hand separation.

Sorting Desiccated and Waterlogged Samples

I have focused up to this point on how to process standard flotation samples, that is, those in which ancient plant remains are preserved by charring. For many regions of the world, these are the most common types of samples encountered. But extremely dry or extremely wet conditions can produce valuable deposits of uncharred botanical remains. Fine-sieving procedures for recovering desiccated or waterlogged botanical materials were discussed in Chapter 2. Procedures for sorting and preliminary classification of such remains are similar to those discussed above, but they must be adapted to accommodate the special needs of these collections. The key point in dealing with waterlogged or desiccated materials is minimizing deterioration.

Waterlogged materials are preserved by the anaerobic condition created by continuous wetting and, in peat bogs or other situations of high raw humus deposition, by the action of polyphenols (Dimbleby 1978; Renfrew 1973). Waterlogged materials swell and become very heavy. Wood, for example, can absorb up to 700% of its oven-dry weight in water. If allowed to dry, wood in this condition shrinks dramatically, especially in the plane perpendicular to wood rays and stem radius, and becomes very hard (Levy 1970). Willcox (1977) reports that, if waterlogged seeds are allowed to dry at room temperature, distortion occurs. Drying also creates conditions for continued decay by fungi and bacteria. This is especially true for remains from pH neutral or basic bogs, where both bacterial and fungal decay take a rapid toll (Dimbleby 1978).

In processing waterlogged botanical remains it is therefore essential that material not be allowed to dry out. Following sieving, samples awaiting sorting should be stored wet in sealed plastic bags or other watertight containers and checked periodically for evidence of drying. Willcox (1977) recommends storing sorted seeds in alcohol (industrial methylated spirits). Kenward et al. (1980) suggest storing sorted material in alcohol-glycerin-formalin solution. The latter authors sort each sieve fraction by spreading the residue in water in a glass or ceramic sorting dish. Glass vials are recommended for storage of sorted material.

Desiccated botanical remains present a problem opposite to that described above; material must not be allowed to absorb too much moisture from the air. Storage under controlled humidity (less than 70% relative humidity) is recommended for such materials in order to retard fungal growth. Botanical material from dry cave sites in Missouri, stored with other artifacts in uncontrolled climatic conditions, was found to suffer from mold growth (Fig. 3.8; Pearsall et al. 1983).

Figure 3.8 Mold growth on basketry fragments from dry cave sites in Missouri: (a, b) before treatment; mold appears as a fine white powder; (c) after treatment (photographs courtesy of M. Cornman).

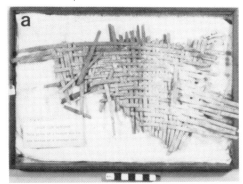

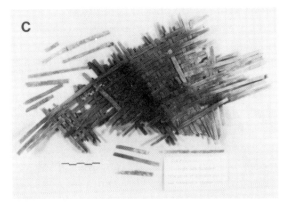

Killing mold and storing materials in clean containers under controlled temperature and humidity arrested further deterioration of those collections. Desiccated materials may also be fragile, especially if some deterioration has already occurred. Careful treatment is indicated. Desiccated grain often darkens while appearing fresh in other characters (Helbaek 1970).

Few special procedures are required for sorting desiccated botanical materials. As is the case with waterlogged materials, richness of samples can increase processing time dramatically. A critical decision to be made at the beginning of a project is what materials should be removed from various sieve fractions. Small-sieve fractions may be full of fine fragments of material also represented in larger sieves. For analysis of materials from El Paraiso de Chuquitanta, in Peru, for example, all botanical remains were removed and classified from the two largest sieve fractions (2.0 mm and 0.85 mm) while seeds only were removed from the two smallest fractions (0.425 mm and 0.212 mm) (Pearsall and Ojeda 1988). Soil residue (that which passed through the sieves) was spot-checked for seeds; none were observed. This strategy was decided upon after a number of samples had been sorted completely and it had become apparent that removing all the abundant grass stems present in all fractions would greatly slow sorting. Removing these and other vegetative material only from the larger sieve fractions speeded work considerably.

Another difficulty caused by the excellent preservation of dry caves or deserts is in distinguishing remains that are culturally introduced from those brought in by rodents or wind. The broader issue of determining the source of seeds in archaeological sites is considered in "Issues in Interpretation" at the end of this chapter.

Building a Comparative Collection

Identification of macroremains is carried out through one-to-one comparisons between known plant specimens and unknown archaeological materials. A key to successful identification is access to adequate comparative material. Because many archaeological botanical remains are charred, fragmented, or otherwise altered from their fresh condition, comparative specimens are most useful if reduced to a similar state. In many cases, this precludes the use of material collected for other purposes (e.g., specimens housed in herbaria). It is also the case, however, that charred seeds and fragments of charcoal do not in themselves form a complete comparative collection. All such "working" specimens must be backed up by identified voucher specimens, ideally two sets—one curated in the lab and the other deposited at an herbarium. A voucher specimen is a dried, identified example of the complete plant.

Collecting plant specimens for use as comparative material serves a second and equally important purpose, that of providing information on vegetation of the study area. Data on plant associations, seasonality and abundance of resources, and distribution of resources (especially useful for catchment analysis) can be gathered as plants are collected. It is usually necessary to do additional observations on vegetation, but much useful preliminary information can be obtained during collecting trips focused on acquiring comparative specimens. Techniques discussed here are intended not only to guide collecting for study of macroremains but also for application to pollen and phytolith analysis.

Plant-Collecting Procedures

Many plant taxonomy textbooks include a section on field and herbarium techniques (e.g., Benson 1959:380–400; Harrington 1957:90–98; Lawrence 1951:234–262; Porter 1967:42–54; Radford *et al.* 1986). Hicks and Hicks (1978) is an excellent source of additional references on plant collection and herbarium curation. Refer to these sources for detail on the techniques summarized here.

Equipment and supplies needed for collecting and pressing comparative materials and voucher specimens include the following items:

1. *Plant press* (Fig. 3.9). Frames are available from biological supply houses in metal or wood, or can be constructed with hardwood slats. Standard size is 12″ × 18″. Riveted models are more durable than those nailed or screwed together. The two press frames are held together around specimens by cloth straps or cord. Blotters, used to absorb moisture from specimens, are available in several grades of absorbancy and thickness, precut to 12″ × 18″.

Figure 3.9 Standard plant press.

Blotters can also be cut from roofing felt. Corrugates, made of either aluminum or double-faced cardboard, can also be ordered cut to standard press size. Corrugates are used to allow passage of air through the press (the flutes run across the short dimension of the corrugate). It is often useful to have two presses, one to carry in the field, the other to hold drying specimens.

2. *Pressing papers.* Old newspapers, or newsprint, are used to hold pressed material. A folded newspaper page, torn to about 12" × 16" (just under press length, to accommodate 11.5" × 16.5" herbarium mounting paper) serves as one pressing paper.

3. *Field notebook.* Water-resistant, hard-cover surveyor's notebooks make excellent plant-collecting notebooks. Using pencil to enter data is an additional precaution against sudden showers that can cause lost notes.

4. *Collecting tools.* A sturdy pocketknife, short-handled pruning shears, hand pick or strong trowel, and folding saw or machete are basic tools for collecting. A saw is often easier to use for wood collecting than a machete, but the latter is handy for clearing brush. For collection focused on tuber- or rhizome-producing plants, a long-handled shovel or garden fork is useful.

5. *Collecting bags.* If field pressing is not carried out (see below), plastic bags are useful for holding plants during a collecting trip. Ziplock bags of varying sizes can be used to hold small plants or extra rhizomes and fruits. A large plastic garbage bag is used to hold larger specimens, with each taxon tagged with a collection number. By adding a small amount of water to the bag, specimens are kept fresh during the collecting trip. Traditionally, a metal collecting can, or vasculum, is used to accumulate botanical material (see Lawrence 1951: Fig. 29). In moist climates, a backpack, rice sack, or similar sturdy porous container can be used. These are also useful for accumulating wood samples. Small coin envelopes are handy for collecting flowers for pollen analysis, extra small fruits, and seeds.

6. *Measuring and recording equipment.* In this category are included tape measure (for plant height), camera, Brunton compass (to accurately record slope and aspect of terrain), and a small tape recorder. The last item is useful when working with non-English-speaking informants for recording data on plant names and uses.

7. *Hand lens.* For viewing ephemeral flower characteristics likely to be lost before specimens can be examined in the laboratory.

8. *Labels and tags.* Metal-rim tags for labeling specimens in the field; paper labels for voucher specimens.

Each specimen in a comparative collection must fill three different requirements. It must be identifiable (the portion collected must be adequate to make a precise

Building a Comparative Collection • 131

scientific determination), it must be usable (the collection must include seeds, fruits, wood, flowers for pollen, leaves for phytoliths, and so on), and it must be representative of a population (plants should be selected to represent both average and range for species).

Collecting identifiable specimens means collecting examples in flower, in fruit, or both. Good flowering specimens are needed for identifying the plant, even if flowers are not needed for the comparative collection. If plants are not in flower, then fruiting specimens should be collected. Plants not in flower or fruit are quite difficult to identify without the help of a botanist familiar with the flora. Few botanical keys can be used on nonflowering material.

Because it often happens that a plant is collected in flower one day but not observed in fruit until some time later, or that additional flowering specimens have to be searched for on later collecting trips, one should compile an illustrated plant catalogue to keep track of collecting progress (Fig. 3.10). Divide the catalogue into sections (trees, shrubs, rosette plants, grasses) and sketch enough of each plant collected to serve for later cross-reference. Accompany this drawing by information on flower color, fruit type, and leaf size, and note whether more material is needed.

Figure 3.10 Page from illustrated plant catalog used during research in Ecuador.

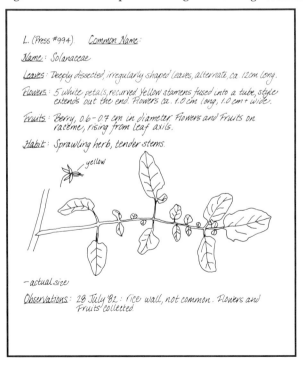

By updating the catalogue after each collecting trip, and carrying the notebook on subsequent trips, one can easily fill gaps in the collection and avoid duplication. The illustrated plant catalogue is also useful when carrying out vegetation survey or plant density observations, since plants can be readily recalled and tallied by collection number if unidentified. One often must do detailed observations on plant communities before obtaining final plant determinations, and such an illustrated notebook greatly increases accuracy of vegetation description.

Making sure that plant collections will be of use in identifying archaeological plant remains means careful attention to collecting potentially useful parts of plants and anatomically different parts of the same useful plant. In wood, for example, twigs, branches, trunk, and roots vary in the number of growth rings and abundance of support tissue. If possible, collecting should follow some preliminary analysis of archaeological botanical remains, or at least familarization with types of materials found in previous studies. Compiling a list of useful plants from ethnographic or ethnohistoric sources is another way of making collecting time more productive. For wood collecting, it is useful to focus first on dominant, common trees and those trees and shrubs said to be good for firewood or construction, rather than attempting to collect all taxa in a diverse forest environment. Getting at least dominant species from all environmental zones in the study area is also important. For seeds and fruits, collections should include weedy annuals as well as berry- and nut-producing taxa. If tuber or root remains are anticipated, collecting should not neglect these plant parts. In short, the more precollection time spent making lists of useful plants (and what parts are used), studying reports of prior paleoethnobotanical analyses from the region, and thinking about the ways the collection will be used, the more productive collecting time can be.

Because identifications of archaeological materials are based in large part on the comparative collection, it is vital that specimens representative of the species be collected. Diseased or insect-damaged plants should be avoided, as should individuals much smaller or much larger than the population as a whole. Range of variation should be noted, however. If possible, fruits should be collected from a variety of different plants, so that the collection represents the natural variation in the population. Wood samples should not be collected from obviously stressed trees.

Finally, here are a few miscellaneous tips to make collecting, and later use of collected material, go smoothly:

1. In advance of collecting plants, make sure that permission to collect is obtained from the landowner and, if required, the local government. Many countries require that foreign botanists obtain collecting permits, deposit examples of some or all plants collected, obey regulations on collecting endangered species, and obtain permission to export collections. No

special permit is required to import most dried, fumigated specimens for scientific use into the United States. Information on prohibited taxa and importing live plants can be obtained from the U.S. Department of Agriculture (Animal and Plant Health Inspection Service, Plant Protection and Quarantine Programs, Federal Center Building, Hyattsville, Maryland 20782). Application for soil importation permits may also be made to this address.

2. Collect enough material initially to make a *minimum* of three complete voucher specimens; this allows one for identification and donation to an herbarium, one for curation in the lab, and one that can be traded, deposited to fulfill collecting permit regulations, or cannibalized at a later time.

3. Collect abundant extra seeds, flowers for pollen, fruits, nuts, and rhizomes for charring and use in the lab. Specimens are lost and fragmented with use, and fellow ethnobotanists may request examples. If these are collected at a later date, they should have different voucher specimen numbers to distinguish them from the original collections.

4. Collect wood specimens at least 2' long from branches at least 2" in diameter. If it is possible to obtain a trunk specimen (from a fallen tree, by salvaging lumbered materials, or by arrangement with forestry officials), do so. Give wood specimens the same collection numbers as vouchers.

5. Be sure to make detailed notes on plant habit (height, whether it is vining or erect, branching pattern) for all collected species. It is also useful to note flower color, type, and arrangement, fruit type, presence of spines or prickles, and similar characteristics. Field notes should also include plant associations—what plants are growing near the collected specimen, what plants are dominant in the area.

6. At the beginning of collecting in an area, make a brief entry in the field notebook describing the collecting area. This should include the date, general location (town, country, or similar), and description of the immediate area.

7. During the course of collecting, make observations on occurrences of previously collected taxa. This gives useful data on what habitats different species occupy, relative abundance of species, and plant associations.

Pressing and Drying Specimens

References cited in the last section for field-collecting procedures also include detailed information on how to press and dry voucher specimens. There are two approaches to plant pressing: field pressing and pressing in camp or lab. Many botanists recommend that plants be pressed as they are collected (field pressing).

The advantages of this approach are several: pressed specimens are made from freshly cut and undamaged plants, excessive material is not accumulated and carried, notes on flower character and color are made before flowers wilt, and all note taking is completed on the spot. Accumulating plants for pressing later in lab or camp has the primary advantage of allowing all field time to be devoted to collecting. Depending on working conditions, it may be impractical to field press (for instance, when working alone under windy conditions). On the other hand, if laboratory time is limited or plants being collected do not save well, field pressing may be called for. Field-pressing procedures parallel those described below, except that to minimize press weight few or no blotters are used in initial pressing.

Lawrence (1951:238–240) gives excellent directions for creating good, usable voucher specimens. Only one species should be pressed per page, although pressing several examples of the same small plant on a page is efficient use of press space. Whenever possible, the entire plant (root, stem, leaves, inflorescence) should be pressed on one page (Fig. 3.11). For plants longer than 16", the stem is folded in a V or N shape. Overlapping leaves or branches are trimmed off, leaving basal portions, so that the original arrangement of parts is preserved. One or more leaves should be turned with lower side uppermost. If flowers cannot be pressed open to show the arrangement of flower parts, some should be split open and spread out. For specimens that cannot be pressed on one sheet, several sheets can be used (e.g., one for the basal portion, one for the middle, and one for the top). Notes should include original size of the specimen and proper ordering of sheets. Very large leaves, such as palm fronds or large compound leaves, are pressed by splitting lengthwise and including as much of the leaf as possible, including the entire petiole and portion of the stem where it was attached. Notes should be included on leaf arrangement and size.

When pressing, one often wishes for two sets of hands. Plants can be held in place while being arranged by lightly taping one part down (taking care not to place tape on a portion that will tear when the tape is removed). Placing a small heavy metal bar on one part of the specimen while arranging the rest, sliding a slotted strip of paper over V bends, or using wet newspaper to hold the specimen in place are helpful techniques. Pressing woody specimens often results in vouchers with wrinkled leaves because leaves close to stems are not pressed flat; this can be remedied by placing pads of newspaper or foam rubber next to bulky portions of the specimen to keep adjacent leaves flat.

Once the plant is arranged in the folded pressing paper and the collection number is written with indelible pen on the outside of the sheet, it is placed between two blotters. Specimens are piled up, the stack placed in the press frame, and press straps tighted. A 24-hour "sweating" period now follows, after which the press is opened and wet blotters removed. Each specimen is examined and rearranged if necessary. Plants are now flaccid and can be easily manipulated.

Building a Comparative Collection • 135

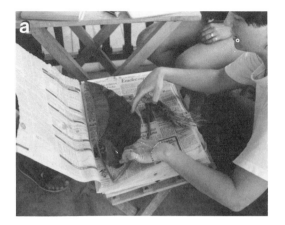
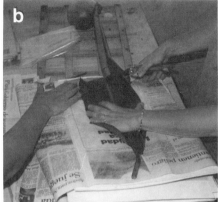
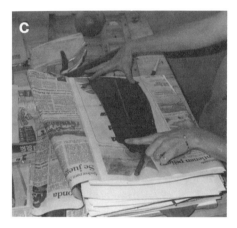

Figure 3.11 Pressing plants: (a) an entire plant is fitted into the pressing paper by folding the leaf petiole or main stem; in this example all but one leaf is removed, and that leaf is cut in half at the mid-rib; (b, c) large leaves are cut into sections, which are pressed individually; (d) detailed notes aid later identification.

There are two techniques for drying pressed voucher specimens: using no heat, and using an artificial heat source. Although there was considerable debate among botanists early this century when the idea of using artificial heat sources was introduced, drying with heat is now the prevailing method. Good voucher specimens can be created either way, however. To dry specimens without artificial heat, "sweated" specimens are returned to the press with dry blotters and the press is closed tightly. After 24 to 36 hours, the press is opened and wet blotters replaced by dry ones. This process continues until specimens are dry, which usually takes one week. Dryness is tested by lifting each specimen by the stem; it should stay rigid, with little droop of branches, flowers, or leaves. Disadvantages of this method of drying are the time required, the number of blotters needed (and the time required to sun dry wet ones), and the possibility of damage to specimens by mold or insects during the long drying process.

Drying specimens over an artificial heat source is faster, requires fewer blotters, and also kills mold and insect pests. In humid climates it is the more practical method. "Sweated" specimens are placed between dry blotters, and each set of blotters between cardboard or aluminum corrugates. The press is tightened, but with somewhat less pressure than is used for drying without heat (to avoid crushing flutes in the corrugates). The press is then placed over a heat source, positioned so that the corrugated air flutes run up and down (Fig. 3.12). Heat used to dry pressed specimens can come from a charcoal fire, kerosene lantern or heater, camping stove, or bank of incandescent light bulbs mounted on a board. In choosing a heat source, remember that too-rapid drying results in brittle, browned specimens. Lawrence (1951) recommends a heat level giving a minimum of 12 hours drying time; Benson (1959) recommends drying at no more than 100° F. Fires, stoves, or heaters must be carefully regulated to avoid overdrying or too-rapid drying of specimens. If electricity is available, incandescent bulbs are probably the safest, least expensive low-temperature option. Drying is most efficient if the heat source and press are enclosed (in a drying cabinet or by covering with canvas, a heat-reflecting blanket, or odd pieces of siding or wood).

Once plant voucher specimens are dried, they should be bundled together (wrapped in newspaper, tied securely so no specimens can slide out of papers) and stored in a closed plastic bag. A tightly closed bag maintains low humidity, avoiding redampening and mold growth. Moth balls repel insects. To kill pests, put a piece of No Pest Strip in each storage bag. Extra fruits, seeds, roots, and the like should be air dried, placed in paper bags or boxes, and stored in a carton. Collections should be checked periodically for insect infestation, and insecticide placed in the storage carton. Root and tuber specimens sometimes mold even after being allowed to air dry. These can be dried over artificial heat (on a screen on the plant press drier frame), but substantial shrinkage occurs for some types of tubers. An alternative is storage in alcohol; this method maintains specimens at their fresh size but leads to discoloration.

Identification of Comparative Materials

Scientific determination of voucher specimens is likely to be a skill outside the areas of expertise of students or professionals whose training is primarily in anthropology and archaeology. However, courses in general botany, plant taxonomy, and field botany allow the anthropological ethnobotanist to determine family and sometimes genus affiliation of specimens. At that point, specimens can be sent to the appropriate taxonomic specialist for species determination. It is also advisable

Building a Comparative Collection • 137

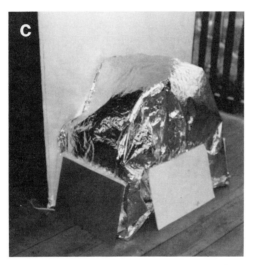

Figure 3.12 Field drying plant specimens: (a) wooden drying rack, enclosing the heat source (three lightbulbs), increases drier efficiency; (b) press with voucher specimens is positioned above the heat source; (c) press and rack are covered with a reflecting blanket to hold heat in.

to check genus-level determinations by consulting herbarium collections, or by asking herbarium staff for help with difficult specimens.

However much the archaeologist is involved with determining voucher specimens, at some point the specimens must be prepared for final checking and deposition in an herbarium. The first step is to contact an herbarium with good collections in the project region and inquire about donating specimens for determination. After an herbarium or specialist has been contacted, voucher specimens are prepared for shipment. If unmounted material is preferred for determination, vouchers can be sent in clean newspapers. Each specimen should be clearly labeled with its collection number and should include labels with provenience and collection information. It is a good idea to include an inventory of material with duplicate information. Bundle the newspapers together securely and ship in a strong box. Be sure to include specimens with good flowers and enough material to make a good mounted herbarium specimen.

Mounting voucher material, either for shipment to an herbarium or for curation in the lab, is relatively easy. Taxonomy textbooks, such as those cited at the beginning of this section, include directions on how to mount specimens (another good source is Knudsen 1966). Supplies needed for mounting specimens are available through biological supply houses. Necessary materials include standard-size herbarium paper (11.5" × 16.5"), paste or glue (use prepared herbarium paste or consult an herbarium on how to make paste), brush (to apply paste), gummed linen tape (to hold bulky materials), and labels (approximately 4.5" × 2.75", varying 0.5" in either direction) (Lawrence 1951:243–248). Do not apply excessive paste to specimens. Labels are usually pasted on the lower right or left of the herbarium sheet. As was the case for labels sent with unmounted specimens, labels for mounted specimens should give full provenience and collection information, typed if possible. Place loose seeds, flowers, or fruits in paper folded into an envelope, and fix it onto the sheet. Let the finished mounted voucher specimen dry thoroughly (with weights placed on areas likely to pull loose during drying). Store dried vouchers in genus covers (folded stiff paper, available through biological supply houses) in a closed cabinet or in a storage box.

Preparing a Working Laboratory Collection

The final step in building a comparative botanical collection is to organize collected materials into a working, hands-on collection easily used by lab personnel. This usually involves charring comparative materials to render them more directly comparable to archaeological remains. Uncharred examples of all materials should be retained, however, both for use in identifying uncharred remains and for replacing lost charred materials. Extra material is also useful for exchange with other labs, or for techniques such as thin sectioning, scanning electron microscopy, and starch grain analysis (see discussion of these techniques in "Specialized Identification Techniques," below).

There are several ways to char comparative materials. The easiest is to heat materials in a muffle furnace or kiln. Such units achieve high temperatures in a short time, allowing rapid processing, and their temperature is easily controlled.

The following is a basic procedure for charring in a muffle furnace (Fig. 3.13). Preheat the furnace, positioned in a fume hood (to draw off smoke created during

Figure 3.13 Charring comparative wood specimens in a muffle furnace: (a) specimens are wrapped in aluminum foil labeled with specimen numbers; (b) specimens are stacked in the preheated furnace, and a diagram of the arrangement of specimens is made; (c) specimens are heated until smoke no longer escapes from the furnace; (d) specimens are unwrapped and allowed to cool completely; (e) charred wood is stored in boxes in the comparative collection.

charring), to 400–500° C. Wrap material to be charred in aluminum foil. Write sample numbers on the foil (the impression of the number generally survives heating). For wood specimens, sample numbers may be written in pencil on the wood itself. Place material of like size in the furnace. For wood, pieces of similar diameters should be charred together; this ensures that all pieces are ready at about the same time. Small seeds char very quickly and should never be mixed with larger materials. Use a lower temperature and extend the charring time (200° C, 12 hours) if moisture content is likely to result in excessive distortion of seeds (Renfrew 1973). Seeds should be placed loosely in foil packets so that they are less likely to fuse together during charring; alternatively, use small covered crucibles to hold seeds. As a safeguard against mixing samples, make a diagram of the arrangement of materials in the furnace; sketches of seeds may also help identification if samples are confused. After charring is completed, remove material in order and check numbers against the diagram. This method is especially useful for charring wood specimens, which can be stacked in pyramid fashion in the furnace, allowing processing in batches. Once materials are in the furnace, close the unit and turn on the fume hood. If the furnace or kiln has an opening to admit oxygen, close it (i.e., creating a reducing atmosphere for charring). Timing charring is something that comes with experience. Wood takes 30–60 minutes, depending on diameter and density. A good indicator to watch for is smoke cessation; samples are usually ready when smoking stops. Nuts and larger fruits require somewhat less time; seeds require only 5–15 minutes and should be carefully monitored. When charring is complete, remove the samples carefully (use long-handled tongs and wear insulated gloves). If foil has torn and exposure to air causes smoking or burning, sprinkle areas affected lightly with water. Specimens should not be placed in storage containers until completely cool. Because smoke emitted during charring contains resins and other residues which become deposited on the outside of the furnace door, the unit should be wiped clean before residues harden. Blackening on the inside of the furnace will burn off.

If a muffle furnace or kiln is not available, charring can also be done in a household oven or by heating materials in a container of sand. Small seeds are the easiest materials to char in this way, although larger specimens can also be charred using long heating times. (I do not recommend using a household oven for anything but small seeds, however, since larger specimens produce large volumes of smoke.) To char materials in a container of sand, place a sturdy pot, half filled with sand, on an electric hot plate or over a bunsen burner, in a fume hood. Wrap materials to be charred in foil, as described above, and bury them in sand. Seeds can be placed in small, covered crucibles, buried to their tops in sand. Charring times vary considerably with this technique, depending on temperature of the heat source, temperature of sand, and material being charred. Run a few test samples to determine approximate charring times before processing comparative materials.

Bohrer and Adams (1977) recommend that uncharred materials be heated gently to 212–250° F to kill insect eggs or larvae before being placed in the comparative collection. They also suggest that charring or parching of seeds be carried out in the same season in which the material is collected so that any distortion that occurs during charring will parallel that of archaeological material. However, because the degree and type of distortion that occurs during seed charring depends on the moisture content of the seeds as well as their structure (Helbaek 1970; Renfrew 1973; Stewart and Robertson 1971), it is difficult to generalize about how to char seeds to achieve distortion parallel to that of archaeological specimens. Renfrew (1973), experimenting with four different species of cultivated grain (100 seeds each), found that length of all charred grain decreased but that changes in width and thickness were more variable. Distortion due to charring is discussed further below for various classes of macroremains.

Arranging charred and dried materials in a collection so that specimens are easy to find and use can be difficult. I agree with Bohrer and Adams (1977:45): there is no entirely satisfactory way of storing and organizing specimens. Missouri lab collections are divided by geographic region and type of material. Wood forms one collection, which is divided into North America, Caribbean, coastal Ecuador, and high-altitude Peru subcollections. Charred working material for each subcollection is housed in a portable desktop storage unit (Fig. 3.14). Taxa are arranged alphabetically by family, then alphabetically by genus within the family. Uncharred wood and extra charred materials are stored in the same way in small boxes or bags in drawer cabinets (Fig. 3.15). Fruits, seeds, and nuts form another collection, which is also divided into regional subcollections. These materials are stored in assorted boxes, vials, and envelopes in cabinet drawers. A working collection of larger North American materials (nuts, large seeds) is stored in a portable desktop unit. This arrangement is not very suitable for small seeds, however. Seeds can be mounted on microscope slides; this is useful for teaching recognition of seed types. But to be used for identification, seeds have to be mounted in a variety of positions, so loose seeds may be preferable. Storing charred seeds in small glass vials, gelatin capsules, or coin envelopes makes access fairly easy. The Missouri lab also has a small root and tuber collection and a collection of mounted phytolith comparative material, all arranged alphabetically by family and genus.

Gunn (1972), describing the Beltsville, Maryland, U.S. Department of Agriculture seed storage system, recommends arranging collections systematically by orders and families. The Engler and Prantl (1889–1915) or Cronquist (1968) systems may be used. Drawer or box units for each family are numbered to facilitate filing collections. Using systematic order rather than alphabetical order has the advantage of placing materials from closely related families together in the collection—but a user who does not know the system takes longer to locate specimens. Gunn (1972) reports that screw-capped or cork-topped glass vials make good storage containers

Figure 3.14 A small desktop storage unit is used to hold comparative wood specimens at the Missouri lab.

Figure 3.15 Most comparative material at the Missouri lab is stored in drawer cabinets arranged by type of material and geographic area. Arrangement within each collection is alphabetical by family. (a) wood collection; (b) seed collection.

Building a Comparative Collection • 143

for seeds, since these permit the collection to be scanned without removing seeds from containers. Additional information on building and maintaining seed collections may be found in this article and in Harrington (1972).

Using a comparative collection involves much searching through specimens. This is how one learns to recognize taxa. It is sometimes useful, however, to develop aids to identification that shorten the search. These aids should never become a substitute for hands-on use of comparative materials, however. At the Missouri lab we have developed several specimen card files that are used in the process of identi-

Figure 3.16 Wood identification card. One side of this card for *Guaiacum* shows a sketch of the cross section. Radial and tangential sections are also very useful. Notes about parenchyma and ray appearance help remind one of distinctive characteristics. Relative pore size (P.S. 4) is noted. The other side of the card summarizes pore, ray, parenchyma, and other characteristics of the charred specimen.

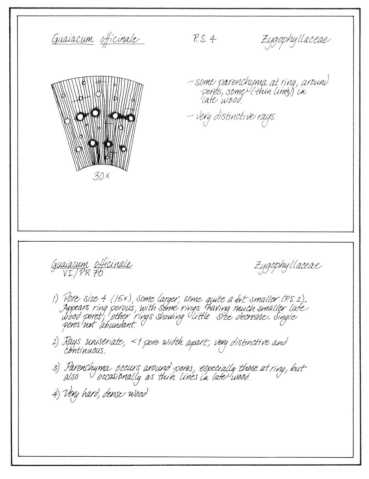

fication. Each of the wood specimens in the Latin American subcollections, for example, has been described and sketched on a 5" × 8" card (Fig. 3.16). These cards can be grouped in a variety of ways to form a desktop key. Using the key allows one to eliminate types from consideration quickly, shortening the identification search process. Making cards for wood, nut, seed, and other comparative materials is an excellent way to learn to identify them. We have also developed a dichotomous key to aid in identification of North American wood taxa. Another card file exists for seeds. Seed drawings are made to scale, attachment areas carefully depicted, and information on size, surface characteristics, and distortion during charring noted. A file of photographs would serve the same purpose. This approach is used for our phytolith comparative collection.

Basic Identification Techniques

In this section I present basic techniques for identifying macroremains including seeds, fruits, nuts, wood, tubers, and other vegetative materials. I also include a short section on the special problems associated with identification of cultivated plants. The goal of these discussions is to teach the reader how to approach identification of materials encountered in archaeological botanical collections rather than how to identify specific taxa.

Seeds

In this and the following section on identification of fruits and nuts the focus is on plant reproductive structures encountered as archaeological macroremains. I discuss seeds separately from fruits and nuts because these two classes of remains often differ greatly in size and degree of fragmentation. An excellent source of taxonomic literature on identification of seeds and fruits, as well as vegetative plant materials, is Delcourt *et al.* (1979).

A seed is a reproductive structure composed of a protected embryonic plant (Shackley 1981:108–124). The major group of seed-bearing plants, the angiosperms, further protect the embryonic plant by enclosing seeds in carpels. Seeds are formed in the ovary of the plant by fertilization of the ovum (female gametophyte) by pollen (male gametophyte). The ovary, protected by an outer wall, the pericarp, and the seeds in their enclosing carpels form the fruit (Fig. 3.17).

Angiosperm seeds have three major parts (Fig. 3.18): embryo, endosperm, and seed coat (testa) (Esau 1977:455–473; Fahn 1982:479–496). The embryo is formed of plumule (first bud), radical (first root), and cotyledon (first leaf). Dicotyledons have two embryonic leaves, monocotyledons only one. In some plants, such as legumes, cotyledons are large and contain stored nutrients to nourish the germinating seed. Other plants contain specialized tissue, the endosperm, which provides nutrients

Basic Identification Techniques • 145

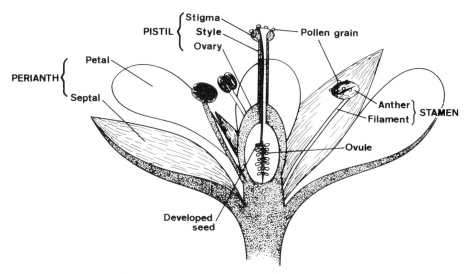

Figure 3.17 Representative flower structures.

for the embryo. Amount of endosperm and its position in the seed relative to the embryo differ among plant groups and are important diagnostic traits in seed identification. Embryo and endosperm, if present, are protected by the seed coat, a two-layer outer covering. Seeds coats may have distinctive coloration, texture, or surface features that can be important distinguishing characteristics. The point where the seed was attached to the ovary is the hilum, or attachment scar. The seed may also show evidence of attachment of the flower style or the point of penetration of the pollen tube.

Only a limited number of characteristics can be used to identify seeds recovered loose in flotation samples. Size and shape are two characteristics usually readily describable for whole, unbroken seeds (but see discussion of seed distortion during charring, below). The seed coat is an important source of diagnostic characteristics, including color for uncharred seeds, texture (whether the testa is smooth or reticulate, and if reticulate, what patterning is present), attachments (spines, bristles, glumes, awns, pappus, etc), and scars (hilum, style scar, pollination scar, etc). If the seed coat is lost during charring or badly eroded in desiccated materials, identifiability decreases markedly. There are also minute characteristics of the seed coat, such as fine reticulation, which are important in identification at the species level. Such characteristics may require use of scanning electron microscopy for study (see Brisson and Peterson 1976).

Other important clues to identification are found in the nature and placement of embryo and endosperm. Number of cotyledons present allows placement of the

146 • Chapter 3 Identification and Interpretation of Macroremains

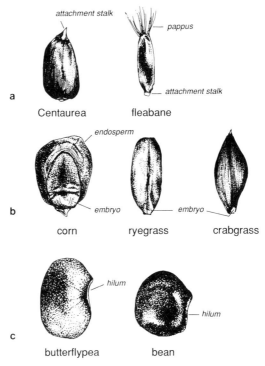

Figure 3.18 Examples of angiosperm seeds, illustrating characteristics that aid identification: (a) Compositae (daisy family); seeds in this family are often oblong and sometimes ridged; a pappus (ring of hairs or scales) may be present; an attachment stalk is often prominent. (b) Gramineae (grass family); seeds have a lateral or basal-lateral embryo; endosperm tends to be abundant; some grasses have elongated, grooved seeds (e.g., ryegrass); others have seeds which are flattened and lightly ridged (e.g., crabgrass). (c) Leguminosae (bean family); many legumes are beanlike, with a notched attachment area or hilum; the seed coat is usually smooth; seed interior consists mostly of broad, thick cotyledons, with little or no endosperm. Not to scale.

seed in either monocot or dicot groups. Abundance, or lack, of endosperm is also an important characteristic. One of the most useful clues is the position of the endosperm and embryo in the seed. This can often be determined without dissection, by observing overall shape of the seed, features on the seed coat, and the nature of seed distortion after charring. One of the most useful guides to embryo and endosperm placement in common North American seeds is Martin and Barkley (1961).

To illustrate how such characteristics are used to identify seeds, I review a hypothetical identification of an unknown seed (Unk A), (which turns out to be the commonly encountered taxon *Chenopodium*).

Basic Identification Techniques • 147

Having encountered Unk A in flotation samples often enough to recognize it as a distinctive seed type, the first step in identifying it is to describe it. This is best done with a detailed sketch of the type. Figure 3.19 illustrates how Unk A appears in flotation samples. In addition to making a sketch, it is important to describe size, shape, and seed coat characters from whole, unbroken examples. Note that we have made the useful discovery that the endosperm tends to swell and pop the seed coat off Unk A, and that loose curved embryos, puffed endosperms, and half seed coats also belong to this type. Waiting to attempt identification until a number of examples of unknowns accumulate often leads to such useful information on taxon variability.

The process of matching characteristics of Unk A to a known seed type can follow one of several routes. It is sometimes possible to recognize immediately the order or family of plants to which the unknown seed belongs. Bohrer and Adams (1977), for example, state that they frequently distinguish seeds of the order Caryophyllales and the families Boraginaceae, Euphorbiaceae, Labiatae, Gramineae, Compositae, and Umbelliferae. Gunn (1972) and Martin and Barkley (1961) are excellent sources of information on family- and order-level seed characteristics. It is also often possible to recognize legume seeds, either at family or subfamily level. The ability to recognize order- or family-level seed groups can greatly speed the identification process. This skill comes from experience, careful study of comparative materials, and training in systematic botany. Unk A is recognizable as a member of the order Caryophyllales (or Centrospermae). This order contains the families Chenopodiaceae, Amaranthaceae, Nyctaginaceae, Phytolaccaceae, Gyro-

Figure 3.19 Archaeological appearances of *Chenopodium* seeds: (a) planar view and cross section of a whole, undistorted seed; (b) planar view of a seed without a seed coat; (c) cross section of a "puffed" seed, illustrating disarticulation of seed coat, embryo, and endosperm. Not to scale.

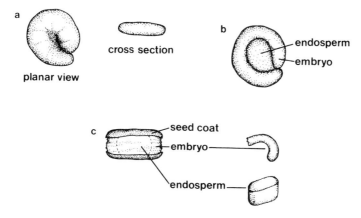

stemonaceae, Achatocarpaceae, Aizoaceae, Portulacaceae, Basellaceae, and Caryophyllaceae. Distinguishing seed characteristics of this group (omitting the obscure families Gyrostemonaceae and Achatocarpaceae) are, (1) presence of a coiled or curved embryo, (2) central endosperm, (3) generally small size and flattened shape, and (4) textured seed coat.

If Unk A had not fallen into a higher taxonomic group with a diagnostic type of seed, we would follow a more time-consuming route toward identification: a search through seed manuals and comparative materials for specific taxa that look like Unk A. During such a search, we compile a list of likely candidates, seed taxa that resemble the unknown in one way or another. In particular, we note what genera, families, or orders contain seeds somewhat like the unknown. We also check plant groups that are known to occur in the project study area, but whose seeds are not familiar. Once the field has been narrowed down to a few groups, we can study these taxa in detail. This is also the next step in determining to which group in the Caryophyllales Unk A belongs.

Sometimes a search of available comparative materials and manuals does not yield any close matches to an unknown. At such a time, a good description and drawing or photograph of the unknown should be prepared and filed, or distributed to colleagues, in hopes that a match will be encountered at some future point.

Once the search for Unk A has been narrowed to the order Caryophyllales, seeds from taxa in families of this order are examined. Some families are quickly eliminated. Because Unk A is not highly reticulate, a match in the Caryophyllaceae is unlikely. The minute seeds of Portulaceae seem too small, although the genus *Calandrinia* has seeds somewhat similar in surface texture. The seeds in Phytolaccaceae are too robust. Through this process, we finally come up with two families that seem especially likely: Chenopodiaceae and Amaranthaceae. Checking which genera in these families are native to our study area, we are able to narrow the likely choice to *Chenopodium* and *Amaranthus*. It is important to remember, however, that eliminating taxa because they do not occur in the study area can lead one astray if climatic and vegetation conditions were different in the past, or if seeds come from plants obtained through trade. Do not force an unknown to fit a local species, since it may not.

Since we have now narrowed the field to a choice between two genera, it is perhaps time to consider the general point of how to know when the process of identification can be taken no further. This depends to a large extent on two factors, the condition of archaeological specimens and the taxonomic group in question. Experience and the state of one's comparative collection also play roles. A seed may lack diagnostic characteristics needed to identify it below family or genus level. Charred seeds often lack attachments, such as awns or spines, needed for more precise identification. Outer seed coats may be missing, or eroded, so that reticulation patterns are not clearly visible. Paleoethnobotanists sometimes indicate their level of confidence in an identification by using the latin abbreviation "cf." (e.g., cf. *Chenopodium*, "looks like" *Chenopodium*), or by giving identifications a "confidence rating" (e.g., * very secure, ** secure, *** somewhat insecure).

Alternatively, the unknown may fall into a taxonomic group for which a multitude of genera and species occur in the study area, many of which produce similar seeds. Even if few species occur, seeds may be so similar among them that only a genus-level identification is possible. Plant species are not defined on the basis of seed character, after all. A large part of good seed identification is knowing when a more precise identification is inappropriate.

Getting back to Unk A, we have decided that *Chenopodium* or *Amaranthus* are the most likely genera. For the purposes of this example, let us assume that enough good specimens are available that we can in fact go further in our identification. Being able to do this is partly skill and partly good luck. By examining local species of the two genera, we eliminate all the *Amaranthus* species, having the good luck to have a *Chenopodium* type of "classic" shape: lenticular in cross section, with no rim, round rather than tear-shape when viewed from the top, and clearly notched.

Once we have determined that Unk A is a species of *Chenopodium*, how do we determine which species it is? If we are working on material from Missouri, Steyermark's (1963) flora tells us that there are eighteen species growing today in the state, half of which are considered native. In addition, cultivated *Chenopodium berlandieri* is known to occur prehistorically in the Midwest (Asch and Asch 1985). Species-level determination must thus proceed very cautiously and may involve scanning election microscopy study of seed coats and use of detailed metric and shape data (for *Chenopodium*, see Wilson 1981 and Asch and Asch 1985). It is not always so difficult to identify specimens to species level, however. One encounters seeds from families with few genera and species, or in which taxa all produce distinctive seeds; in such cases, species can be distinguished for good specimens.

Seed identifications are sometimes made difficult by charring distortion. In the mock identification of *Chenopodium* discussed above, reference is made to "puffed" specimens, seeds in which the central endosperm expanded, popping off or fracturing the seed coat. Many ethnobotanists have noted this phenomenon; it is perhaps more common to find distorted *Chenopodium* than intact specimens. Many seeds are altered by the process of charring. Heating and post-depositional processes tend to reduce seeds and fruits to a simpler structure. In corn, for example, kernel endosperm and embryo may separate; cobs may break up into glumes and single cupules. Several researchers have reported experiments in seed charring designed to study the nature of distortion and size changes under various conditions. Stewart and Robertson (1971), for example, investigated effects of moisture on degree and type of size change in *Triticum* and *Hordeum* species. They found less size change in drier grain (11% moisture) than in moister specimens (15% moisture). Renfrew (1973), as reported above, conducted similar experiments using *Triticum, Hordeum, Avena,* and *Secale.* Both studies were concerned primarily with size changes, since to compare charred archaeological grain to modern cultivated species, the nature and magnitude of size change during charring must be understood. As a result of their study, Stewart and Robertson (1971) caution against using size as a diagnostic characteristic of charred grain. Similar experiments have been carried out for maize (*Zea mays*) (Cutler 1956; Cutler and Blake 1976; Pearsall 1980a).

The most comprehensive study to date on effects of charring on seed size and shape is that of Wilson (1984). Rather than examining effects of charring on individual seeds, Wilson took the approach of investigating effects of carbonization on a seed assemblage. He was interested primarily in what effect loss of species during carbonization would have on interpreting seed assemblages. This study contains much interesting information on effects of charring seeds of varying chemical compositions (oily, starch, protein-rich) and moisture levels, as well as the way charring temperature, duration, and presence or absence of soil affects seed distortion and destruction. The implications of this study for quantifying and interpreting seed assemblages is discussed below. It is sufficient here to note that seeds varied widely in the way they were distorted by charring; degree and nature of distortion were markedly affected by moisture and heat of charring. Wilson concludes that "it is not yet possible to deduce the original size of fossil carbonised seeds, because the heating environment cannot at present be deduced" (1984:206). Whether or not this is overly pessimistic, at least for some seed types it should serve as a warning against indiscriminate "correcting" of seed size. Distortion caused by charring, whether in the form of "puffed" or extruded endosperm, cracked seed coats, or fused masses of seeds, often limits the ability of the investigator to identify specimens.

The following references are useful aids for seed identification: Bertsch (1941), Brouwer and Stählin (1975), Corner (1976), Delorit (1970), Gunn (1972), Gunn and Selden (1977), Gunn et al. (1976), Isley (1947), McClure (1957), Martin (1946), Martin and Barkley (1961), Montgomery (1977), Musil (1963), Renfrew (1973), Schopmeyer (1974), and USDA (1948). Refer to Delcourt et al. (1979) for additional general references and a listing of sources by botanical family. Remember: seed manuals are no substitute for a comparative collection.

Fruits and Nuts

The other type of plant reproductive material commonly encountered archaeologically is whole or fragmented fruits. A fruit is a ripened ovary and its adhering parts—the seed-bearing organ of the plant. The pericarp, or ovary wall, can be soft and fleshy, leathery, hard, or thin and paperlike, depending on the form of the fruit. Internal arrangement of seeds in the fruit is also variable. Some fruits have only one seed, which is fused to the ovary wall; other fruits have several seeds. Although finds of whole fruits are reported archaeologically, especially from desiccated and waterlogged sites, it is more common to find fragments of inedible fruit parts, discarded and accidently charred. I include in this discussion identification of very large seeds, such as palm and other tropical fruit seeds, since remains of these are more similar in size and appearance to fruit fragments than to small seeds.

Figure 3.20 illustrates a number of common fruit types. Nuts are indehiscent, one-celled, one-seeded hard and bony fruits. Some nuts, such as those in Juglandaceae, are covered with a leathery husk or exocarp. Husks, nut shell (hard pericarp), and nut meats (embryo) are all found archaeologically. There are two major types of succulent fruits: drupes and berries. Drupes are fleshy fruits with one or more seeds, each enclosed in a hard, stony portion of pericarp. Drupe pits (seed with fused pericarp), such as the peach pit illustrated in Figure 3.20, are frequently encountered archaeologically. Berries are similar to drupes, but their seeds are surrounded by a hardened seed coat rather than a portion of the pericarp. The archaeological *Achras* seed illustrated in Figure 3.20 is from a berry in Sapotaceae, a tropical family with edible fruits. Legumes are dry, one-carpel fruits that dehisce along two sutures. Legume valves (half the opened fruit) are common finds in desiccated sites where cultivated or wild beans were utilized. The cotton fruit is an example of a capsule, a multicarpel fruit that dehisces along one or more sutures.

Identification of fragmented fruit remains such as nut shell, legume valves, or husk and rind pieces presents the analyst with a set of problems somewhat different from those encountered in identification of seeds. One rarely encounters whole specimens. Published manuals or drawings are therefore less useful as identification aids, since fragmented remains do not resemble illustrated whole specimens. Photographs or drawings often do not illustrate internal portions of fruit, the nature of tissues, or details of attachment areas. A comparative collection thus becomes of primary importance for correct identification of such remains. Large seeds, such as those of *Achras* or common North American taxa such as persimmon (*Diospyros virginiana*) and pawpaw (*Asimina triloba*), present similar difficulties, since they are often recovered in fragmented form.

Identifying fragmented fruit, nut, and large seed remains is therefore largely a matter of detailed study of comparative material. For identifying charred macroremains, comparative materials should be charred and broken up into pieces of about the same size and composition as the archaeological materials. For each comparative taxon, features such as pericarp thickness, texture, and surface (internal and external) characteristics, nature of embryo and endosperm, presence of sutures, attachment scars and the like, and nature of exocarp (thickness, texture, and curvature) should be described. Figure 3.21 is an example of an identification aid developed for black walnut (*Juglans nigra*) from lab comparative materials. The descrip-

152 • Chapter 3 Identification and Interpretation of Macroremains

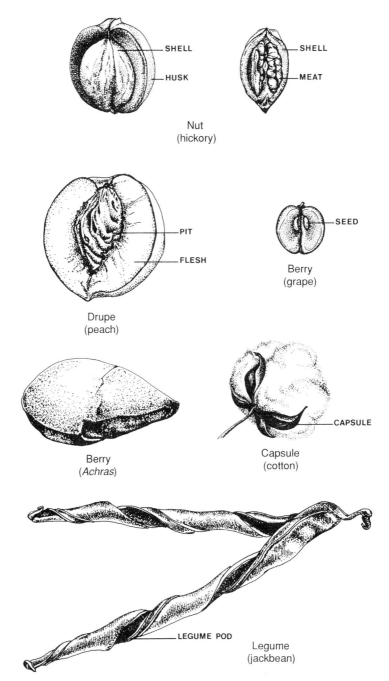

Figure 3.20 Examples of common fruit types, illustrating characteristics that aid identification. Not to scale.

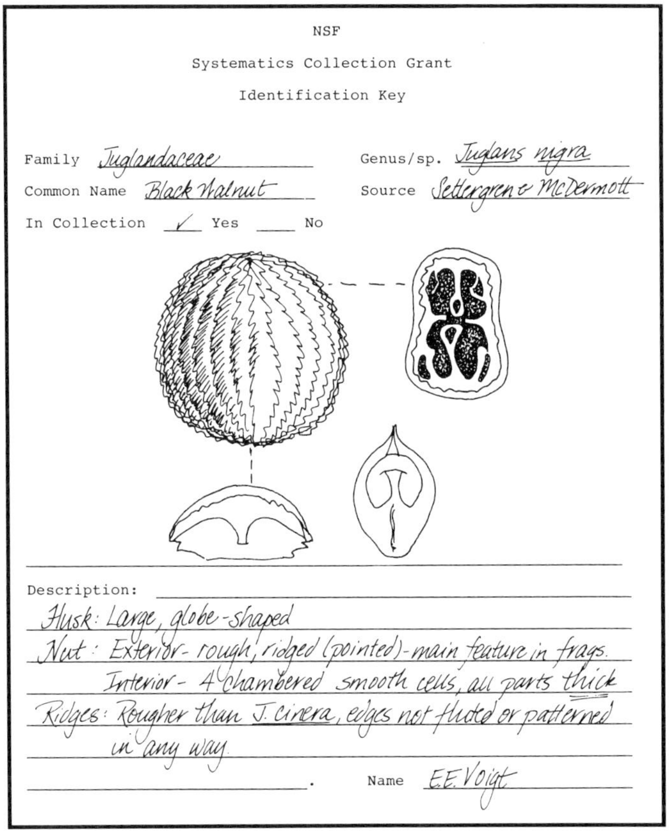

```
                        NSF
              Systematics Collection Grant
                   Identification Key

Family  Juglandaceae          Genus/sp.  Juglans nigra
Common Name  Black Walnut     Source  Settergren & McDermott
In Collection   ✓  Yes  ____ No
```

Description:
Husk: Large, globe-shaped
Nut: Exterior- rough, ridged (pointed)- main feature in frags.
      Interior- 4 chambered smooth cells, all parts thick
Ridges: Rougher than J. cinerea, edges not fluted or patterned
        in any way
                                    Name E.E. Voigt

Figure 3.21 Laboratory guide to species-level identification of walnut shell.

tion includes notes on characteristics that distinguish black walnut from butternut, *J. cinerea*. Figure 3.22 shows similar notes for distinguishing valves of *Phaseolus vulgaris* (common bean, *frejol*) and *P. lunatus* (lima bean, *pallar*). Such identification aids can be grouped into a desktop, informal key for each major class of material (nuts, rind, legume valves). Developing a formal dichotomous key may also be useful in some situations, if the full range of taxa likely present in an assemblage can be determined. It is also useful to describe in detail the appearance of fragmented cultivated materials that may be mixed with fruit and nut remains, and how to distinguish these from wild taxa that may resemble them. I return to identifying cultivated material below.

Identification of fragmented fruit and nut remains begins during sorting, when the non-wood category is removed from flotation large splits. This material is accumulated from a number of samples until the full range of material is present. In

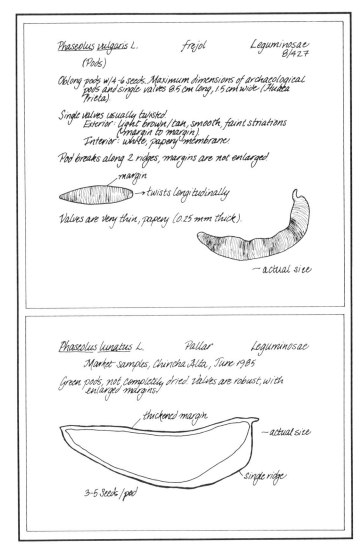

Figure 3.22 Laboratory guide to species-level identification of valve (pod) fragments of common and lima bean.

temperate regions, many dense, curved pericarp fragments are likely to be nuts of some kind. With the comparative collection, fragments are compared to local nut taxa for determination of genus or family. One might, for example, recognize walnut, hickory, and hazelnut fragments in an archaeological assemblage. For each genus-level group, one must then decide whether sufficient characters are present to distinguish among species. If material is highly fragmented, only groups of species may be distinguished. In the hickories, for example, it is often possible to separate thin-shelled from thick-shelled species (using pericarp thickness), but large fragments with pericarp ridging patterns, showing overall size and shape of the nut, are needed for more precise identification.

The non-wood material will also often include other distinctive remains. It is not unusual, for example, to find fragments of large seeds. Some, such as those of legumes and persimmon, are mostly dense cotyledon, usually lacking seed coats. Location of the hilum helps distinguish the different taxa. Other large seeds, such as grape and peach pits, may be well preserved and readily recognizable. Another type of material to watch for in the non-wood category are fragments of squash and gourd rind. The more flattened curvature of the fragments makes them easily recognizable from nuts.

The end result of examining non-wood material in flotation samples may be disappointing: perhaps several nut taxa, including lots of fragments not identifiable except as walnut family (e.g., hickory or walnut), persimmon seed fragments, and numerous pieces of dense and porous amorphous tissue that does not look like much of anything. With luck, a larger fragment, or whole example, of the fruit, large seed, or other plant part contributing amorphous material to the assemblage may be found, allowing eventual identification of fragmented material. When working on material from Krum Bay, for example, I was able to identify numerous seed coat fragments only after a single whole *Achras* seed had been identified (Pearsall 1983a). Similarly, it was only after identification of several whole specimens of *Canavalia plagiosperma* from the Real Alto site that numerous dense cotyledon fragments were recognized as the same taxon (Damp *et al.* 1981). Unfortunately, fragmented fruit, nut, and large seed remains often remain unidentified because of lack of specimens with sufficient diagnostic characteristics or a lack of material in comparative collections.

Although illustrated manuals often prove to be less useful for identifying fruit and nut remains than similar texts on seeds, a number of the sources listed in the previous section include descriptions of fruits and nuts. I have found Schopmeyer (1974) to be especially helpful. Refer to Delcourt *et al.* (1979) for a listing of references by botanical family.

Wood

Fragments of charred wood are often the most abundant macroremain recovered by flotation. Identification of wood can give insight into patterns of selection of firewood, timber for house construction, or material for tool manufacture. Data useful for reconstructing local vegetation may also be provided through study of wood remains.

Wood is xylem tissue of the secondarily thickened plant. The term "wood" is used broadly here to include all plant tissues found in secondarily thickened stems. A number of texts describe the structure of wood in detail, among them Harlow (1970), Jane (1970), and Panshin and de Zeeuw (1980). Before attempting wood identification, consult a basic text and familiarize yourself with wood anatomy. General descriptions of the structure of wood can be found in Dimbleby (1978), Shackley (1981:93–107), and Western (1970). The basic reference used here is Panshin and de Zeeuw (1980). Minnis (1987) is an example of an archaeological wood key. University forestry departments have comparative wood collections and personnel trained in wood identification. The structure of wood is also covered in plant anatomy courses.

Wood, whether charred, dried, or waterlogged, is identified by comparing the structure of unknown pieces to known comparative material or keys. Features of wood commonly used for identification include (1) vessels and their arrangement, (2) size and arrangement of rays, (3) abundance and nature of parenchyma, (4) physical characteristics such as color, texture, and hardness, and (5) microanatomical features such as vessel pit arrangement. These and other features of wood can be viewed from three different planes or sections, the transverse or cross section (plane across the stem), the radial longitudinal section (plane paralleling the radius of the stem), and the tangential longitudinal section (plane at a right angle to the radius of the stem) (Fig. 3.23). Most features used for identification are viewed in transverse

Figure 3.23 Transverse, tangential, and radial sections of a typical hardwood.

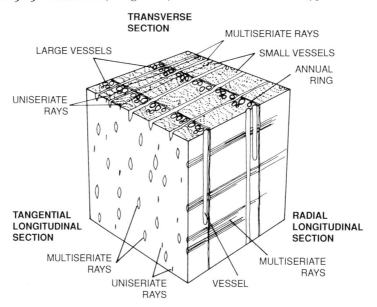

section at 10–45×. Radial and tangential longitudinal sections are used primarily for studying ray length and depth and require higher magnification (100×).

The nature and arrangement of vascular tissue as viewed in transverse section are key features for distinguishing woods. Presence or absence of vessels, for example, can be used to distinguish softwoods (gymnosperms) from hardwoods (angiosperms). Vessels are specialized cells in hardwoods for conducting water and dissolved salts. Vessel cells are arranged end to end, running parallel to the direction of stem and branch growth. End walls of vessel cells eventually dissolve, resulting in the formation of long tubes of dead tissue. In transverse section, vessels appear as circular holes referred to as pores (Fig. 3.23). Unlike hardwoods, softwoods have no vessels; instead water is conducted through trachids, passing through perforated end walls of trachid cells (trachids also occur in hardwoods, where they comprise much of the tissue around the vessel elements).

One of the easiest features to recognize is the annual growth ring. Annual rings are formed if growth stops and starts on a seasonal basis. In the temperate zone, for example, trees lay down new xylem throughout spring and summer but are dormant during winter. Beginning and end of growth is marked by narrow bands of thick-walled trachids (Shackley 1981:93–107). If vessels formed during spring growth of the tree (early wood) are larger than those formed during the summer growth period (late wood), the wood is ring porous (Fig. 3.24). In some taxa, change between early and late wood pores is dramatic. In other taxa there is a gradual decrease in vessel size; this is semi-ring porous wood. Both variants are ring porous, as can be seen by comparing the latest pores of one year to the first pores of the next year. As is illustrated by Figure 3.24, semi-ring porous wood shows a clear size difference. If vessels formed during growth of early and late wood are the same size, wood is classified as diffuse porous.

It is not always as easy as the above description may suggest to determine if a piece of wood is ring, semi-ring, or diffuse porous. In the warm tropics, for example, some trees do not undergo a dormant period. Evergreen taxa may grow throughout

Figure 3.24 Ring porous, semi-ring porous, and diffuse porous wood seen in transverse section.

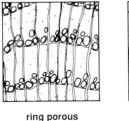

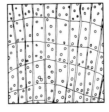

ring porous   semi-ring porous   diffuse porous

the year, laying down wood with no discernable growth rings. It is difficult to characterize vessel patterning in such taxa. Deciduous trees do occur in the warm tropics, usually in areas with seasonal dry periods. These taxa undergo seasonal dormancy related to rainfall. It is also sometimes difficult to determine whether a piece of wood is diffuse or semi-ring porous. Depending on circumstances of growth of the tree and changes in wood due to charring, drying, or waterlogging, actual vessel size within and between rings may vary. It is often necessary to key the same piece of unknown wood as both semi-ring and diffuse porous.

There is much more variety in vessel arrangement than that defined by the relation of vessels to growth rings. Vessels may occur singly, in groups, or in chains (Fig. 3.25). Chains may run radially (parallel to rays), obliquely (at an angle across ring), or tangentially (parallel to ring). Groups or clusters may vary from several pores grouped together to large flamelike groupings. Density of vessel occurrence also differs among wood taxa. Vessels may be sparse or quite numerous in relation to trachids and other tissue. Relative abundance of vessels in early and late wood also varies among tree species. Vessels may also have spidery inclusions, called tyloses.

A final feature of vessels which can be useful in identification is absolute size. Examination of wood plates or comparative wood collections generally reveals variation among taxa in largest vessel diameter, as viewed from transverse section. As is discussed further below, vessel size can change because of charring or post-depositional conditions, so absolute vessel size cannot be used as a final criterion in

Figure 3.25 Common vessel arrangements.

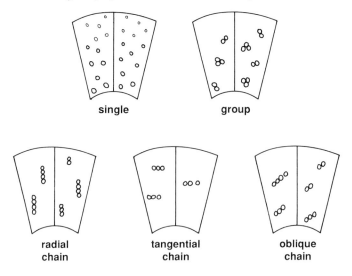

identification. It is, however, useful to note vessel size, using a qualitative scale, and to use this information as part of the total description of comparative or archaeological specimens (see Fig. 3.16). During identification of wood from the Real Alto site, for example (Pearsall 1979), charred comparative specimens were arranged in a desktop key by vessel size, using a scale of 1 to 6 (size 1 vessels being smallest).

Size and arrangement of rays, which are composed of parenchyma tissue and trachids, are additional features very useful for wood identification. Viewed from transverse section, rays appear like spokes of a wheel, radiating out from the center of the stem (Fig. 3.26). Using this section, size and arrangement of rays can be described as they appear when "sliced" through one horizontal plane. In some taxa, all rays are very narrow, only one cell-width wide. These are uniseriate rays. In other taxa, rays in transverse section appear variable in width, perhaps 2–6 cells wide. These are multiseriate rays. When a taxon with multiseriate rays is viewed from the tangential longitudinal section (ray cross section), rays are seen to change in width from top to bottom, being widest in the middle (giving a boat-shape cross section; Fig. 3.23). Ray height can be seen in tangential longitudinal section, as well as in radial longitudinal section. Multiseriate and uniseriate rays can occur in the same taxon, as is the case for oak.

In addition to determining width and height of rays and mix of sizes present, it is useful to characterize overall spacing and abundance of rays. Using transverse section, ray spacing may be described in terms of number of pore widths between rays. Regularity of spacing can also be noted. Abundance of rays can be characterized in terms of number present in a standard area.

Charred wood may fissure along rays, forming radial splits. Although formation of fissures is dependent to some degree on charring temperature and wood moisture content (see below), some wood taxa are more prone to radial splitting than others, making this a potentially useful identification criterion.

Abundance and nature of parenchyma tissue are the third set of features commonly used for wood identification. Parenchyma cells are storage cells with undifferentiated walls. There are three types of wood parenchyma: ray, epithelial, and axial or longitudinal. Parenchyma cells make up much of the tissue of rays. Epi-

Figure 3.26 Uniseriate and multiseriate rays viewed in transverse section.

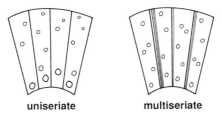

thelial parenchyma surrounds resin canals. Axial parenchyma occurs in strands running longitudinally through wood. Axial parenchyma is most easily seen in transverse section and can be very useful in identification. Arrangements of axial parenchyma that commonly occur in hardwoods have been described by Panshin and de Zeeuw (1980) and include the following types (Fig. 3.27):

1. *Paratracheal.* parenchyma associated with vessels. Paratracheal parenchyma can be scanty (a few cells around vessels, not easily visible), vasicentric (a row of cells completely surrounds vessels), aliform (cells form winglike projections on the sides of vessels), or confluent (irregular tangential or oblique bands are formed from winglike projections around vessels).
2. *Apotracheal.* parenchyma occurring independently of wood vessels. There are two basic types, diffuse (single parenchyma cells scattered in the wood, not easily visible) and aggregated (cells arranged in short tangential lines between rays, forming what looks like the steps in a ladder).
3. *Banded.* parenchyma arranged in more or less regular bands or lines, running tangentially. Thin-banded parenchyma differs from aggregated apotracheal parenchyma in forming bands that cross rays. In some wood taxa, bands include vessels; in others, parenchyma bands occur independently of vessels.
4. *Marginal.* parenchyma occurring at either the beginning or the end of annual growth rings. It is often difficult to distinguish between these

Figure 3.27 Common arrangements of parenchyma tissue.

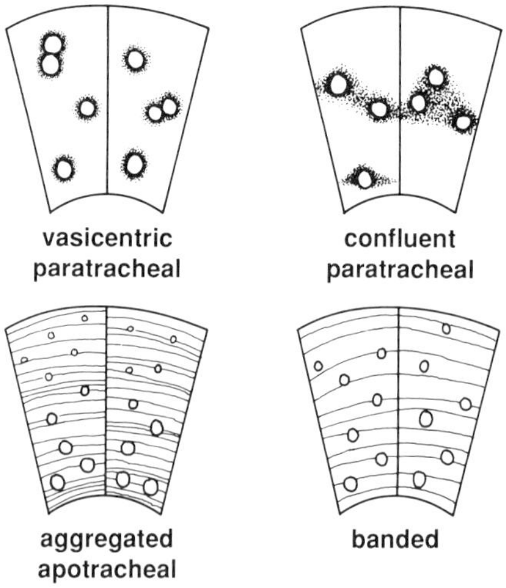

two types of marginal parenchyma, so they are grouped in most presentations. Marginal parenchyma can appear as a distinct band or as occasional cells.

Besides the anatomical features reviewed thus far, which are apparent under low to medium magnification (15–60×), there are also a variety of physical characteristics visible at low magnification or with the naked eye which are useful in identifying material, especially remains that are not charred. These include color of heartwood, luster (degree of shininess or dullness of the wood), odor and taste, grain and texture, weight, and hardness. Some of these characteristics are also useful in identifying charred wood. Different species of wood charred under similar conditions often retain luster differences, for example. Texture and relative hardness may also be assessed from charred specimens. Dense, heavy wood produces heavier charcoal than light wood, making this characteristic useful in some circumstances.

There are also a variety of useful microanatomical features that are visible only at higher magnification. Since studying these features often means examining thin sections or using electron microscopy, they are used only to distinguish among very similar taxa. Alternatively, a metallurgical microscope with top lighting and 50–900× can be used to examine wood at higher magnification. Included among minute structures of wood useful for identification are cell structure of rays (homocellular or heterocellular), presence of sheath or oil cells in ray tissue, nature of vessel perforations (simple or scalariform), arrangement and size of vessel pits, and size and nature of trachids. Wood anatomy texts such as Panshin and de Zeeuw (1980) describe in detail these and other microanatomical wood features and illustrate how these features aid identification.

Actual procedures for using the characteristics of wood described above to identify an unknown archaeological wood specimen depend on the nature of the material (charred, dried, or waterlogged), its state of preservation, and composition of the woody flora of origin. How precisely wood can be identified (or whether it can be identified at all) is limited by these same factors, as well as by availability of comparative materials. I discuss these points by considering in turn the process of identifying charred, dried, and waterlogged wood.

Having to identify charred wood is a common circumstance for many ethnobotanists. Charring eliminates or alters a number of features useful for identifying fresh wood. Color, odor, and taste are of course lost, and hardness and luster must be redefined. During the process of charring, wood shrinks an average of 40–45% by volume (McGinnes et al. 1974). This causes distortion, because shrinkage is not even for each dimension or for all types of tissue. McGinnes et al. (1971) discovered, for example, that charring oak in a commercial kiln resulted in an average shrinkage of 25.68% in tangential dimension, 15.45% in radial dimension, and 11.43% longitudinally. This uneven shrinkage caused circular spring pores to become elliptical and folds to develop in cell walls. Other changes following charring included occurrence of severe radial splits, deposition of residual tars in cells, and fusing of adjacent cell walls to form thick, dense double walls. Controlled laboratory charring experiments (McGinnes et al. 1976) confirmed observations of charring in commercial kilns and demonstrated that charring at higher temperature (800° C vs. 400° C) produced greater shrinkage. There are three important implications of these experiments for charcoal identification: (1) conditions of charring of archaeological wood must be considered unknown and unequal among specimens; (2) comparisons between comparative charcoal and archaeological charcoal should deemphasize traits affected by shrinkage; and (3) all comparative wood used for identification should be charred under similar temperature and moisture conditions to minimize variability in shrinkage among specimens.

To examine charred wood, simply snap the specimen or cut it with a sharp razor blade to create a fresh transverse surface. View this surface under low magnification (15×) with a dissecting microscope and intense, obliquely angled top lighting; note characteristics such as vessel arrangement, abundance, and size, ray spacing and abundance, and physical characteristics of the wood (Fig. 3.28). Using higher magnification (30–60×), note parenchyma characteristics and ray widths. Next snap or cut radial longitudinal and tangential longitudinal sections to give additional views of rays. Index cards for each distinctive archaeological wood type which include a description of the type, sketches of sections, a list of all possible match-ups in the comparative collection, and characteristics that distinguish the type from very similar ones are useful at this stage. Punch cards may also be used to record anatomical features and physical characteristics of archaeological specimens and comparative material. Leney and Casteel (1975) report good results examining charcoal specimens with incident light microscopy, including dark field, utilizing an instrument such as the Zeiss Epi microscope.

Charred specimens may be too fragile to examine by snapping or cutting a clean surface. Before such specimens can be examined, they must be strengthened. This is done by impregnating the specimen with an agent that infiltrates air spaces and then hardens, binding tissue together. The specimen can then be cut for microscopic examination. Because embedding and thin sectioning are techniques useful for studying a variety of botanical materials, I describe them in detail below along with scanning electron microscopy, starch grain analysis, and other specialized techniques.

Dry ancient wood presents a number of problems in identification (Western 1970). Dry wood specimens may be too hard and solid to snap for a clean surface, or so brittle and powdery that they disintegrate when handled. Wood may become impregnated with mineral crystals, which obscure features and make specimens

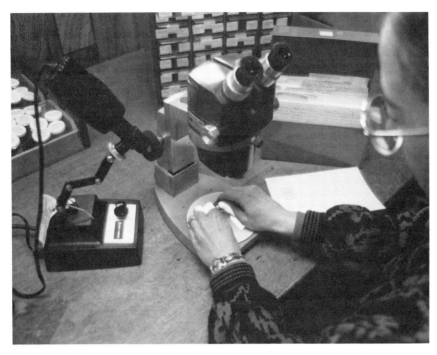

Figure 3.28 Identification of charred archaeological wood. Pieces are snapped to give fresh sections and viewed with side lighting. Comparisons are directly to known wood specimens (in desktop storage unit) and to identification cards.

difficult to view. Although specimens appear well preserved, partial decay may have taken place, altering or destroying features needed for identification. The process of gradual drying, like charring, causes shrinkage in wood tissues. Western (1970) reports shrinkage up to 25% in the radial and tangential directions, and less in length.

When identifying dry ancient wood, overall condition of specimens must first be assessed. Choose a sample with numerous specimens and snap (if possible) a number of pieces. Note the quality of the transverse section. If the wood is soft enough to snap but gives a ragged section, cut specimens with a sharp razor blade to give a better section. For larger, sturdy specimens, it may be possible to obtain a good transverse surface by sawing a thick section and sanding one surface. If snapping, cutting, or sawing does not yield good results, either because the wood is too hard or too fragile for such treatment, or because the pieces are too small, then identification must entail embedding and sectioning or grinding selected specimens (see "Specialized Identification Techniques," below).

The condition of waterlogged wood can also vary considerably and affect identification. Sometimes waterlogged specimens are soft enough to be cut easily by hand with a sharp razor blade or scalpel. Structural waterlogged wood is sometimes identified in situ this way at historic period sites in Great Britain (Williams 1975). After cutting into a specimen 10 mm or more to expose good structure, a section is cut out for examination. If wood is so soft that cut sections disintegrate, Williams recommends wetting the area to be cut with alcohol, which hardens wood slightly. Specimens to be transferred to the laboratory or permanently curated must not be allowed to dry out. Waterlogged wood shrinks dramatically and becomes very hard if allowed to dry (Levy 1970).

Waterlogged wood is sometimes too spongy to give a good transverse surface when hand-sectioned. In this situation, specimens are embedded in aqueous wax to stabilize them and permit microtome sectioning (Dimbleby 1978; and see below). As is sometimes the case for dry ancient wood, waterlogged wood can become infiltrated with mineral or salt deposits, which cause it to harden. Such specimens must also be cut or ground for examination.

Identification of waterlogged wood may be plagued by partial or total deterioration of diagnostic features. Dimbleby (1978) notes, for example, that cell walls often disappear in waterlogged specimens. Although decay by aerobic bacteria is inhibited by waterlogging, anaerobic bacteria and fungi can cause deterioration. In a study of a large collection of wood and other vegetal materials preserved under peat at the Monte Verde site in Chile, a variety of post-depositional changes were documented in specimens, some of which rendered correct identification problematic (Thomas Dillehay, personal communication). Although the analyst can prepare comparative collections of fresh wood or charcoal, it is difficult to prepare dried or waterlogged collections that duplicate the preservation state of archaeological specimens. It is important that changes caused by processes of decay, desiccation, or waterlogging be recognized and distinguished from taxonomically valid features.

As is the case for identification of all archaeological botanical materials, it is important that precision of identification of wood specimens reflect not only limitations imposed by the innate structure of wood but also those resulting from charring or post-depositional changes. For paleoethnobotanists working in regions where detailed studies of wood are available, limits of wood identification may be precisely known. For example, in North America numerous studies have shown that wood from most oak species is very much alike and can often only be divided into white oak and red oak groups (Panshin and de Zeeuw 1980). Similarly, willow and poplar wood can only be distinguished at high magnification and are therefore usually identified only to the family level, Salicaceae. It is more difficult to determine the correct level of identification when working only with locally collected comparative material from a region with a poorly studied wood flora. My advice in this situation is to be conservative and not attempt to distinguish taxa that appear very similar. As was discussed under the topic of sampling, one usually identifies only a random sample of wood from flotation or sieve samples, rather than entire samples. Identification time may also be reduced by not identifying wood from all samples, but only from a selection of samples in different context categories.

The following sources are useful in learning the basics of wood identification: Core et al. (1979), Dimbleby (1978), Harlow (1970), Harrar (1957), Jane (1970), Kribs (1968), Leney and Casteel (1975), Panshin and de Zeeuw (1980), Miles (1978), Schweingruber (1978), Shackley (1981), and Western (1970). Refer to Delcourt et al. (1979) for numerous references to specific taxonomic groups.

Roots and Tubers

Human populations throughout prehistory have utilized underground storage organs of plants for food. These include cultivated plants such as manioc or yuca (*Manihot esculenta*), potato (*Solanum tuberosum*), and yam (*Dioscorea* spp.) as well as a multitude of wild plants from many botanical families. Concentrations of starch in fleshy underground storage organs make these valuable food resources.

In spite of the important role roots and tubers must have played in prehistoric subsistence, macroremains of underground storage organs are sparse for any region of the world, and for any crop or wild plant. This is the result of both problems of preservation of these remains and difficulty in identifying them. Although I focus here on roots used as foods, some roots are deposited archaeologically through use as fuel (e.g., through burning of wood roots or sod).

Fleshy underground storage organs can be divided into two types: those originating in stem tissue, and those that are actually roots. Since there are a number of differences between stem and root tissue which aid in identifying remains, it is useful to start with this basic distinction. A root is the underground portion of the main axis of a plant or branches of the axis (Benson 1959:649–666). A stem is the portion of the main axis or branch which is leaf bearing and flower bearing (Lawrence 1951:737–775). When stems occur underground, they retain anatomical structures of leaf-bearing organs: nodes (areas that bear leaves), buds, and sometimes remnant, scalelike leaves. By contrast, roots do not have nodes, internodes (portions of stem between two nodes), or remnant leaves (Fig. 3.29).

Crops with storage organs that are true roots include carrots, beets, and sweet potatoes. The carrot is a good example of a tap root enlarged through ordinary secondary growth. Secondary growth is formation of secondary vascular tissues (xylem and phloem) from a vascular cambium, and of periderm (protective tissue replacing the epidermis) from a cork cambium. Formation of secondary root tissue is a characteristic of gymnosperms and many dicotyledons. Monocotyledons, by contrast, rarely exhibit secondary root growth. Figure 3.30 and 3.31 illustrate transverse

166 • Chapter 3 Identification and Interpretation of Macroremains

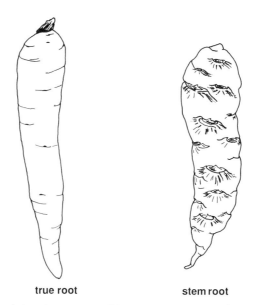

true root   stem root

Figure 3.29 True root (a) and stem root (b). In a stem root such as mashua, leaf nodes, the "eyes" of the tuber, indicate its origin as stem tissue. The carrot, a true root, has no eyes, only inconspicuous secondary root scars.

sections through tap roots enlarged through ordinary secondary growth. Note the displacement of the vascular cambium outward, so that it appears like a ring in transverse section.

Not all secondary root growth takes place in the pattern illustrated in these figures. Secondary root development in beets and sweet potatoes are examples of anomalous secondary growth (Esau 1977:243–256). In these roots, secondary vascular cambium is located out of its "ordinary" position. In beets, for example, many cambial layers develop concentrically. Sweet potato roots are characterized by cambial development around individual vessels or vessel groups. Many patterns of anomalous secondary growth exist.

There are a variety of kinds of fleshy underground storage organs which originate in stem tissue; they occur both in monocotyledon and dicotyledon families. Common types are rhizomes, corms, bulbs, and tubers (Fig. 3.32). Rhizomes are horizontal underground stems that can be distinguished from roots by the presence of nodes, buds, or scalelike leaves. If the rhizome gives rise to a new plant at its tip, it is a stoloniferous rhizome. A corm is an enlarged, usually subterranean, stem base. Corms are usually covered by a membranous or fibrous tunic, or outer skin. A bulb is an underground bud covered by fleshy scales formed from leaf bases; this gives the characteristic layered, or tunicated, appearance of bulbs such as onion.

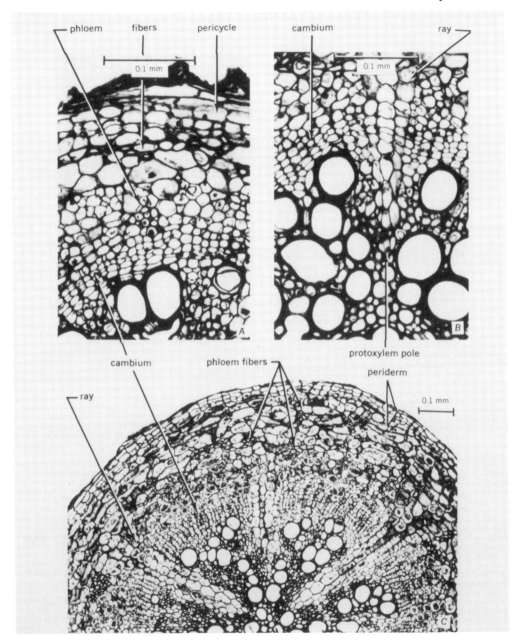

Figure 3.30 Cross section of root of alfalfa (*Medicago sativa*) with details of secondary growth: (a, b) early stage of secondary growth; (a) arrangement of tissues along radius through phloem; cortex (outside pericycle) is collapsed; (b) region opposite a protoxylem pole; (c) root in advanced stage of secondary growth (from Esau 1977).

168 • Chapter 3 Identification and Interpretation of Macroremains

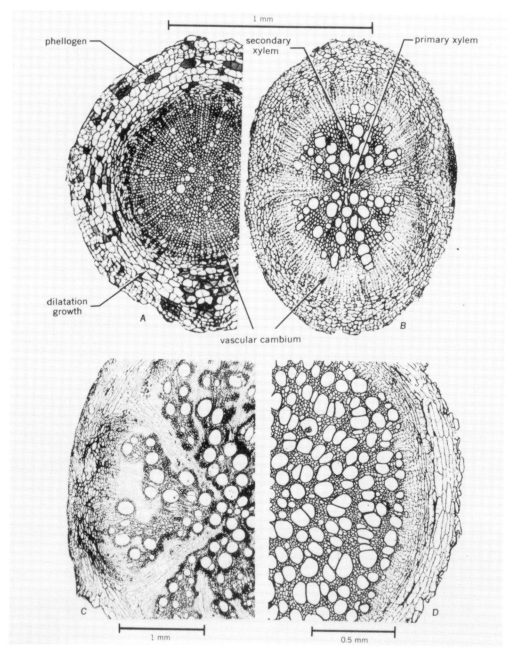

Figure 3.31 Cross sections of roots of herbaceous plants in secondary state of growth: (a) tomato (*Lycopersicum esculentum*); (b) cabbage (*Brassica oleracea*); (c) pumpkin (*Cucurbita* sp.); (d) potato (*Solanum tuberosum*) (from Esau 1977).

Basic Identification Techniques • 169

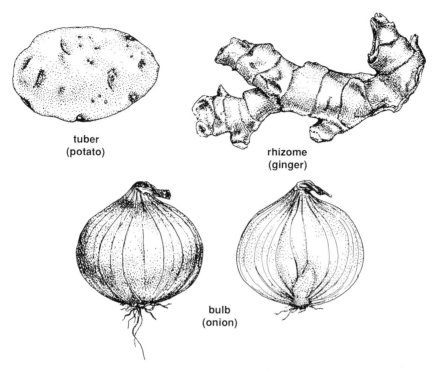

Figure 3.32 Various types of stem roots.

Tubers, like rhizomes, are subterranean stems with nodes and buds. Tubers are shorter and thicker than rhizomes, however, being formed from underground branches of the stem.

Paucity of root and tuber macroremains in archaeological sites is closely related to how these resources are processed and used. Whatever the preservation situation, the only part of a root or tuber which often remains after use is the peel. Although some root peelings are tough and fibrous (e.g., manioc or *Canna*), other roots or tubers have thin periderms that quickly decay (if they are not eaten with the root). In a situation of charred preservation, recovering tuber or root peelings is very unlikely, since such remains are quite fragile. Finds do occur in dry preservation settings; for example, it is not uncommon to recover dried achira (*Canna edulis*) and jicama *(Pachyrrhizus tuberosus)* peelings in early coastal Peruvian deposits (Pearsall 1985a; Ugent et al. 1984).

It is also unlikely that whole roots and tubers end up in an archaeological deposit. The most likely sources of whole roots or tubers are those discarded as spoiled (and preserved in a dry or waterlogged setting) and any accidentally charred during roasting. The quantity preserved in these ways at a site might be small, but such remains provide a good opportunity for identification.

Although there is sufficient variation in form and structure of many roots and tubers to permit identification, success may be limited by state of preservation of remains. Identification may also be difficult, and time consuming, because techniques such as embedding and sectioning, scanning electron microscopy, and starch grain analysis are often necessary for precise identification. Before discussing the role of these techniques, I review the basics of identifying root and tuber material.

In a situation of dried preservation, both peelings and whole or partially complete roots or tubers may be present. The first step in identification of such material is to sort remains into like groups based on overall form and external characteristics visible at low to medium magnification. Peelings, for example, can be sorted by presence or absence of leaf nodes (true root vs. rhizome or tuber) or thickness and presence of fibrous or corky periderm. If peelings have leaf nodes, these can be further divided into tubers (shortened internodes) and rhizomes (elongated internodes). Whole or partially complete roots are sorted in a like manner. After initial sorting, it may be clear that several different kinds of roots or tubers are present, and peelings can be associated with whole examples. One might make 5" × 8" cards that include a sketch and description of remains for each archaeological type.

After preliminary sorting and description of tuber and root remains, each type should be studied carefully and compared to known voucher material. This detailed study may include measuring longitudinal internode distances on rhizomes, describing node shape, size, and patterning (whether nodes are arranged in an opposite, alternate, or whorled pattern; how many ranks of nodes are present), noting striations, folds, and other surface characteristics, describing any branching pattern present, and sketching and measuring stem or leaf base attachment scars or remnants. Comparing unknown types to known comparative material using these characteristics as well as any preserved physical characteristics such as color, tissue weight, and texture may allow identification, even if only to family or genus.

It is often necessary to go beyond the use of overall form and external characteristics to arrive at precise identification of root or tuber material. For example, when attempting to identify small charred tap roots in upper strata of Panaulauca and Huashumachay caves, Peru (Fig. 3.33), I discovered that many local plants produced secondarily thickened tap roots. Although most could be eliminated as very unlike the archaeological type in overall shape and nature of leaf attachment scars, a group remained which overlapped in form. The anatomical structure of the archaeological remains was used to make the final determination as *Lepidium* sp. (Pearsall 1985b).

To use anatomical features such as pattern of secondary growth to aid identification, it is necessary to view the unknown tuber or root remains in section, es-

Figure 3.33 Charred roots found in upper strata of Panaulauca and Huashuamachay caves, Peru: (a) variation in shape of archaeological roots; (b) well-preserved roots, showing intact epidermis; (c) top view of root, showing circular leaf-attachment area.

pecially transverse section, and to compare them to known materials. Figure 3.34 shows such a comparison for the Panaulauca root remains. This viewing can sometimes be done by simply cutting a thick section of specimen with a razor blade or scalpel. The section surface is then viewed at high magnification using incident lighting or prepared for scanning election microscopy. If it appears that a specimen cannot be hand-cut (being too fragile or too hard), it may be possible to obtain a good transverse section by freeze fracturing. In this procedure, the specimen is quick-

172 • Chapter 3 Identification and Interpretation of Macroremains

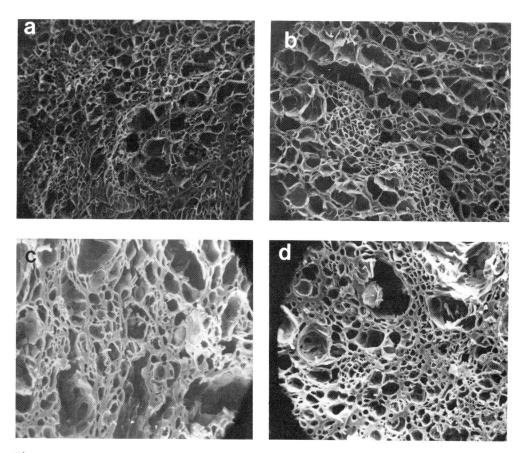

Figure 3.34 Scanning electron microscope photographs of cross sections of charred archaeological and comparative roots from Peru: (a, b) comparative specimens of cultivated maca, *Lepidium meyenii*; (c, d) archaeological specimens showing similar anatomical features; after comparison to other material, these archaeological specimens were identified as *Lepidium* sp., probably early cultivated maca.

frozen in liquid nitrogen and then broken in the desired section. Freeze fracturing is described in more detail below with other specialized techniques.

It is sometimes impossible to obtain a clean section without first stabilizing root material in an embedding medium. Because embedding media infiltrate small pieces more readily than large, broken or partial examples of each type can be used. After embedding, specimens can be sectioned for either light or electron microscopy.

Once transverse sections of archaeological root or tuber types and relevant comparative materials (i.e., taxa similar to the unknown in overall form and external characteristics) have been prepared, the structure of the root tissue should be described and compared to known specimens. Features useful in comparing roots or tubers of different taxa include nature of secondary growth (ordinary or anomalous; if the latter, what form does it take), abundance of parenchyma (storage cells), abundance and arrangement of vascular tissue, arrangement and thickness of rays, abundance of phloem fibers, presence of sclerenchyma in phloem, and thickness of periderm. Anatomy texts such as Esau (1977) and Fahn (1982) serve as guides to identifying these and other features.

By studying the overall form of root or tuber material, external characteristics, anatomical structure, or a combination of these, it is often possible to identify archaeological material. Precision of identification is limited by the innate structure of roots and stems and by changes that occur during charring, drying, or waterlogging. It is important not to force too precise an identification on root material in a poor state of preservation.

Another approach useful in identifying dried tuber remains is starch grain analysis (Fig. 3.35). This technique was used by Towle (1961) in her early work in coastal Peru. In a series of articles on analysis of dried tuber and root remains from sites in the Casma valley, Peru, excavated by Shelia and Thomas Pozorski (Ugent et al. 1981, 1982, 1983, 1984), Ugent and coauthors successfully identified cultivated sweet potato (*Ipomoea batatas*), achira (*Canna edulis*), and potato (*Solanum tuberosum*) using extant surface features and starch grains. Each of these species was found to produce starch grains distinctive from other cultivated Andean tubers. In addition, Ugent and Verdun (1983) found that 19 wild Mexican tuber-producing species of *Solanum* could be identified using starch grain shape, length, width, position of hilum, and percentage occurrence of eleven defined grain types. Cultivated potato could also be distinguished from wild species studied. Unfortunately, starch grain structure is not preserved in charred root specimens (grains burst upon heating) or in all dried specimens. For example, Smith (1980) reported no success in extracting starch from dried tuber and root material from Guitarrero Cave. Techniques for starch grain extraction and analysis are presented in "Specialized Identification Techniques," below.

I have focused thus far on identification of edible underground storage organs. Root, bulb, corm, and rhizome material can also be deposited from plants used for other purposes. Burning grass sod or woody root material as fuel is one source of deposition of nonedible root and stem material. The identification of woody root material follows procedures outlined above for wood identification. Root wood is not identical in form to stem wood of the same taxon, however. Roots have a higher proportion of parenchyma tissue and a smaller proportion of cells with lignified walls than stem wood. This may mean, for example, that root wood has higher, wider rays (which consist mostly of parenchyma) than stem wood. Comparative root material is needed to ensure correct identification.

174 • Chapter 3   Identification and Interpretation of Macroremains

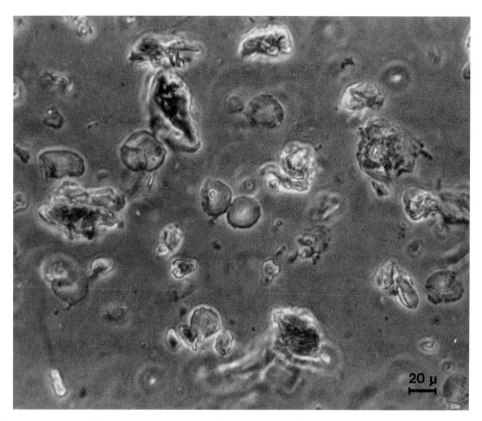

Figure 3.35   Starch grains from cultivated sweet potato (*Ipomoea batatas*). Photographed at 400×.

Fibers, Leaves, and Non-Woody Stems

This final grouping of macroremains consists of plant parts recovered archaeologically which are not reproductive structures (seeds, fruits, nuts), underground storage organs (roots, tubers, corms, bulbs, rhizomes), or wood. Paleoethnobotanists who work in regions where only charred material is preserved may never encounter cordage, textile fragments, bundles of coca or tobacco leaves, or cane, but finds of aboveground, vegetative portions of plants do commonly occur in sites with waterlogged or dried preservation. In those settings, a major problem may be what to do with masses of grass, fibers, stems, leaves, and twigs which may constitute most of the excavation matrix.

I divide this discussion into two parts, (1) fibers, and (2) leaves and stems, because some specialized techniques have been developed for fiber identification, whereas identification of leaves and stems is basically a matter of having a broad knowledge of appearance of vegetative portions of plants and comparing archaeological speciments to comparative material.

Fibers

Three types of fibers are of interest here: true fibers (flax, hemp, ramie, among others), seed hairs (such as cotton), and animal hair. I include animal hair in the discussion because it may be confused with vegetable fiber on first examination.

True fibers are long, slender sclerenchyma cells, many times longer than wide (Esau 1977:71–82). Sclerenchyma cells have a secondary wall deposited over the primary wall after the cell has completed its extension growth. Sclerenchyma functions to give support, hardness, and rigidity to plant tissue. Fibers can be found in many parts of plants but are particularly common in the vascular tissue. Stem phloem fibers of dicots are the "soft" fibers of commerce: hemp, jute, and flax, among others. Figure 3.36 illustrates location of fibers in hemp. Phloem fibers usually occur in strands, bundles of overlapping fibers. Leaf fibers of monocots are the "hard" fibers of commerce: Manila hemp and sisal, among others. These fibers are strongly lignified, hard, and stiff.

Cotton "fiber" is actually epidermal hair from the seed. It is soft and characteristically flattened and convoluted. It does not occur in strands; each hair is attached individually to the seed epidermis.

Analysis of fibers, whatever their origin, generally involves microscopic examination of individuals as well as observation of texture and color of fabric, cord fragment, or fiber mass (Appleyard and Wildman 1970; Catling and Grayson 1982; Cook 1968). It may also be necessary to cross-section fiber specimens to observe the cross-sectional shape and thickness of cell walls. In their discussion of identification of vegetable fibers, Catling and Grayson (1982) suggest the following general procedures.

1. Make general observations of color and texture.
2. Make a mount of a whole fiber strand (bundle of fibers) by boiling the specimen in water to remove air and mounting it on a microscope slide in 50% glycerin. Note arrangement of fibers in the strand and inclusions of tissues other than sclerenchyma.
3. To view individual fibers, fiber strands must be macerated. Place specimens in a 50:50 solution of glacial acetic acid and 20% hydrogen peroxide and heat them in a water bath for 7 or 8 hours. Remove specimens and place them in water; shake vigorously until the strand separates into individual fibers. Mount in 50% glycerin. If desired, the mount can be stained with a 1% aqueous solution of chlorazol black. Note shape of

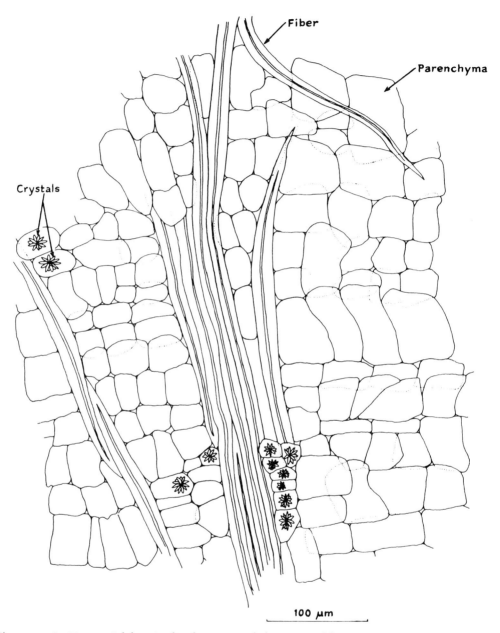

Figure 3.36 Tangential longitudinal section of the stem of hemp, *Cannabis sativa* L, illustrating location of fibers as well as crystal inclusions and parenchyma tissue (from Catling and Grayson 1982:21).

fiber cell end, thickness of cell wall and lumen, appearance of dislocations and cross markings, inclusion of crystals, and fiber cell diameter and length.
4. For stems, pseudostems, and leaves (sources of true fibers), prepare transverse sections by microtomy. Examining sections allows study of fibers in situ.

Microscopic examination of individual fibers in whole mount allows easy separation of animal and vegetable fibers. Animal hair is composed of cuticle and medulla. Cuticle scales are diagnostic characteristics of animal hair which segregate it from vegetable fiber. Because of the differing chemical characteristics of animal and vegetable fibers, staining can also aid in distinguishing these fibers. Animal hair can often be identified to the family level or below using characters such as distribution of pigment, thickness of cuticle, type of medulla, and pattern of scale margins (Appleyard and Wildman 1970).

Stephens and Moseley (1974) reported success distinguishing fiber of domesticated cotton from that of wild cotton. By measuring fiber diameter of wild and primitive cultivated *Gossypium barbadense* from the coasts of Peru and Ecuador, these authors demonstrated that mean fiber diameter is narrower in wild than in domesticated cotton. Using these data, as well as measurements of seeds and bolls (cotton fruit), they studied archaeological remains of cotton from four preceramic sites in the Ancon-Chillon area of Peru. They determined that mean seed sizes and mean fiber diameters increased through time from older to more recent sites, suggesting that the process of domestication of cotton was under way.

Leaves and Stems

Identifying whole leaves and other vegetative plant material is a difficult task. Success depends entirely on the presence of features that are distinctive enough, and of limited enough distribution in a flora, to allow determination. And regardless of how distinctive leaf or stem material might be, successful identification depends on an adequate comparative collection.

There are a variety of features of stems and leaves which may be useful in identification. In the leaf epidermis, for example, features like stomata guard cells, trichomes (epidermal appendages, such as hairs), cells with inclusions (tannin, oils, crystals), and sclerenchyma can be studied (Esau 1977:351–374) (Figs. 3.37 and 3.38). Leaf form or outline, margin characteristics, shape of base and apex, surface characteristics, and venation pattern can be described (Lawrence 1951:737–775) (Fig. 3.39). Rury and Plowman (1983) and Plowman (1984), for example, have used detailed study of leaf venation to identify archaeological specimens of coca (*Erythroxylum* spp.).

178 • Chapter 3 Identification and Interpretation of Macroremains

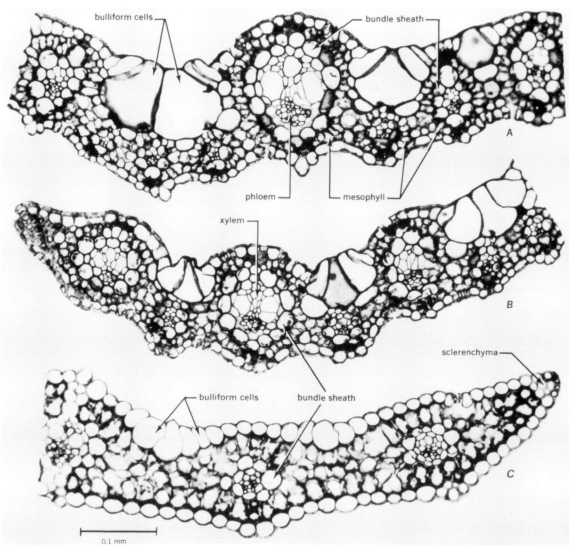

Figure 3.37 Grass leaves in transections: (a, b) *Saccharum* (sugar cane) showing structure associated with C_4 photosynthetic cycle; (c) *Avena* (oat), a C_3 plant. Double bundle sheaths as in wheat. Spatial association between mesophyll and vascular bundles closer in sugar cane than in oat (from Esau 1977).

Stem material can be analyzed for leaf arrangement (whorled, distichous, alternate), bud type and bud-scale arrangement, and arrangement of vascular tissue. The last named feature can be used to distinguish stem tissue of many dicots from most monocots (Esau 1977:295–319).

The following are useful references for identifying vegetative plant parts: Ap-

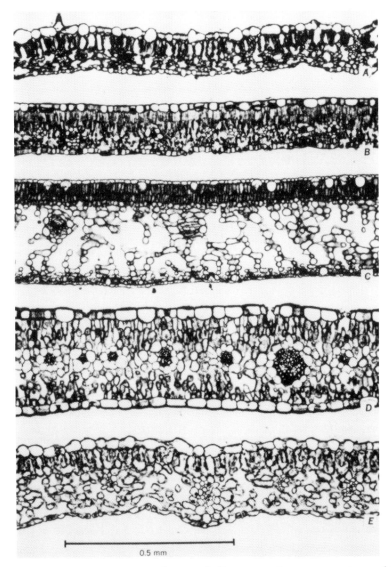

Figure 3.38 Variations in structure of mesophyll as seen in cross sections of leaves: (a) parsnip (*Pastinaca sativa*); (b) peach (*Prunus persica*); (c) lemon (*Citrus limon*); (d) *Dianthus*; (e) *Lilium* (from Esau 1977).

pleyard and Wildman (1970), Catling and Grayson (1982), Cook (1968), Dilcher (1974), Martin and Juniper (1970), Ogden (1974), and Palmer (1976). Refer to Delcourt et al. (1979) for additional references to specific taxonomic groups. Detailed anatomical studies, such as Metcalfe (1960) for the grasses, contain much valuable information for identifying vegetative plant material.

Table 3.3 • Size Changes in Archaeological Sunflower (*Helianthus annuus*) Achenes

	No. of Measurements	Corrected achene dimensions (L × W)			Index (L × W) (mm)
		Mean (mm)	Std. dev. (mm)	Range (mm)	
Middle Archaic					
Koster, Horizon 9A	1	4.4 × 2.5	—	—	11
Koster, Horizon 6	1L,2W	6.0 × 2.4	— × 0.1	— × 2.3–2.4	14
Late Archaic					
Napoleon Hollow, Titterington	4	6.1 × 2.7	0.2 × 0.3	6.0–6.4 × 2.5–3.0	16
Middle Woodland					
Crane	5L,8W	6.6 × 3.3	2.2 × 0.9	4.2–9.4 × 1.9–4.6	21
Loy	2L,4W	6.5 × 3.6	2.4 × 0.7	4.8–8.2 × 2.7–4.2	23
Massey	8L,11W	6.8 × 3.2	0.9 × 0.7	5.5–8.7 × 2.2–4.8	22
Smiling Dan	3L,15W	7.2 × 3.4	0.8 × 0.4	6.2–7.8 × 2.8–4.1	24
Macoupin	1	8.5 × 3.8	—	—	33
Early Late Woodland					
John Roy (large)	2	11.3 × 6.7	1.3 × 0.6	10.4–12.2 × 6.2–7.1	75
(small)	1	5.9 × 3.3	—	—	19
Newbridge	3	8.3 × 4.0	1.3 × 0.5	6.7–9.1 × 3.4–4.4	33
Carlin	1	7.9 × 4.1	—	—	32
Crane	64L,114W	7.1 × 3.3	1.2 × 0.6	5.2–11.1 × 2.2–5.6	24
Late Late Woodland					
Koster East	3W	— × 3.9	— × 0.6	— × 3.4–4.5	—
Deer Track	1	6.1 × 3.6	—	—	22
Carlin	3L,4W	9.7 × 4.5	0.6 × 0.4	8.8–10.1 × 4.1–4.9	43
Kuhlman	1	3.9 × 2.7	—	—	10
Loy	1W	— × 3.6	—	—	—
Healey	2	8.7 × 4.1	1.7 × 0.4	7.5–9.9 × 3.9–4.5	37
Mississippian					
Audrey North	2	10.3 × 3.4	0.6 × 0.5	9.9–10.8 × 3.0–3.8	35
Hill Creek	3L,9W	9.0 × 4.5	1.8 × 0.9	7.4–10.9 × 3.2–5.7	41
Historical/Protohistoric					
Oak Forest	1	9.4 × 4.2	—	—	39
Zimmerman	7L,10W	7.0 × 3.3	0.5 × 0.3	6.5–7.7 × 2.9–3.8	23

Source: Asch and Asch (1985:168).

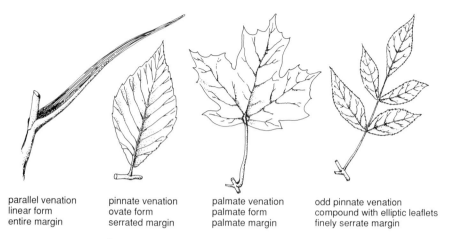

parallel venation	pinnate venation	palmate venation	odd pinnate venation
linear form	ovate form	palmate form	compound with elliptic leaflets
entire margin	serrated margin	palmate margin	finely serrate margin

Figure 3.39 Variation in leaves. Not to scale.

Cultivated Plant Material

This chapter on identifying macroremains is not intended as a guide to identification of particular taxa. This limitation also extends to remains of cultivated plants. General techniques for examining and analyzing plant remains, presented above, apply to cultivated as well as wild materials.

During the process of domestication, plants undergo genetic change. Genetic changes may be manifested in characters of the fruit and seed which are discernible archaeologically, such as increased size of seeds and presence of mechanisms in the fruit which inhibit seed dispersal. To study cultivated material adequately, it is important to take precise measurements, draw or photograph remains, and compare fragmented materials carefully to known materials. Size of seeds or fruiting structures and overall morphology can often be used to distinguish cultivated plants from wild relatives, or to separate remains of closely related species or varieties. Table 3.3 illustrates use of seed size to trace the development of a cultivated plant, sunflower, over time. In other cases it is necessary to rely on detailed anatomical study, utilizing thin sectioning or scanning electron microscopy.

Techniques for identifying archaeological cultivated plant material have been developed and published by botanists and paleoethnobotanists specializing in various crops. These studies can serve as the starting point for analysis of materials. It is often advisable, however, to have identifications of cultivated materials confirmed by a specialist in that crop; one not only avoids mistakes but can contribute to the specialists' knowledge of existing variation in, and geographic distribution of, the crop. References to works describing how to identify macroremains of cultivated plants are given below. This is not an exhaustive list, but it should give the beginner enough to get started.

Squashes and Gourds
  Cutler and Whitaker (1961)   "History and Distribution of the Cultivated Cucurbits in the Americas"
  King (1985)   "Early Cultivated Cucurbits in Eastern North America"

Cotton
  Smith and Stephens (1971)   "Critical Identification of Mexican Archaeological Cotton Remains"
  Stephens (1971)   "The Botanical Identifications of Archaeological Cotton"
  Stephens (1975)   "A Reexamination of the Cotton Remains from Huaca Prieta, North Coastal Peru"

Tobacco
  Haberman (1984)   "Evidence for Aboriginal Tobaccos in Eastern North America"

Roots and Tubers
  Leon (1964)   "Plantas Alimenticas Andinas"
  Martins (1976)   "New Archaeological Techniques for the Study of Ancient Root Crops in Peru"
  Pearsall (1985b)   "Paleoethnobotanical Data and Procedures"
  Towle (1961)   *The Ethnobotany of Pre-columbian Peru*
  Ugent et al. (1981)   "Prehistoric Remains of the Sweet Potato from the Casma Valley of Peru"
  Ugent et al. (1982)   "Archaeological Potato Tuber Remains from the Casma Valley of Peru"
  Ugent et al. (1984)   "New Evidence for Ancient Cultivation of *Canna edulis* in Peru"
  Ugent and Verdun (1983)   "Starch Grains of the Wild and Cultivated Mexican Species of *Solanum*, Subsection *Potatoe*"

New World Grains
  Bird (1980)   "Maize Evolution from 500 B.C. to the Present"
  Bird (1984)   "South American Maize in Central America?"
  Callen (1967)   "The First New World Cereal"
  Cowan (1978)   "The Prehistoric Use and Distribution of Maygrass in Eastern North America"
  Cutler (1956)   "Plant Remains"
  Mangelsdorf (1967)   "Prehistoric Wild and Cultivated Maize"
  Mangelsdorf (1974)   *Corn: Its Origin, Evolution, and Improvement*
  Nabham and deWet (1984)   "*Panicum sonorum* in Sonoran Desert Agriculture"

Old World Grains
  deWet et al. (1983)   "Domestication of Sawa Millet"
  Helbaek (1969)   "Plant Collecting, Dry-farming, and Irrigation Agriculture in Prehistoric Deh Luran"
  Renfrew (1973)   *Palaeoethnobotany: The Prehistoric Food Plants of the Near East and Europe*

Legumes
  Burkart (1952)   *Las Leguminosas Argentinas Silvestres y Cultivadas*
  Damp et al. (1981)   "Beans for Valdivia"
  Kaplan (1965)   "Archaeology and Domestication in American *Phaseolus* (Beans)"
  Renfrew (1973)   *Palaeoethnobotany: The Prehistoric Food Plants of the Near East and Europe*
  Sauer and Kaplan (1969)   "*Canavalia* Beans in American Prehistory"

Native North American Cultigens
  Asch and Asch (1977)   "Chenopod as Cultigen: A Reevaluation of Some Prehistoric Collections from Eastern North America"

Asch and Asch (1985)	"Prehistoric Plant Cultivation in West-Central Illinois"
Asch and Asch (1978)	"The Economic Potential of *Iva annua* and Its Prehistoric Importance in the Lower Illinois Valley"
Smith (1984)	"Chenopodium as a Prehistoric Domesticate in Eastern North America: Evidence from Russell Cave, Alabama"
Wilson (1981)	"Domesticated *Chenopodium* of the Ozark Bluff Dwellers"
Yarnell (1972)	*Iva annua* var *macrocarpa*: Extinct American Cultigen?
Yarnell (1978)	"Domestication of Sunflower and Sumpweed in Eastern North America"

Coca
Hastorf (1987)	"Archaeological Evidence of Coca (*Erythroxylum coca*, Erythroxylaceae) in the Upper Mantaro Valley"
Plowman (1984)	"The Origin, Evolution, and Diffusion of Coca, *Erythroxylum* spp., in South and Central America"
Rury and Plowman (1983)	"Morphological Studies of Archeological and Recent Coca Leaves (*Erythroxylum* spp.)"

Fruits
Pickersgill (1969)	"The Domestication of Chili Peppers"
Pickersgill (1984)	"Migrations of Chili Peppers, *Capsicum* spp., in the Americas"
Renfrew (1973)	*Palaeoethnobotany: The Prehistoric Food Plants of the Near East and Europe*
Smith (1967)	"Plant Remains"
Smith (1980a)	"Plant Remains from Guitarrero Cave"
Smith (1980b)	"Plant Remains from the Chiriqui Sites and Ancient Vegetational Patterns"
Towle (1961)	*The Ethnobotany of Pre-Columbian Peru*

Specialized Identification Techniques

Although much of the work of identifying charred, dried, or waterlogged archaeological macroremains can be carried out at low magnification (10–45×) with a standard dissecting microscope or a research microscope with incident lighting, I have discussed a number of instances where this approach is not practical. Wood or tuber remains may be too fragile to be broken cleanly or cut by hand for examination. Minute characters of seed coats necessary for species-level identification may not be clearly visible. Examination of starch grains or microanatomical features may be necessary to confirm identification of fragmented materials. In this section, I review specialized analysis techniques useful for identifying macroremains, including embedding, sectioning and grinding specimens, preparation for electron microscopy, starch grain analysis, electrophoresis, isotopic analysis, and morphometric analysis. For a review of another new technique, lipid analysis, refer to Rottlaender and Schlichterle (1980).

Embedding, Sectioning, and Grinding

The basic principle involved in embedding archaeological plant specimens is to permeate specimen tissues with a liquid medium that eventually hardens and binds them together. The stabilized specimen can then be cut into thin sections for viewing by transmitted light microscopy or transmission electron microscopy (TEM). TEM requires sections between 0.01 and 0.2 microns in thickness (Wischnitzer 1981). Sections for light microscopy can be thicker, although best results are obtained with sections less than 10 microns in thickness. Materials are embedded and sectioned when it is necessary to view features in the interior of specimens. Scanning electron microscopy (SEM) is used for viewing exterior surfaces, or interior features of specimens strong enough to be hand snapped or cut.

The first step in embedding specimens for examination is to decide on an embedding medium. There are a great variety of media available, many of which have been adapted for TEM (Glauert 1965; Wischnitzler 1981). The ideal embedding medium should (1) have a low viscosity so that it penetrates specimens easily, (2) harden evenly, causing little tissue distortion, (3) be hard enough so that thin sections can be cut, and (4) be stable enough to hold up under electron bombardment (Glauert 1965). Epoxy resins, polyester resins, and water-soluble media (such as Aquon, JB-4) are the most widely used embedding media. Embedding materials in paraffin wax does not allow thin enough sections for TEM examination, although sections adequate for transmitted light microscopy can be produced. Resins have largely replaced paraffin even for this purpose, however, since they are easier to use and can be mixed in a variety of hardnesses. Use of water-miscible waxes such as Carbowax or polyethylene glycol is appropriate for waterlogged materials, which cannot be allowed to dry out (Dimbleby 1978; Western 1970).

In most cases, it is best to chose an embedding medium soft enough to be sectioned with a sliding microtome. Many resins are appropriate for microtome sectioning. Most archaeological botanical material is soft enough that it can be supported by soft embedding media. By contrast, a hard embedding medium is necessary for bone sectioning, since the tissue to be supported is very hard. Sections are then cut by saw and ground to appropriate thinness. It is also possible to use hard resins for embedding botanical specimens, but such specimens can be cut and sanded only as long as they are completely permeated with resin. If one needs fairly large sections, or if archaeological botanical material is embedded with sand or hardened in other ways, cutting and grinding surfaces for viewing may be preferable to microtome sectioning (Smith and Gannon 1973).

Whether one chooses microtome sectioning or sawing, the key to successful thin sectioning is good permeation of embedding medium into specimens. Specimens not adequately permeated fall apart when microtomed or sawn. To ensure good penetration of the embedding medium, follow carefully infiltration procedures recommended for the medium. Placing specimens in a vacuum desiccator often aids permeation. Another way to enhance penetration is to place specimens first in a very viscous solution of the medium (e.g., 25% resin, 75% ethanol) and then after several hours of soaking transfer specimens to less viscous solutions (50% resin, 50% ethanol; then 100% resin). Starting with small pieces (5–10-mm cubes) which have openings directly into the interior of the specimen (e.g., epidermis or periderm removed) also aids penetration.

The following is a general procedure for embedding and sectioning botanical specimens. It follows Glauert (1965), Smith and Gannon (1973), and my own experiences and assumes use of a soft resin embedding medium unless otherwise indicated. Consult the manufacturer's instructions for specific embedding media. In addition to the sources cited above, O'Brien and McCully (1981) is a very useful general text on laboratory techniques for the study of plant structure.

1. *Sample preparation.* For most organic materials fixation and dehydration steps precede actual embedding of specimens. Fixation is the process of stabilizing the fine structure of organic tissue after death of an organism. Since archaeological materials are long dead, this step is not necessary.

In most cases it is also unnecessary to dehydrate charred or dried specimens prior to embedding. Dehydration is the process of removing all free water from specimens and replacing it with ethanol (or acetone, in the case of polyester resins). Dehydration is carried out by placing specimens in a series of solutions of increasingly pure ethanol. Since the process of charring drives moisture from specimens, removing water from charred specimens is not necessary unless samples have become rewetted during deposition. Soaking charred remains in 100% ethanol (or acetone) for a short time prior to embedding helps resin penetration, however, since the more viscous liquid penetrates air spaces in specimens more easily than does resin. Smith and Gannon (1973) recommend subjecting soaking specimens to a vacuum to ensure that all air is removed. With porous materials, use of a vacuum may not be necessary.

Dried archaeological material also contains little free water and therefore should not require a full dehydration series. A short soak in 100% ethanol (or acetone) prior to embedding may, however, remove air and improve resin penetration. Waterlogged material should be embedded in a medium that does not require removal of water prior to embedding.

The first step in preparing most archaeological materials for embedding is to cut small specimen pieces for study. If material is to be soaked in ethanol or acetone, cut pieces before soaking. Small pieces are easier both to embed and to section. Use of cubes of tissue measuring 5–10 mm on a side generally gives good results. A separate piece should be prepared for each view (i.e., transverse, radial longitudinal, and tangential longitudinal sections). Place each piece in proper orientation, on a slip of paper on which sample number and view is recorded, in an embedding boat.

It is often possible to cut pieces for embedding by hand. A razor blade or scalpel works quite well. Blades should be very sharp; cut material in one firm downward motion. Cut wood along the transverse section, then trim it to expose other sections. You may have to saw dried wood. I have found that charred tuber specimens are often too fragile for cutting or snapping. Such material may be broken into small pieces by freeze fracturing (see "Electron Microscopy," below).

2. *Embedding specimens.* Once samples have been cut to appropriate size and subjected to either a dehydration series or soaking in 100% ethanol or acetone, they are ready to be embedded. Following manufacturer's instructions for the embedding medium, mix a fresh batch of resin. Some resins can be stored after mixing if refrigerated or frozen (Smith and Gannon 1973). Others, such as JB-4 water-soluble resin, have a long shelf life if unmixed but should be fresh-mixed for each embedding series. This resin has three components: a water-soluble monomer, an activator, and a catalyst. Once the resin is catalyzed, it must be used.

For resins such as JB-4, which have one solution for infiltration and another for actual embedding, specimens are first placed in infiltration solution. Other resins are activated by heating (see Smith and Gannon 1973), so the same solution is used for infiltration and embedding. Specimens may be placed immediately in a 100% solution of resin or moved from a dilute resin solution into solutions of increasingly higher concentrations of resin until a 100% resin solution is reached. As discussed above, gradually replacing ethanol (or acetone) with resin, and placing soaking specimens in a vacuum desiccator, may improve penetration of resin.

Specimens may float to the surface of the infiltration solution when it is poured into the embedding boat. Although most sink eventually, this can be speeded by placing specimens in vacuum during soaking. If a specimen does not sink, it will not become well permeated with resin. If this happens, attach the specimen to its label and fix both to the bottom of the embedding boat. A small amount of activated resin can be used for this purpose. Infiltration takes from several hours to several days.

Once specimens have been completely infiltrated with resin, remove them from the infiltration solution (or drain off solution if specimens are attached to embedding boats), return them to boats, and cover them with embedding resin. Specimens must be completely submerged in resin or blocks will not microtome well. If a resin with an activator is used, polymerization (hardening) begins immediately at room temperature, taking about 2 hours for complete curing. Cover samples, or place in a vacuum, to exclude air from the solution. Curing can be speeded by heating in a low-temperature oven. If the resin requires heat for polymerization, place boats in a vacuum desiccator and heat.

3. *Sectioning: microtomy.* Once resin blocks are cured, pop them out of the embedding boats and trim excess resin so that the blocks will mount easily on the microtome. It is especially important to trim excess resin from the surface to be sectioned. Resin should be cut away with a sharp razor blade until the face of the section is almost exposed.

Detailed instructions for microtomy and descriptions of equipment can be found in electron microscopy texts such as Wischnitzer (1981) and general technique texts such as O'Brien and McCully (1981). Cutting good sections—sections that are thin, uniform, and uncurled—is a matter of practice. When beginning sectioning of a specimen, first cut thick sections (7–10 microns), until all resin covering the face of the specimen is removed and full sections obtained. Then reduce thickness gradually.

Lift cut sections from the microtome blade with forceps or a brush, place them in mounting medium on microscope slides, and cover with cover slips. For water-soluble resins, place sections in water on slides and heat on a drying unit. As water evaporates, resin dries and serves as the mounting medium. Stain sections, if desired, before cover slips are added.

Problems in obtaining good microtome sections usually relate to incomplete permeation of specimens with resin. It may be possible to obtain somewhat thick but usable sections of such specimens. If specimens are not centered in the resin block (i.e., one side is not totally covered with resin), that edge of the section will be crushed during microtomy or will curl when cut.

4. *Sectioning: sawing and grinding.* If a microtome is not available or if specimens are too hard to cut by microtomy, thin sections can be produced by sawing thick sections from specimens and grinding them down until a thin section is obtained.

After specimens have been embedded and trimmed as described above, grind flat faces desired for viewing with a graded series of sandpapers (beginning with roughest paper, ending with the smoothest) or screens impregnated with different size particles of silicon carbide (Smith and Gannon 1973). Cement flat, polished surfaces face down onto microscope slides.

Once specimens are firmly attached to slides, use a jigsaw or rotating saw to cut most of the resin block away from the slides, leaving as thin a

188 • Chapter 3 Identification and Interpretation of Macroremains

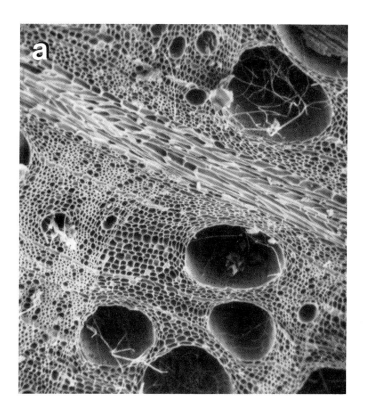

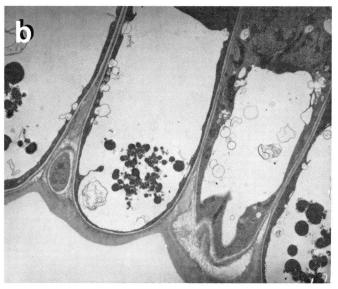

section as possible attached to it. Then grind this section, as described above, until it is approximately 10 microns or less in thickness. Specimens are now ready for viewing with light microscopy or TEM.

Electron Microscopy

The electron microscope is a valuable tool for study of biological materials (Fig. 3.40). Examination of thin sections using TEM allows high-resolution study of cell structure at a wide range of magnifications. The scanning electron microscope permits detailed examination of surfaces, and the excellent depth of field possible with SEM gives a vivid three-dimensional quality to photographs. Applications of SEM in paleoethnobotany can be found in Damp *et al.* (1981), Haberman (1984), Hastorf (1983), Pearsall (1985b), Miller (1982), and Wilson (1981), among others. Examining seed coat characteristics using SEM often allows the detailed, high-magnification study necessary for species-level identification. For wood, tuber, or stem material, internal "surfaces" can be created without embedding and sectioning and then examined by SEM.

The process required to prepare archaeological macroremains for examination using electron microscopy is determined by the nature of the material and by whether TEM or SEM is to be used.

Preparation of materials for TEM follows the procedure for embedding and sectioning outlined above. Sections between 0.01 and 0.2 microns in thickness are required. General procedures for examination of materials using TEM can be found in electron microscopy texts such as Wischnitzer (1981) and vary for the specific hardware used. Since many electron microscopy facilities require either that all work be done by a facility technician or that the user go through a training course on the microscope, the beginner is advised to make inquiries at the microscopy facility while still in the planning stage of research.

Preparation of samples for SEM does not include embedding and sectioning of materials, although fracturing material to obtain a clean internal "surface" for examination is necessary. It is usually not necessary to pretreat archaeological materials for SEM examination. Fixation is not necessary, for example, and a complete dehydration series often is not required. It is necessary to dry specimens before

Figure 3.40 Applications of electron microscopy to study of plant materials: (a) transverse section of oak (*Quercus* spp., red group) viewed by scanning electron microscope; rays, pores, and ring transition are clearly visible (photographed at 100× by Dave Pickles, provided by B. Cumbie); (b) secretory cells in the nectories of crown of thorns (*Euphorbia millii*) viewed by transmission electron microscope; the thick cuticle (wall) of the cells is clearly visible (photographed at 3650× by Gary Brown).

examination, however. This is done by air drying, which is usually adequate for charred materials, or by critical-point drying. Critical-point drying removes liquid from specimens without altering their structure. Surface tension forces may collapse structures as liquid evaporates (Wischnitzer 1981). There is a "critical point," in terms of temperature and pressure, however, at which liquid is converted to gas and removed from the specimen without surface tension forces damaging structure. Because this critical point is lower for ethanol than for water, free water in specimens is replaced by ethanol prior to critical-point drying. This requires a short dehydration series for dried materials and a full series for waterlogged specimens.

Following drying, specimens are mounted on standard SEM specimen holders, or stubs. Material is attached to the stub with either a conductive liquid adhesive, such as copper print (powdered copper in suspension), or double-sided copper tape. Blocky specimens are mounted with surfaces to be viewed at a 45° angle to the plane of the stub; this improves reflection of the electron beam. Additional copper print is put around the specimen, to "smooth" sharp angles; this also helps reduce charging (absorption of electrons) of blocky specimens.

The final step before viewing with SEM is to coat specimens with a thin layer of conducting material. Although charred material tends to conduct electrons, and thus can be viewed without coating, charging can occur rapidly in uncoated charred specimens. To avoid the problem of too rapid charging, coat charred as well as uncharred materials. Coating specimens with a conducting material, such as gold, gold-palladium, palladium, carbon, or an antistatic agent, sets up an electrical connection between specimen stage and coating which carries away the charge. Two basic methods exist for coating specimens: vacuum evaporation and sputter-coating. Each coats the specimen surface with a very thin layer of conductive material (0.01–0.1 micron thick) (Wischnitzer 1981). Because wood is very porous, coating may require more time than expected.

Although small archaeological macroremains such as seeds can be mounted whole for SEM examination, it is possible to examine only small pieces of wood, stem, tuber, nut shell, or other larger materials. I have obtained good results examining pieces cut from charred or dried tuber and root specimens with a sharp razor blade. Although some crushing of cells occurs, in general minute structure is not obscured. Successful results preparing charred tuber, legume, and maize tissue this way have been reported.

Freeze fracturing is an alternative that produces excellent viewing surfaces, especially in specimens too fragile, or too hard, to be cut. For freeze fracturing, specimens are first soaked in ethanol to replace free water and air with that liquid. Each specimen is then placed in a small cylinder made of parafilm which has been crimped shut at one end. Cylinders are then submerged in an ethanol bath and shaken gently so that they completely fill with liquid. Open ends are then crimped shut. Each ethanol-filled cylinder is quick-frozen by emersion in liquid nitrogen. The frozen specimen is then fractured by striking the cylinder sharply with a razor blade clamped into a hemostat. A number of small pieces should fracture, so that a good selection for mounting is obtained. Once the pieces have warmed and dried, they are examined under the dissecting microscope, and specimens are selected for mounting.

Starch Grain Analysis

As mentioned above, starch grain characteristics have been used to identify dried tuber and root remains from Peru (Ugent *et al.* 1981, 1982, 1984). It is not just in identification of underground storage organs that this technique can be useful, however. Plants store more starch than any other food material. Starch storage occurs in seeds, roots or tubers, and other fleshy structures. The structure, shape, and size of starch granules varies considerably among plant species, so much so that it is often possible to distinguish starch of different plants (Dean 1978). The common occurrence of starch, and the distinctiveness of starch granules, make these structures useful for identification of macroremains. Because starch structure is destroyed by heating, the utility of starch grain analysis is limited to dried and waterlogged botanical materials.

Extraction and examination of starch grains is a simple process. The following procedure comes from Dean (1978:65–71). Soak dried seeds or fruits in water overnight, or until pericarp or seed coats are completely loosened. Remove seed coats, pericarp, embryos, and other non-starchy structures. Using mortar and pestle, pound up starch-bearing structures (cotyledons and endosperm) in a small amount of water to make a smooth paste. Let the paste dry, then regrind it into a fine powder.

Preparation of starch samples from tuber, root, and other fleshy starch storage organs is similar. First peel or debark these materials and then pound a portion of the fleshy interior into a smooth paste (adding water, if necessary). After this paste dries, regrind it into a fine powder. Ugent *et al.* (1984) recommend mounting starch extract for examination using light microscopy in 50:50 glycerin-water. Iodine potassium iodide solution can be used to stain starch granules. Dry extract can be mounted for SEM examination on double-backed tape. Ugent *et al.* (1984) note that it is often possible to scrap enough material for examination from the insides of root or tuber peels.

Electrophoresis

Starch-gel and gel-acrylamide electrophoresis are used to separate and isolate protein molecules on the basis of their electrical charge and size (Boulter *et al.* 1966). Electrophoresis has broad applications in plant taxonomy, since protein patterns reflect genetic makeup of plants. Shifts in protein patterns reflect genetic differences among taxa (Maurer 1971:169–171). Plants can be distinguished on the basis of protein pattern at a number of taxonomic levels, including population, species, and genus, depending on the type of protein present. Protein patterning has been used, for example, to determine the relations among soybean cultivars in Asia (Hymowitz and Kaizuma 1981). Data from electrophoresis were used to define three gene centers for soybeans, and to aid in defining paths of dissemination of the crop from China to the rest of Asia. Smith and Lester (1980) used protein patterning as one technique to investigate relations among maize and related species and genera. They found, for example, no differences between maize and teosinte from Mexico and northern Guatemala.

Johannessen (1985) reviews the application of electrophoresis in paleoethnobotany and summarizes protein studies on archaeological materials. The potential of this technique for identifying very fragmentary macroremains, for determining which cultivar of a crop is represented archaeologically, and for studying relations among wild, weedy, and cultivated varieties of plants may be considerable, if it can be demonstrated that proteins are sufficiently long-lived to survive in ancient plant remains. Data on this point are inconclusive, since few studies using archaeological materials have been carried out. Derbyshire *et al.* (1977) report obtaining protein profiles from archaeological maize kernel remains from a site in Arizona dating to the thirteenth century or earlier. Their analysis showed protein patterning in archaeological maize to be very similar to modern examples, except in the proportion of glutelin present (archaeological maize contained less than modern). Shewry *et al.* (1982) were not able to detect protein patterning in grain samples dating from 1000 to 3000 B.C., although amino acid analysis did show great similarity between ancient and recent samples. Proteins were degraded sufficiently so that banding did not occur.

As Johannessen (1985) suggests, archaeological application of electrophoresis must be systematically tested, using well-preserved, securely dated, identified plant remains and modern examples of identified taxa, before its full potential as an identification tool can be evaluated.

Isotopic Analysis

There is considerable interest in archaeology in the use of nutritional assessment of bone to reconstruct past diet. Chemical analyses of bone, including carbon and nitrogen isotope and trace element analyses, are the usual approaches. The impact of this research on paleoethnobotany is considered in Chapter 6. Of interest here is a new application, the use of isotopic analysis to identify highly fragmented macroremains and cooking residues. This work has been pioneered by DeNiro and Hastorf (1985; Hastorf and DeNiro 1985). Different isotopes of elements exist in nature.

Plants do not necessarily incorporate these isotopes in the same ratios in which they occur in nature, however. This differential accumulation among plant groups is the basis for isotopic analysis.

Hastorf and DeNiro have determined that it is possible to discriminate among nonleguminous plants that follow the C_3 photosynthetic pathway, nonleguminous plants that follow the C_4 or CAM pathways, and legumes by determining ^{13}C and ^{15}N isotope ratios of charred plant material. Dried ancient plant remains can only be discriminated using ^{13}C, that is, only as C_3 or C_4/CAM plants, because of shifts in ^{15}N values during the initial stages of diagenesis. The cause of this postburial alteration in N isotope ratios of dried material is unknown.

This research has several implications for identification and analysis of macroremains. By determining carbon pathways for food plants common in a given region, using standard sources such as Downton (1971), and compiling a list of common legumes, an isotopic analysis of cooking residues or very fragmentary remains may lead to a fairly short list of possible identifications. If isotopic analysis indicates a legume or C_4/CAM plant, for example, possibilities may be quite limited. The ability to identify very charred material to at least legume, C_4/CAM, or C_3 plant could give a less biased picture of macroremain occurrence in preservation settings where only sturdy materials such as large seeds, pits, or nut shell are readily recognizable. The frequency of certain types of plants in cooking residues, as compared to their frequencies in archaeological deposits, might give some insight into biases present in dietary reconstructions based on macroremain occurrence frequencies.

Morphometric Analysis

Video and computer technologies that permit rapid, precise morphometric analysis—the detailed study of the external form of objects—are now available. Using computer image analysis, objects can be studied in all dimensions and in any rotation. This technology constitutes a major advance in both measurement and the statistical evaluation of metric data.

Basically, a video camera makes an image of a specimen, such as a seed. This image is then analyzed by computer. Any number of measurements can be taken and converted into areas, volumes, ratios, and so forth. Not only can many more measurements be made by computer than are possible with calipers (e.g., partial widths), but measurement is precise and rapid.

Applications of this technology in paleoethnobotany are clear. For example, precise measurement of seeds is an essential component of identifying certain cultivated plants. As Decker and Wilson (1986) demonstrated in their study of cucurbit seeds, morphometric analysis utilizing multiple measurements, ratios, and areas permits segregation of groups of cultivars. Morphometric analysis of very small remains, such as phytoliths, can also be easily achieved through computerized image analysis (Russ and Rovner 1987). Because of advances in video and computer technology, morphometric analysis will play an increasingly important role in paleoethnobotanical analyses of cultivated plants, and in other situations in which taxa must be segregated by fine distinctions of size and form.

Presenting and Interpreting Results

When flotation, sorting, and identification of samples are accomplished, the most challenging and interesting part of macroremain analysis remains: presenting and interpreting results of research. There are no easy step-by-step instructions for the beginner to learn how to make sense of macroremain data. One learns by reading the work of experienced paleoethnobotanists and trying and accepting or rejecting their approaches, and by thinking about how research issues can be addressed (and resolved) using botanical data. In addition, interpretation must always be tempered by consideration of biases in data: depositional bias (what gets into the site in the first place), preservation bias (which deposited materials survive), and recovery bias (what comes out of the site). Critical evaluation of macroremain studies demands similar evaluation of results of analysis.

In this section I discuss approaches to presenting, quantifying, and interpreting macroremain data. To say that these are "standard" approaches would be misleading; rather, they are approaches that have worked for various researchers and have generated results that can be compared and replicated. I begin by considering qualitative and quantitative approaches to presenting data. Every macroremain data set has both qualitative, or descriptive, and quantitative aspects. Simple lists of recovered taxa are useful, but limited. Quantitative analysis, ranging from counts and weights to multivariate statistics, adds another dimension to interpretation. I also present minimum standards for reporting results. Finally, I consider several important issues of interpretation, including the impact of depositional, preservation, and recovery biases on data interpretation and the criteria for proof and falsification in paleoethnobotanical interpretation.

Qualitative Presentation

The qualitative aspect of data is the information gained from determining that certain plants are present at sites. To trace the spread of a crop outside its hearth area, for example, one is interested in where securely identified remains are found, and at what time periods. A report of presence of wheat or barley at an early site in Europe and a detailed description of the appearance and size of its grains help us discover when those crops spread from the Near East. Determining which plants co-occur in the archaeological record of a region is another important qualitative aspect of data. Plants from different ecological zones found in the same deposits may give insight into patterns of trade or seasonal population movements.

To get the most out of rosters of plants recovered at sites, it is necessary to understand their ecology and potential utility for human populations. Information of this kind should be researched and presented for all recovered taxa. The work of C. Earle Smith, Jr., provides an excellent model for this aspect of data presentation. In his botanical analysis for the Tehuacan project, for example, Smith (1967) presented a summary of modern-day vegetation and agricultural practices in the study area, discussed the ecology of each site, presenting a summary of plant remains found in each, prepared a data table detailing remains of food and fiber plants from all sites by zone and phase, described each taxon, including comments on identification problems, possible uses, occurrence in the region, and material recovered, and discussed significance of data for documenting vegetation change and development of agriculture. This study illustrates how much can be learned from considering ecological and behavioral implications of macroremain assemblages. It also shows how much work is involved in getting the most from macroremain data.

To summarize, documenting the occurrence (presence) of macroremains can be an important source of information on seasonality of site occupation, past vegetation and ecology, diet, subsistence practices, trade, and domestication, among other issues, without quantitative analysis of data. Seasonality of occupation is difficult to document without use of biological indicators, including botanical remains. Plotting monthly availabilities of plants recovered from sites allows one to pinpoint the most likely season of occupation or to argue for year-round occupation. Insight into diet can be gained by compiling nutritional data on plants considered foods. To determine which plants may have been foods, fuels, accidental weedy inclusions, or medicinal plants, one relies on ethnographic analogy, historical records, and biochemical characteristics of plants. Determining present-day habitats of wild taxa found in macroremain assemblages allows modeling of resource zone use. Similarly, recovery of cultivated plants or agricultural weeds may give insight into agricultural practices such as irrigation and field construction. Careful description (including drawings or photographs) and measurement of cultivated plant remains permits comparisons of materials among sites, leading to increased understanding of crop evolution and dispersal.

Quantitative Analysis

Increasingly, paleoethnobotanical analyses include quantitative data (see Hastorf and Popper 1988). Quantitative analysis may be anything from tabulation of counts or weights to presentation of a principal components analysis; there are many different approaches in use today. I summarize these in two groups: non-multivariate and multivariate approaches. I recommend quantifying macroremain data, but with three qualifications: (1) do not use any statistical technique you do not fully understand, (2) begin with simple tabulations and then apply more complex techniques, and (3) do not use approaches that require more rigor than the data are capable of sustaining.

Non-Multivariate Approaches

Into the broad category of non-multivariate approaches fall those ranging from nominal and ordinal measures to species diversity calculations. All macroremain analyses should include counts and/or weights of recovered remains. These may be presented in tabular form for every sample analyzed or summarized for each taxon, stratigraphic level, or other relevant grouping. Table 3.4, for example, shows data summed by level at Pachamachay Cave (Pearsall 1980b). It is probably more common to present counts or weights for groups of samples, rather than for individual samples; raw data tables are often too lengthy for publication. Putting large tables

Table 3.4 • Seed Counts, Summed by Level, from Pachamachay Units 2–10 (from Pearsall 1980b:224)

Level	Provenience	Opuntia whole	Opuntia frags	Chenopodium	Amaranthus	Poaceae	Ephorasma	Plavia	Malvaceae	Freucaceae	Stipa	Calamagrostis	Sporobolus	Muhlenbergia	Grass mystery	Silenums	Graminae sps	Scirpus	Ratiluncs	Siliuis	Smooth land	Luzula	Ruppiaceae	Reticulate Opuntia	Leguminosae	Compositae	Others	Umbellifereae	Lepi/Brasicaceae	Unknown	Total
2		1		3	3	2	3	5	2	3	1		3		6		5 6		11		2	1	3	2		14	76
4				14	11	5	8		1	2	5	2	66		56	1	8		9	4		4	20	1		14	231
9		1		19	1	1	7	5		1	6		42		15		2		4 12	1	6 13	1	27	1		27	192
10			4	5		1	2	5			2		176		2	1	4		17 2	2		1	6			6	242
11				1				3				3	18		33		4						5 1				73
12		5		4	3	2	1	1	3		2		152		73				5	2		3	8			6	290
13		8	2	46	59	32	2	5	1 10	8			23		1272		14 27			5 3		12	25 7			8	1544
14		11	6	15	26			10	2				1400		2		1					1					1483
15		1		5		1	3	1		14	1	3	140		65				10 3	1		1 5	33 2			12	304
16		3		7	2	2	1	3	1	1	3	2	95		9	1	9		11	1 2		1 2			1	12	169
17		3															1										6
18		4		7		1	4	1					31		7		12 1		9			3	2	3		2	86
19		9	13	13	1		8	4	2	1	14		87		13		23 1		23	2 1		12 3	3			12	246
20		10	12	21	14	1	11	5	1		8		27		44	2	22		10	3 11		4				5	211
21		17		10		1	3	7	1		1		4		4		15		11 7			4	2			20	107
22			3	2		2	1	3	2	1			6		3		1					2	1			8	34
23		14		7		2	2	1	1	6	9		45		14		10		7	1 5		3				5	132
24		14	11	11	1		3		1 11		3		38		10	1	2	1	11	2		2	2			8	129
25		9	16	18	4	1	1	4	4				39		1				4	2 1		1 2	2	1		3	111
26		12	2		1		1						11							1		1					30
27		6	14	1	5		4						11		7		4			4			1			6	63
Total		128	86	209	135	53	40	76	18	51	68	21	2414		1636	1	20 159	9	122 62	33	10 43	55	136	5	1	168	5759

on microfishe or computer diskette allows interested researchers easy access to raw data.

Counts or weights may add little new information to an analysis until converted into ratios, used in calculating measures such as diversity or biomass, or analyzed using multivariate statistical techniques. Patterning in data (e.g., changes in taxon frequency over time, differences in occurrences among features) is often obscured by unevenness in grouped data. Level summaries, for example, may differ in total number of samples analyzed or total amount of soil floated per level. Such unevenness obscures real differences in wood or seed assemblages. Also, assemblages may be so complex, with many types of plants represented, that differences cannot be easily recognized from raw data.

The simplest measure used by paleoethnobotanists to standarize macroremain data (i.e., to "even out" unevenness in data) is the ratio. Miller (1988) presents an overview of different types of ratios used in paleoethnobotanical analyses. Although a ratio is a simple statistic, the comparison of two values by division, choosing which two values to compare is not so simple. I return to this point below.

Miller divides ratios commonly used by paleoethnobotanists into two types: (1) those in which material in the numerator is included in material in the denominator (e.g., percentage of seeds which are *Chenopodium*), and (2) those in which materials quantified in the numerator and in the denominator are mutually exclusive (e.g., seed-to-wood ratio).

Density ratios are a common example of the first type. In a density ratio, the denominator is the total volume of soil processed to obtain the count or weight of botanical material of interest. In flotation analysis, for example, it is not uncommon to express seed, nut, or wood occurrence by count or weight per liter of soil. Table 3.5, from Johannessen's (1984) analysis of floral materials from the American Bottom, summarizes wood, nut, and seed data for cultural components by mean weight or count per ten liters of floated soil. For sites where only charred materials are preserved, density ratios may be used to compare amount of burning activity among different contexts in one component or between different cultural components. Demonstrating that the level of burning activity is similar among contexts, components, or sites reduces concern about preservation biases in data. For dry or wet deposits, density measures indicate uniformity of deposition. Miller (1988; Miller and Smart 1984) also reports using quantity of seeds per liter of soil as a seasonality indicator; the practice of burning seed-laden animal dung as fuel gave higher seed densities in winter than summer deposits for sites in Iran.

Even if all flotation samples are the same size, quantity of material varies. Percentages allow one to standardize for different quantities of material per sample, regardless of the cause of uneven density. For example, if quantities of seeds recovered from flotation samples vary widely, directly comparing the number of seeds

Table 3.5 • Summary of Floral Data for Emergent Mississippian Components

Phase	Dohack		Range	Merrell	Edelhardt
Site	Dohack (uplands)	Joan Carrie (uplands)	Range (flood plain)	Robinson's Lake (flood plain)	BBB Motor (flood plain)
Total features in component	61	25	42	58	149
Total features analyzed	25	14	42	24	26
Total liters fill analyzed	720	184.3	720	435.5	687
Total wt. charcoal (g)	334.9	36.4	193.9	35.0	32.5
Mean wt. charcoal/10 l (g)	4.7	2.0	2.7	0.8	0.5
Mean wood frags./10 l	154.5	68.2	175.1	46.1	27.0
% bottomland taxa[a]	3.8	2.6	3.6	13.1	5.9
% upland taxa[b]	73.0	89.3	90.4	53.2	56.4
Mean nut frags./10 l	78.9	10.1	5.6	0.3	2.9
% Juglandaceae	20.6	25.3	21.2	9.1	3.0
% Carya spp.	47.3	46.2	28.4	90.9	10.3
% Juglans nigra	0.1	2.2	1.7	—	—
% Quercus spp.	5.3	24.7	41.2	—	83.7
Mean seeds/10 l	83.2 (516.9)[d]	104.7 (216.1)	47.5	35.1	26.3
% starchy seeds[c]	93.6 (99.2)	97.2 (96.6)	94.7	88.2	90.8
Nut : wood ratio	0.51	0.15	0.03	0.01	0.11
Features with cucurbits, %	0	21.4	38.1	0	3.8
Features with maize, %	60.0	50.0	78.6	70.8	92.4

[a]Defined as Ulmaceae + *Fraxinus* spp. + *Morus rubra* + *Gleditsia triacanthos* + *Gymnocladus dioica*; % of identifiable fragments.

[b]Defined as true *Carya* spp. + *Quercus* spp.; % of identifiable fragments.

[c]*Phalaris caroliniana* + *Polygonum erectum* + *Chenopodium* sp.; % of identifiable seeds.

[d]Numbers in parentheses denote mean if single features with seed concentrations are included in average.
Source: adapted from Johannessen (1984).

of one species between one sample and another is not valid. Note how seed totals vary in Table 3.6, for example. This table compares occurrence of seeds from some hearths at Panaulauca Cave (Pearsall 1985b). As is discussed further below, interpreting differences in percentage occurrence of seed taxa among samples may be difficult, since individual species differ in "preservability" or likelihood of being deposited. Additionally, because percentages are relative rather than absolute measures of abundance, it is impossible to determine with certainty which taxa are actively changing in occurrence and which only appear to be changing because all must sum to 100%.

If preservation conditions, as measured by density of material per volume of soil, are reasonably even for a series of samples, changes in occurrence of individual species can be examined independently of other taxa occurring in samples. Figure 3.41 shows distribution of seeds from three food plants over five occupation levels at

Table 3.6 • Percentage Occurrence of Seeds from Some Panaulauca Hearths (from Pearsall 1985b)

	Pan 17 L 10 F 28	Pan 16 L 11B F 34	Pan 21 L E F 57	Pan 34 L GG F 103
Festuca				
Gramineae	5.20		10.00	
Cf. Gramineae	5.20	P		
Scirpus	1.70			2.60
Cyperaceae				
Cyperaceae, thin				
Juncus	12.10	P	46.70	7.90
Luzula	1.70			
Sisyrinchium	20.70	P	23.30	2.60
Polygonum				
Chenopodium	8.60			60.50
Cf. Chenopodium				
Amaranthus				
Portulaca				
Calandrinia	1.70			5.30
Cf. Cerastium				
Lepidium				
Cf. Trifolium	1.70			
Lupinus				
Leguminosae				2.60
Cf. Leguminosae				
Geraniaceae				
Malvastrum				
Malvaceae				
Malvaceae, 2				
Opuntia floccosa	32.80		10.00	13.20
Echinocactus				
Umbelliferae, R.				
Umbelliferae, P.				
Labiatae	3.40		10.00	
Solanum				
Solanaceae				
Plantago	1.70			2.60
Relbunium	1.70			2.60
Galium				
Compositae	1.70			
Compositae, 2				
⋮	⋮	⋮	⋮	⋮
Seed Total	58	8	30	38
Taxa Total	14	3	5	9

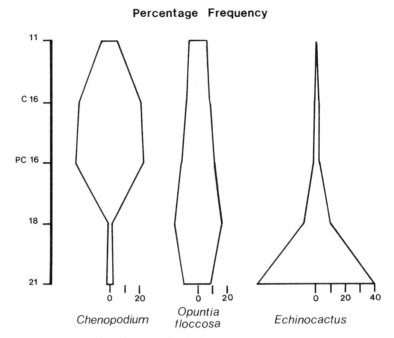

Figure 3.41 Occurrence of seeds from food plants on five occupation floors (preceramic 21, 18, 16; ceramic 16, 11) at Panaulauca Cave, Peru. Level 21 is the earliest (from Pearsall 1988b).

Panaulauca Cave. For each plant, the number of its seeds on each floor were added together to give a total count for that species over the five floors. Occurrence on *each* floor was then expressed as a percentage of that sum; most *Echinocactus* occurred on level 21, and most *Chenopodium* occurred on level 16 preceramic or level 16 ceramic. If preservation conditions had been dramatically different for each occupation level, changes in seed abundances could not be securely attributed to changes in use.

Another way to compare occurrence of species through time independently of each other is to convert raw counts into standard scores (Pearsall 1983b). In this approach, mean and standard deviation of each species through time are calculated, then each occurrence (count) is converted into the number of standard deviation units, standard score, of each away from the mean. Figure 3.42 is a graphic representation of standard scores of five taxa from a deep test pit at Panaulauca Cave. Again, preservation conditions must be reasonably even throughout the sequence being compared.

The second type of ratio described by Miller, that in which numerator and

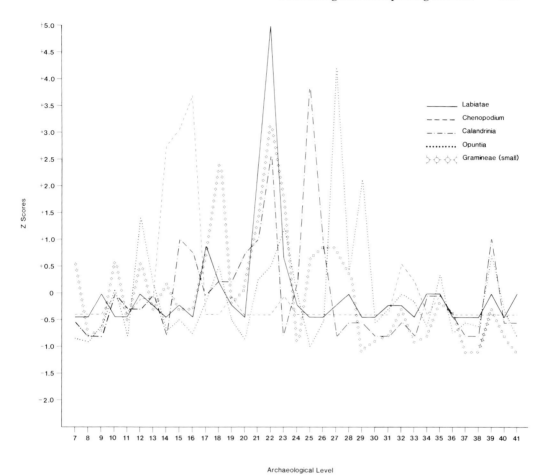

Figure 3.42 Standard scores of five seed taxa in a deep test column at Panaulauca Cave, Peru. Level 7 is the most recent (from Pearsall 1985b).

denominator are mutually exclusive, compares quantities of two different items (individual taxa or material categories). Comparison ratios have been used in two different ways in paleoethnobotany—to demonstrate that one taxon has replaced another over time or space, and to control for likelihood of preservation. Using the first approach, one can test an idea that one fuel type replaced another, or that one food plant increased in use at the expense of another, by comparing the occurrence of one to the other through time. Seed–nut ratios were used by Asch *et al.* (1972), for example, to compare occurrence of seeds and nuts used as foods between the Archaic Koster and Middle Woodland Macoupin sites. A seed–nut ratio of 0.22 for Koster suggested that seeds were much less important there than at Macoupin, where a ratio of 8.32 was obtained. Johannessen's (1984) data on seed and nut occurrence at sites in the American Bottom can also be transformed into seed–nut ratios (Table 3.7); although site-by-site variation exists, a pattern of increasing seed use, at the expense of nut utilization, over time is exhibited. Replacement comparisons should be made between equivalent items (e.g., food–food, fuel–fuel). Nuts may be overrepresented in the examples presented, since hickory and walnut shells could have been used as fuel, enhancing the likelihood of their preservation.

One way to control for likelihood of preservation in assessing macroremain assemblages is to use charcoal as the denominator in comparison ratios (although this does not correct the type of bias mentioned above). At sites where remains are only preserved if charred, there are two avenues (Fig. 3.43) by which botanical materials find their way into the archaeological record: by accidental or by deliberate burning (a more detailed discussion of the sources of seeds deposited in archaeological sites appears in "Issues in Interpretation," below). If no houses burned at a site, abundance of wood charcoal should reflect deliberate burning activities (cook-

Table 3.7 • Seed–Nut Ratios, FAI-270 Project, American Bottom[a]

Period/site	Seed–nut ratio[b]	Period/site	Seed–nut ratio[b]
Late Archaic		Emergent Mississippian	
McLean	0.01	Dohack(3)	1.05
Go-Kart N	0.01	Joan Carrie	10.37
Mo Pac	0.05	Range(1)	8.48
Dyroff	0.03	Robinson's Lake	117.00
Early Woodland		BBB Motor	9.07
Tep	0.01	Mississippian	
Carbon Monoxide(1)[c]	—	Carbon Dioxide(2)	1.00
Fiege	0.04	Lohmann	8.40
Florence Street	0.30	BBB Motor	11.21
Carbon Monoxide(2)	0.07	Range(2)	1.33
Middle Woodland		Julien(1)	11.29
Mund(1)	0.15	Turner	3.43
Truck #7	0.44	Julien(2)	0.11
Late Woodland		Julien(3)	0.16
Carbon Dioxide(1)	0.01	Oneota	
Alpha 1/7	0.56	Range(3)	1.35
Leingang	0.02		
Steinberg	0.31		
Dohack(1)	0.32		
Columbia Quarry	0.61		
Mund(2)	1.03		
Dohack(2)	0.20		
Julien	0.34		

[a]Source: compiled from Johannessen (1984).
[b]Mean seed count per 10 l ÷ mean nut count per 10 l.
[c]No nut recovered

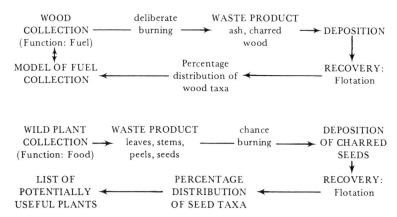

Figure 3.43 Two pathways for charred preservation: deliberate burning (upper loop) characterizes fuels such as wood and dung; chance or accidental burning (lower loop) characterizes foods, medicines, and other nonfuels (from Pearsall 1983b:122).

ing, heating, hide smoking, pottery firing). Where ancient peoples built many fires, there were also more opportunities for accidental burning of other things such as food or medicinal plants. Using charcoal abundance as the denominator in comparison ratios allows one to factor out variation in abundances of accidentally preserved taxa which are due to differences in "amount" of burning activity. The patterning that remains should reflect the relative importance of species. For example, the depiction of seed and nut occurrences at American Bottom sites in terms of nut–wood and seed–wood ratios (Fig. 3.44) gives a clear picture of the changing importance of these resources.

A number of problems can arise with comparison ratios. As mentioned above, it may be difficult to determine if taxa or categories of remains being compared are functionally equivalent. If nut hulls are used as fuel, seed–nut ratios may be low because there are enhanced opportunities for nut preservation. It may be difficult to distinguish a change in use of nuts as food from a change in fuel pattern. If more than wood is being deliberately burned, then expressing species abundances relative to abundance of wood charcoal alone does not fully correct for differences in amount of burning activity. If seeds are introduced in dung burned as fuel, then grouping seeds for seed–wood or other comparison ratios mixes seeds resulting from accidental and from deliberate burning.

Some of these problems can be avoided by careful grouping of taxa into analytical categories (Miller 1988). Categories should be as homogeneous as possible. To compare occurrence of two types of food plants (e.g., small seeded species and nuts), only taxa that fall unambiguously into each category and have the same avenue of

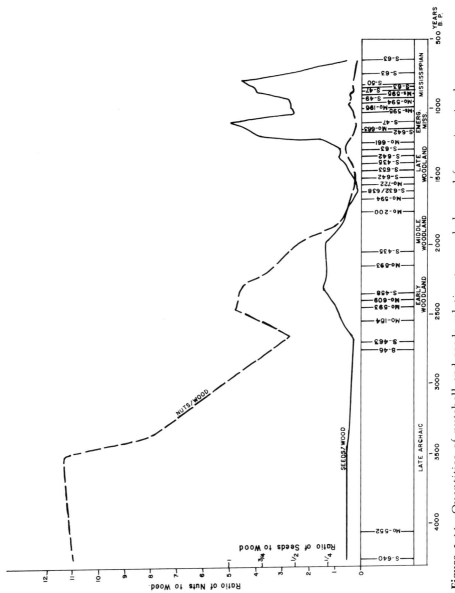

Figure 3.44 Quantities of nutshell and seeds relative to wood charcoal from sites in the American Bottom (from Johannessen 1984:209).

preservation (as in Fig. 3.43) should be included. If important taxa do not fit these criteria but must be included, discuss possible biases this may introduce into the comparisons.

Although ratios are the measures most commonly used by paleoethnobotanists to standardize count or weight data and to demonstrate patterning within them, other univariate measures are also used. Among these are two measures used in zooarchaeology: conversion of counts or weights into estimates of food value and calculation of species diversity.

In his summary of subsistence data from the Tehuacan caves, MacNeish (1967:290–309) converted counts or weights of both animal and vegetable remains into liters of food produced. This approach was also used by Pozorski (1983) in her analysis of floral and faunal remains from early sites on the north coast of Peru, and by Flannery (1986:303–317), who converted raw plant and animal data from Guilá Naquitz cave into grams of edible portion. In each case, food value figures were calculated to determine the relative importance of plant and animal taxa in the diet.

To calculate food value figures from faunal or floral remains, one must know how many individual organisms are represented and how much food each individual provides. In the case of faunal material, this involves calculating the minimum number of individuals (MNI) of each species present in each unit of analysis (level or phase), then converting these figures to available meat weight for each species. Available meat may be calculated in several ways (see Wing and Brown 1979, for a discussion of calculating meat values); in the examples cited here, individuals of average size and use of whole animals were assumed. Table 3.8 shows weights of animals eaten at Guilá Naquitz cave.

Table 3.8 • Weights of Animals Eaten at Guilá Naquitz

Animal	Average weight of 1 adult (kg)	Percentage of weight that is usable meat	Usable meat on 1 adult (kg)	Amt. needed for 100-g portion	
Deer (*Odocoileus*)	40.0	50	20.00	.005	deer
Peccary (*Dicotyles*)	20.0	70	14.00	.007	peccary
Raccoon (*Procyon*)	5.0	70	3.50	.03	raccoon
Cottontail (*Sylvilagus*)	1.2	50	0.60	.17	rabbit
Hawk (*Buteo*)	1.0	70	0.70	.14	hawk
Owl (*Tyto*)	1.0	70	0.70	.14	owl
Pigeon (*Columba*)	0.3	70	0.20	.5	pigeon
Dove (*Zenaidura*)	0.15	70	0.10	1.0	dove
Quail (*Colinus*)	0.15	70	0.10	1.0	quail
Turtle (*Kinosternon*)	0.50	40	0.20	0.5	turtle
Lizard (*Sceloporus*)	0.03	50	0.015	6.6	lizards

Source: Flannery (1986:307).

In the case of floral materials, one must first determine what portion of each utilized plant is edible, then calculate how much food (edible portion) is represented by preserved remains. In the Guilá Naquitz study, Flannery determined the weight of one edible part of each food plant and calculated the amount needed to give a 100-g edible portion (Table 3.9). This in turn allowed him to determine the quantity of archaeological material needed to indicate presence of a 100-g portion. For acorns, for example, 31 nuts yielded one 100-g portion; the archaeological indication of a portion was then 31 nuts. The presence of 6.15 (count) *Agave* quids indicated one edible portion of *Agave* heart. Flannery used these data to calculate grams of edible portion of each species recovered from the site (Table 3.10) and to rank taxa by grams of edible portion represented (Table 3.11).

Methods used by MacNeish (1967) and Pozorski (1983) are similar, except that they expressed plant food quantities in liters rather than 100-g portions. MacNeish interpreted food value data by summing total liters of food for each zone and phase of the Tehuacan caves and then calculating the relative contributions of wild and

Table 3.9 • Relationship between Archaeologically Recoverable Plant Remains and 100-gram Edible Portions for Various Plants Used at Guilá Naquitz

Plant	Edible part of plant	Weight of one edible part (g)	Amt. yielding 100-g edible portion	Archaeological evidence for 100-g portion	
Piñon (*Pinus*)	nut	0.167	600 nuts	600	nuts
Onion (*Allium*)	bulb	11	9 bulbs	9	dried bulbs
Agave potatorum	heart	2,795	100 g	6.15	quids
Agave marmorata	heart	13,000	100 g	6.15	quids
Oak (*Quercus*)	acorn	3.2	31 acorns	31	acorns
Oak (*Quercus*)	gall	5	20 galls	20	dried galls
Hackberry (*Celtis*)	fruit	0.18	550 fruits	550	seeds
Acacia farnesiana	pod	1.67	60 pods	60	dried pods
Guaje (*Leucaena*)	seed	0.125	800 seeds	80	dried pods
Wild bean (*Phaseolus*)	seed	0.125	800 seeds	800 320	seeds or pod valves
Mesquite (*Prosopis*)	pod	1.67	59.9 pods	59.9 809	dried pods or seeds
Nanche (*Malpighia*)	fruit	3.5	28.6 fruits	28.6	seeds
Susí (*Jatropha*)	nut	0.5	200 nuts	200	hulls
Lemaireocereus	fruit	50	2 fruits	2	dried fruits
Prickly pear (*Opuntia*)	nopal	(roasted) 20 (boiled) 32.5	5 nopales 3 nopales	4	dried nopales
Prickly pear (*Opuntia*)	fruit	50	2 fruits	2 100	dried fruits or seeds
Wild *Cucurbita*	seed	0.167	600 seeds	600 24	dried seeds or fruits

Source: Flannery (1986:304).

Table 3.10 • Major Food Plants Recovered in Zones B–E of Guilá Naquitz, Converted to Grams of Edible Portion

Plant	Total recovered in Zones B–E	Weight of edible portion (g)
Piñon nuts	366	61.1
Wild onions	3	33.0
Agave quids	368	5983.7
Acorns	8424	26,956.8
Hackberries	3112	560.2
Acacia pods	3	5.0
Guaje pods	417 (= 4170 seeds)	521.3
Bean pod valves	471 ⎫ 1194.5 seeds	149.3
Bean seeds	17 ⎭	
Mesquite pods	31 ⎫ 372 pods	621.2
Mesquite seeds	4603 ⎭	
Nanche fruits	368	1288.0
Susí nuts	455	227.5
Nopales	158	4147.5
Opuntia fruits	10 ⎫ 46 fruits	2300.0
Opuntia seeds	1819 ⎭	
Cucurbit rinds	55 ⎫ ±150 seeds	±25.0
Cucurbit seeds	17 ⎭	

Source: Flannery (1986:305).

Table 3.11 • Food Plants Recovered in Zones B–E of Guilá Naquitz, Ranked in Order of Weight of Edible Portion Represented

Plant	Grams of edible portion represented	Percentage of total	Rank
Acorns	26,956.8	62.9	1
Agave	5,983.7	14.0	2
Nopales	4,147.5	9.7	3
Opuntia fruits	2,300.0	5.4	4
Nanches	1,288.0	3.0	5
Mesquite pods	621.2	1.4	6
Hackberries	560.2	1.3	7
Guaje seeds	521.3	1.2	8
Susí nuts	227.5	0.5	9
Beans	149.3	0.3	10
Piñon nuts	61.1	0.1	11
Wild onions	33.0	0.08	12
Cucurbit seeds	25.0	0.05	13
Acacia pods	5.0	0.01	14
	42,879.6		

Source: Flannery (1986:305).

208 • Chapter 3 Identification and Interpretation of Macroremains

domesticated resources in the diet (Fig. 3.45). Pozorski tabulated separate food volume totals for plant and animal resources at each of her sites, expressing occurrence of individual taxa as a percentage of either the plant or animal diet. Table 3.12 illustrates converted plant data from a site in Pozorski's study.

MacNeish acknowledges that "estimating the bulk food represented by our archaeological remains on the basis of these rough calculations involves some degree of error" (1967:297). I agree. The major sources of error in this early study (besides use in some cases of rough estimates for quantities of remains representing one liter) stem from use of preservation factors to "correct" quantities of remains in certain contexts, and from combining plant and animal data into one sum for percentage calculations (e.g., Fig. 3.45). Because preservation of plant remains was not

Figure 3.45 Changing trends in the importance of the principal sources of food in the Tehuacan Valley, Mexico (from MacNeish 1967:301).

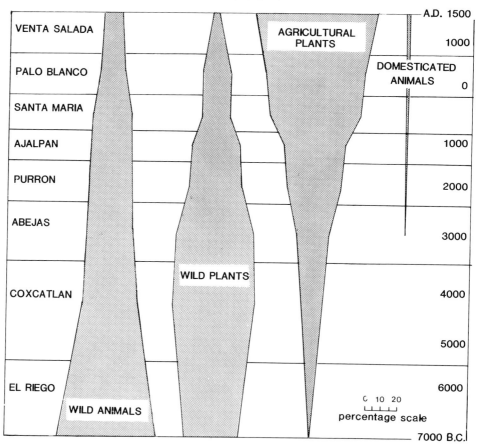

Table 3.12 • Cotton Preceramic Food Plant Remains, Alto Salaverry Site

Species	Cut 1			Cut 2		
	Diagnostic elements	Food vol. (cm³)	% of plant diet	Diagnostic elements	Food vol. (cm³)	% of plant diet
Phaseolus lunatus (lima bean, pallar)	5.5	7.3	0.1	4	5.3	0.1
Phaseolus vulgaris (common bean, frijol)	2.5	1.0	+	2	0.8	+
Capsicum sp. (pepper, ají)	1 stem	20.0	0.2	3 stems	60.0	1.1
Cucurbita sp. (squash, calabaza)	9 stems	9000.0	96.0	5 stems	5000.0	92.1
Persea americana (avocado, palta)	0.5	62.5	0.7	2	250.0	4.6
Inga feuillei (pacae)	0	0	0	0.5	2.5	+
Bunchosia armeniaca (cansaboca)	3	30.0	0.3	11	110.0	2.0
Psidium guajava (guave, guayaba)	2 stems	66.0	0.7	0	0	0
Lucuma obovata (lúcuma)	1.5	187.5	2.0	0	0	0
		9374.3	100.0%		5428.6	99.9%

Source: adapted from Pozorski (1983:27); + = less than 0.1%.

uniform over all occupation floors, quantities were extrapolated for entire floors based on amounts of plant material in squares where remains were present. For example, if remains were present in only two of ten squares, then quantities were multiplied by 5. Such data "correction" assumes that food debris was in fact deposited everywhere on floors, and in the same relative quantities as in areas where it was preserved. One should hesitate to make such assumptions. Combining these data into one total with food values from faunal material, which is subject to a completely different set of preservation, recovery, and analysis biases, may only compound the error.

Flannery's (1986) application of this methodology is both more sophisticated and more conservative. Food value estimates are more accurate because they are based on edible portions rather than on sheer volume. Only remains actually present are used in calculations; no assumptions are made about lost quantities. Flannery notes one pervasive problem with this approach to quantifying floral materials: calculations are based both on edible materials which were not eaten (nut meats, bulbs, edible seeds) and on inedible by-products (nut hulls, quids, fruit pits). Processes of deposition and preservation may have varied considerably between these classes of remains. How did the quantity of acorns deposited and not eaten compare to that processed and consumed at Guilá Naquitz? Acorns were the most common plant food remain at the site, but how important were they in the diet? By calculating harvesting areas needed to procure the mixture of plant and animal resources recovered at the cave, rather than focusing on food rankings, Flannery was able to demonstrate that abundant resources were easily available to support the small site population, whatever the real dietary mix.

To summarize, converting counts or weights of macroremains into food value estimates is not a common method of quantifying macroremain data. The major constraint on applying this method is preservational: it is applicable only at sites where most food refuse is preserved, perhaps by dry or waterlogged conditions. Even in these settings it would be a mistake to assume that deposited remains represent the sum total of plant food consumed by site inhabitants (Cohen 1972–1974). Foods may be consumed totally, leaving no refuse, or eaten away from the site. Refuse dropped in high-traffic areas may be degraded; bulky or messy garbage may be removed altogether from habitation areas. In settings where only accidentally charred material is preserved, too many complications are introduced to attempt reconstructing food values from macroremains. An exception could be made for certain classes of very sturdy remains, like nut hulls, which are subject to enhanced preservation by use as fuel (Lopinot 1984). In a "good" preservation setting, one should attempt food value reconstructions only if (1) preservation of remains can be assumed to be reasonably even over the horizontal and vertical dimensions of a site, (2) there is no evidence for change in food-preparation techniques (i.e., changes that might affect refuse production or disposal patterns), and (3) there is no indication of population replacement (i.e., possible change of concept of "edibility"). And while a direct relationship may exist between quantity of inedible refuse (nut hull, fruit pit, legume pod) and amount of food consumed, a clear relationship between food product not eaten (deposited nut meats and the like) and amount consumed should not be presumed.

Species diversity is a measure that takes into account both total number of species or taxa present in a population and abundance of each species (Pielou 1969:221–235). High diversity results when a large number of species are evenly distributed, that is, when it is difficult to predict what a randomly selected item would be. Low diversity results when the number of species present is low, or when one or a few species account for most of the population. Since macroremain data deal in both numbers of species and counts of each, combining these into one index, diversity, is useful in some situations (Pearsall 1983b).

Yellen (1977) used the Shannon-Weaver information index (Shannon and Weaver 1949) as a diversity measure and demonstrated that a !Kung base camp had a higher diversity index than a specialized activity area. In the !Kung study, each kind of debris encountered in an abandoned campsite (porcupine bones, nut-cracking stones, etc.) was equated to an individual species; amount of each kind of debris was thus equal to abundance of that species. Diversity has also been used by Wing (1975) to characterize faunal exploitation at archaeological sites, and as a means of comparing sites. I applied the Shannon-Weaver index to macroremain data recovered from Pachamachay Cave, Peru, to evaluate the usefulness of diversity as a measure of stability of resource use over time (Pearsall 1983b). The species diversity indices calculated for botanical data by phase at Pachamachay (Fig. 3.46) correlated well with independent data on intensity of occupation at the site. Especially significant was the drop in diversity in phase 7, which corresponds to a change of the cave from a habitation site to a special activity camp.

Other measures may be used to calculate species diversity. Gumerman (1985), for example, used the Simpson index as a diversity, or niche width, measure to test predictions derived from optimal foraging theory and a subsistence model developed for the Great Basin Cosa Junction Ranch site. Developed prior to analysis of remains, the model predicted that through time diet would become slightly more generalized but then narrow to become more specialized. Gumerman used faunal and floral data from the site to test the model; lack of fit between data and some aspects of the model led him to reevaluate the model and examine impact of data-collection methods on results.

Although there have been few applications of species diversity measures to macroremain data, the approach appears to have merit in certain situations. Specifically, calculating diversity for a series of levels or phases at a single site where

212 • Chapter 3 Identification and Interpretation of Macroremains

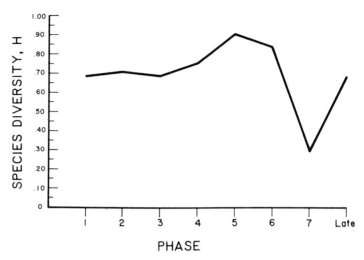

Figure 3.46 Species diversity by phase for Pachamachay seed taxa (from Pearsall 1983b:132).

preservation biases may be assumed to be reasonably constant may give insight into changing patterns of plant exploitation or reveal change in site function. I caution against using the measure for site-to-site comparisons of macroremains, however. Since absolute diversity values are dependent on quantity of remains preserved, differences in preservation conditions among sites or components of a site affect the measure. See Cruz-Uribe (1988) for further discussion of diversity measures.

Before turning to a review of multivariate techniques for manipulating macroremain data, I consider a very simple approach that has gained popularity in recent years: presence, or ubiquity, analysis. In this method, occurrence of a taxon is expressed by the percentage of the total number of units analyzed that contain that taxon (Hubbard 1980). If peach pits occur in 6 of 10 samples analyzed, presence is 60%. How many pits occur in a sample, whether 2 or 200, is immaterial. Popper (1988) presents an excellent review of the application of this approach in paleoethnobotany.

As discussed above, stating that a plant species is present at a site can convey important information in some circumstances. Stating that a plant is ubiquitous at a site, that is, that it occurs widely in site deposits, conveys even more information. Take the case of maize remains in sites in the American Bottom, for instance (Johannessen 1984) (Fig. 3.47). The presence of maize in 5% of the features from a Late Woodland site illustrates that the crop was under cultivation in the region but implies that use was low. Presence in 50% to nearly 100% of the features in Mississippian period sites implies that maize production and use were substantially higher. Percentage presence not only conveys the fact that a taxon is present but

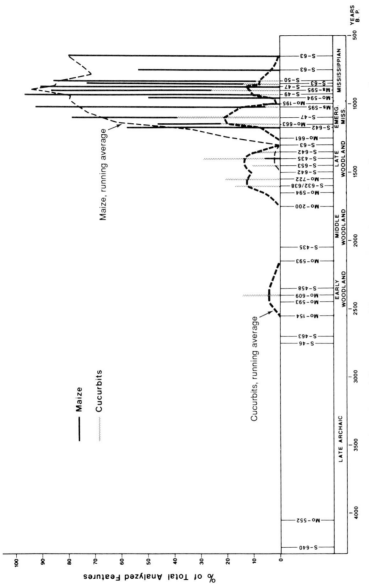

Figure 3.47 Percentage of analyzed features yielding maize and cucurbits from sites in the American Bottom (from Johannessen 1984:210).

also gives an indication of how commonly—in how many discrete locations—it occurs. Occurrence of a plant widely around a site implies that many households had access to the plant—that it was common enough to be frequently charred and preserved. The presence measure minimizes the impact of absolute quantity on evaluation of taxon importance.

In Johannessen's (1984) study, percentage presence was calculated by counting the number of features at a site which contained maize (numerator), dividing by total number of features examined (denominator), and multiplying by 100. Individual flotation or sieve samples can also be used as a basis for calculating presence. Table 3.13 illustrates the occurrence of wood, maize, root or tuber, and large seed remains at the Nueva Era site, Ecuador, using flotation samples as the basis of analysis. Presence was calculated level by level, as well as sitewide (Pearsall 1988a). I used these data to argue that maize played a minor role in subsistence at Nueva Era, a Formative period site in the Ecuadorian Andes, in comparison to root and tuber resources.

Presence analysis is often used as a way of comparing occurrence of taxa among analysis units. Scale of the units compared varies. Hubbard (1980), for example, traced occurrence of eleven cultivated plants in sites throughout Europe and the Near East. Johannessen's (1984) study compared occurrence of taxa on a regional basis, in the American Bottom of midwestern United States. My application of the measure at Panaulauca Cave, Peru, compared occurrence of taxa over time at one site (Pearsall 1988b). Figure 3.48 illustrates occurrence of *Lepidium* roots over nineteen levels; Figure 3.49, occurrence of seeds from food plants for five occupation floors. Is the measure appropriate at these differing scales of comparison?

Hubbard (1980) cautions that the presence calculations on which his study is based were influenced by the quality of raw data available. He is more confident in discussing overall trends in taxon occurrence than in the significance of the absolute value of any individual point. This caution is well taken. If an analysis includes

Table 3.13 • Percentage Presence of Macroremains, Nueva Era Site, Ecuador

Formative period level	Wood	Maize	Root/tuber	Large seeds	No. samples
Level 1	81.0	14.3	71.4	33.3	21
Level 2	86.4	22.7	68.2	50.0	22
Level 3	81.0	14.3	47.6	33.3	21
Level 4	87.0	4.3	39.1	30.4	23
Level 5	33.3	0	0	66.7	3
Sitewide	82.2	13.3	54.4	37.8	90

Source: Pearsall (1988a).

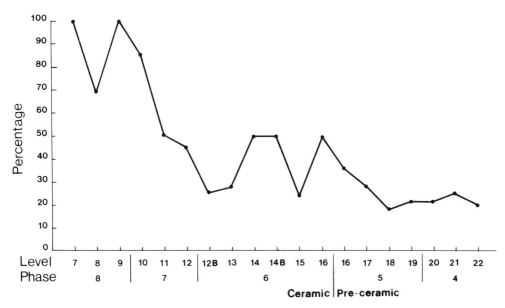

Figure 3.48 Percentage presence of *Lepidium* roots at Panaulauca Cave, Peru (from Pearsall 1988b).

sites excavated by different researchers, in widely varying climatic areas, or with different methods of recovery and analysis of remains, depositional, preservational, and recovery biases will vary widely. Since presence analysis minimizes the impact of absolute numbers of each taxon recovered, it is the most appropriate measure to use in such cases. Hubbard's study makes an important contribution to our understanding of the major trends in the spread of cultivated plants in Europe.

In both Johannessen's (1984) and my (Pearsall 1988b) applications of the presence measure, the analysis units being compared had been subjected to the same recovery and analysis biases. In each case, the same recovery, sorting, and identification procedures were used on all samples. In the Panaulauca analysis, presence calculations for *Lepidium* roots were based on screen and in-situ materials, and calculations for seeds were based on flotation samples. Separate presence comparisons were made for each type of material. There was no evidence in either case for change in depositional biases; that is, accidental, charred preservation was the rule throughout. In both research cases there was a solid basis for quantitative comparisons of macroremain assemblages over time. Presence analysis was an appropriate, in fact a conservative, approach. Note the similarities in the shapes of the curves in Figure 3.49 (percentage presence of seeds from plant foods at Panaulauca) and Figure 3.41 (percentage frequency of the same taxa). Both measures, presence

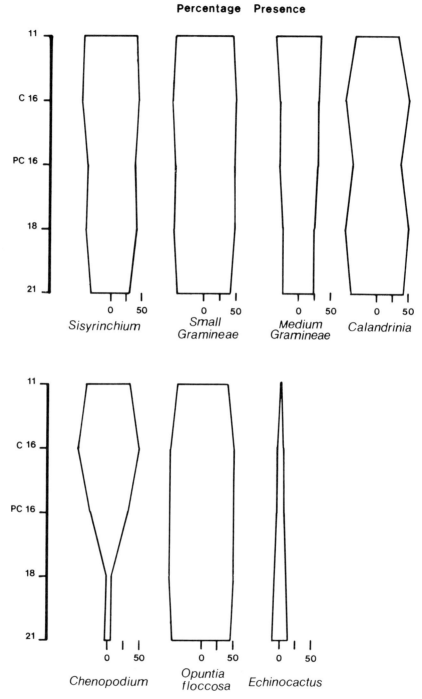

Figure 3.49 Changing seed occurrences on five occupation floors at Panaulauca Cave as measured by percentage presence: above, seeds from dung; below, seeds from food plants (from Pearsall 1988b).

and ratio, show changes in the occurrence of *Chenopodium* (abrupt increase in later levels) and *Echinocactus* (virtual disappearance in late levels) and the relative stability of *Opuntia floccosa*. Comparing the two *Echinocactus* curves shows that, although most of these cactus seeds occurred in the lowest level, the taxon was far from ubiquitous at any time.

Application of Multivariate Techniques

There have been relatively few studies in which multivariate statistical techniques have been used to analyze macrobotanical data. As becomes clear in the chapters that follow, this is the case with paleoethnobotanical studies in general. Even in palynology, where multivariate approaches are used in Quaternary studies, applications in archaeological palynology are rare. Kadane (1988) discusses application of statistical techniques in paleoethnobotany. Lack of familiarity with multivariate approaches, and concern that some may be inappropriate for macroremain data, are probably behind the paucity of studies utilizing such approaches. Before attempting any of the approaches outlined below, seek the aid of an experienced statistician and discuss frankly the nature of the archaeological data.

Simply stated, multivariate statistical approaches aid the researcher in identifying significant patterning in data when the units of analysis are characterized by several variables. In Marquardt's (1974) quantitative analysis of botanical material in coprolites from Mammoth Cave, for example, coprolites were units of analysis; variables were botanical taxa contained within the coprolites (*Chenopodium*, hickory shell, sunflower, etc.). Each unit, or datum, was characterized by the values of its variables.

There are three main reasons for using multivariate approaches in macroremain interpretation: (1) to group or classify previously ungrouped data, (2) to determine the best way to distinguish among known groups of data, and (3) to reduce the number of variables needed to account for differences among ungrouped data.

Classification groups together previously ungrouped data according to defined criteria. The members of each resulting group are more similar to each other than to members of other groups. Classification may be done by eye (intuitively) or by mathematical manipulation of measured values of variables.

To understand how classification, intuitive or numerical, works, consider a lab table covered with corn cobs. Suppose that the cobs come from an old collection; all detailed provenience data have been lost. We know only that all the cobs belong to local maize races from Mexico. Our goal is to classify the cobs in the collection, to determine how many races are present. With the help of someone who has studied thousands of corn cobs, we could quickly sort the collection into major variants. But the ability to classify by eye, by the unconscious consideration of combinations of multiple variables, is a skill that is difficult to teach, or to learn, except through experience.

An alternative is to take a set of measurements of each cob—to measure the values of multiple variables—and to use these to classify the collection. In the case of corn, various measurements of the cob (row number, length, diameter, cupule size and depth) and kernels (length, width, thickness, hardness) have been found useful to characterize maize races. It may be possible to see distinctive trends in certain variables (e.g., by plotting pairs of variables and observing the resulting clusters) and to arrive at a classification this way. To consider all variables together, however, we turn to cluster analysis. Numerical classification, using measured values of multiple variables, is carried out by cluster analysis. A number of different approaches exist, including agglomerative or polythetic strategies, divisive strategies, and reallocation procedures. Refer to Doran and Hodson (1975:158–186) or to similar texts on quantitative analysis for detail on these approaches.

There are advantages to classifying by numerical methods, among them that classes, or clusters, may be precisely defined by the midpoints (centroid) of the range of each variable. Distances between cluster centroids can be calculated to see which groups are closest. A disadvantage is that a clustering program puts every item into some cluster, no matter how different an item might be from the others. In the end, the analyst must interpret the classification and evaluate the validity, or usefulness, of the clusters the computer has generated. Classification procedures are sometimes carried out in conjunction with other approaches, as is illustrated below.

The second way multivariate statistical approaches are useful in macroremain interpretation is in identifying criteria that distinguish previously defined groups. In our corn example, we have a known number of collections of different maize races, and we know which cobs belong to which race. Our question is this: what is the best way to describe the differences among the races so that we can put cobs of unknown origin into the right groups? Original groups may be either given (a botanical species, a maize race) or the result of a prior cluster analysis.

Discriminant analysis is a quantitative approach designed to discover and emphasize those variables that best discriminate between two groups, thus allowing ungrouped units to be assigned to them (Doran and Hodson 1975:187–217). Canonical variate analysis is a form of discriminant analysis used for discriminating among three or more groups. In each type of analysis, a series of linear functions of variables are calculated. These involve multiples of each variable which are used to assign a new unit to a group. In other words, the analysis seeks to reduce the variables to a small number of composite functions which give maximum separation among the groups.

In developing ethnographic models of crop processing, Jones (1984) used discriminant analysis to determine if seed and crop residue assemblages from four processing activities could be distinguished statistically. In this example, groups were of processing products (1, winnowing by-products; 2, coarse-sieve by-products;

3, fine-sieve by-products; 4, fine-sieve products). The variables were seed species and crop residue materials present in the processing products. The value of the variables (i.e., seed counts) varied among the groups. If groups could be discriminated based on the values of the variables, then unknown units (i.e., archaeological seed and residue assemblages) could be assigned to a group at a determined likelihood of misclassification.

To carry out this analysis, Jones reduced the number of variables under consideration by eliminating species present in less than 10% of the samples and standardized and transformed the data to make them more normally distributed. Discriminant analysis was then carried out, using a program available on SPSS, to reduce the discriminating variables to three composite discriminant functions that maximized the statistical separation of the groups. To measure the discriminating value of the calculated functions, all samples were reclassified to groups based only on the value of the functions. Jones found that 93.5% of the original samples were correctly reclassified. This shows good discrimination among groups based on three functions. The first function separated winnowing and coarse-sieve by-products from fine-sieve by-products and products (i.e., separated groups 1 and 2 from groups 3 and 4). The second function separated fine-sieve by-products from products (i.e., separated groups 3 and 4); the third separated winnowing from coarse-sieving by-products (groups 1 and 2).

The results of this exploratory analysis have important implications for interpretation of macroremains, especially if one considers the characteristics of the taxa used to discriminate among processing groups rather than the specific species in the case study. For example, the first function divides groups by seed size, the second by "headedness" (tendency of seeds to remain in heads, rather than loose), and the third by lightness. These characteristics should sort in an analysis, whatever the specific taxa. In other words, ethnographic and archaeological samples could be compared directly, since even if weeds or crops were different, remains of like character should sort the same way. Jones (1987; Jones et al. 1986) has subsequently applied the model to archaeological interpretation of crop processing and storage.

Another example of discriminant analysis is Kosina's (1984) analysis of seeds of five wheat species or subspecies. The goal of this study was to find a limited number of seed measurements which would permit good separation of groups, that is, which would permit species or subspecies identification of archaeological wheat seeds. Kosina found that several measurements of the caryopsis "crease" were especially useful in separating taxa, including those represented by charred specimens.

A third way multivariate approaches are useful in macroremain interpretation is in reducing the number of variables needed to describe differences among units by discovering underlying variables that account for most of the observed differences. This process is similar to discrimination among groups, but in this case units are not grouped but considered individually. Factor analysis is used to discover the underlying variables.

Factor analysis assumes that observed variables (e.g., seed frequencies in a data set, measurements of a corn cob) are linear combinations of some underlying source variables (referred to as factors or components) (Kim and Mueller 1978). The first step in a factor analysis is to examine correlations among observed variables. This is done by calculating correlation coefficients to determine the extent of covariance among variables (i.e., the extent to which values of one variable tend to vary with values of another variable). Factor analysis is then used to determine whether observed correlations can be explained by the existence of underlying source variables (factors or components).

As part of his ongoing research into interpreting archaeological plant remains using ethnographic models of crop processing, Hillman (1984) presented a principal components factor analysis of plant remains from the Cefn Graeanog site in northern Wales. In this analysis, principal components analysis was used to help interprete complex archaeological plant assemblages in terms of residues produced by specific crop-processing activities. The units of analysis were flotation samples; the variables, plant taxa or cropping residue occurring in varying quantities in samples.

The first step in Hillman's analysis was to reduce the number of variables under consideration by grouping taxa in each sample which had the same mode of arrival at the site, associated crop product, growth form and height, and habitat of growth (based on ethnographic observations). Classes of taxa represented in only a few samples were eliminated to avoid overemphasis on rare occurrences.

In each sample, counts of remains were then converted to ratios selected to provide insight into specific questions of husbandry and processing. For example, to determine whether the site was a primary producer settlement or merely a consumer settlement, Hillman compared occurrence of straw waste (B1, threshing floor waste) to fine sievings (B3), that is, the ratio B1/B3. High levels of this ratio indicated a primary producer settlement. Hillman's work is an excellent example of how to choose ratios to fit research questions.

A series of principal components analyses were then run on ratio data from groups of flotation samples (i.e., samples from each cultural phase, samples from like contexts in each phase). This simultaneously compared patterns of variation in the values of ratios from samples tested. All related samples in a phase were then grouped, and these groups were compared between phases. The result was a series of graphs of high-order principal components which showed sample clusters (Fig. 3.50). Interpretation of groups (fine cleanings, fine cleanings mixed with fodder or bedding species, fully cleaned grain) was based on ethnographic observations of residues of various crop-processing activities.

A similar approach was used by Marquardt (1974) in his reanalysis of seed

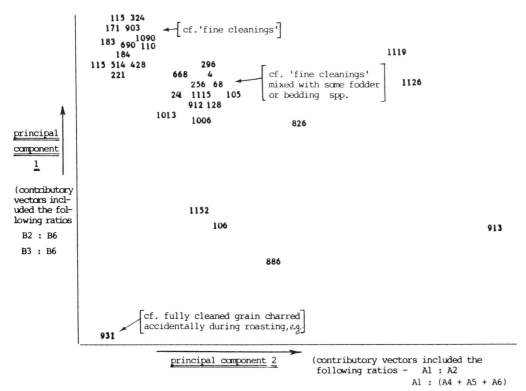

Figure 3.50 Graphic printout from a principal components analysis of ratios between major classes of plant remains in some samples from the site of Cefn Graeanog (from Hillman 1984:36).

remains from Mammoth Cave coprolites. In this case, weights of material were used rather than ratios of types. The principal components factor analysis confirmed observations of a positive correlation among hickory nut, chenopod, sumpweed, and sunflower occurrence and a negative correlation of maygrass with other items in coprolites. Hastorf (1983) used factor analysis to help determine which archaeological taxa in flotation samples from a series of sites in the Jauja region, Peru, covaried. In this case, data were tabulated in percentage presence form, and a factor analysis was adapted for bimodal presence/absence data.

Reporting Results

As illustrated in the last section, a variety of quantitative approaches can be used to aid interpretation of macroremain data. Analyses differ, not only in what quantitative approaches are used, but in sizes of flotation or fine-sieve samples processed to retrieve remains, sizes of screen mesh used in field and laboratory processing, and

222 • Chapter 3 Identification and Interpretation of Macroremains

how samples are chosen for analysis. An important component of presenting and interpreting results of analysis is discussing exactly how the analysis was accomplished, including presenting sufficient basic data for the reader to judge the strength of the conclusions drawn. This section is a brief discussion of minimum standards of reporting results of analysis.

A complete macroremain analysis study should include the following components:

> 1. *Introduction to the site(s) and study area.* This should include the names and official numbers of sites investigated, their locations, the time period(s) and cultural affiliation(s) of the sites, and the nature of vegetation and terrain surrounding them. Whether the analysis is a chapter in a larger work, an article, or a monograph, there should be sufficient information on sites for the botanical analysis to stand alone.
>
> 2. *Context of the paleoethnobotanical analysis.* What questions are addressed through the analysis? What research issues guided excavations? Include discussion of pertinent background information, such as prior studies on the topic or the region. In some cases discussion of hypotheses to be tested and criteria for accepting or rejecting them are appropriate.
>
> 3. *Field and laboratory methods.* A brief discussion of methods should always be included in a report. Choice of methods not only influences results but may limit comparisons between sites. Discussion of field recovery methods should include description or literature citation of flotation system used, method employed to collect in situ samples or bulk screen samples, screen sizes used, and information on sampling strategy (e.g., composite, column, or pinch sampling; all or selected contexts and units). Be sure to include information on volume of soil floated or fine sieved and whether samples were bulk screened prior to flotation. Discussion of laboratory methods should include basic information on how samples were sorted and identified (i.e., what materials were pulled and identified from various fractions), whether subsampling was employed (i.e., either splitting individual samples or selecting a subsample of all samples for analysis), and how data were quantified (i.e., counts, weights, presence/absence, explanation of ratios used, description or literature citation of statistical techniques used). If recovery efficiency tests such as the poppy seed test were run, results should be summarized.
>
> 4. *Presentation of results.* The presentation of results of an analysis differs according to the complexity of the research and to the preference of the analyst. I usually present results of an analysis in tabular form, using counts or weights for taxa in each sample or sample group, with separate

tables for seeds, wood, nuts, and cultivated and other special materials. I arrange taxa either alphabetically by family or in systematic order. Ordering by habitat may also be appropriate. I try to arrange samples in tables in stratigraphic order, grouped by site area or cultural component. I present results of flotation analysis separate from identifications of in situ or bulk screen remains. If space is limited in a final report, such basic data tables may appear as an appendix or on microfishe or be omitted altogether. I recommend including basic data whenever possible, however, at least in summary form (e.g., samples grouped by level or context). This gives added support for conclusions.

5. *Discussion of results.* There are two components to discussion of results of analysis: consideration of taxa recovered and consideration of patterning in occurrence of those taxa, over time, among contexts or site areas, or both. I often begin discussion by taking each recovered taxon in turn and presenting a brief description of its habitat and common uses and summarizing its occurrence at the site. A table summarizing sitewide (or component-by-component) occurrence of all taxa using percentage presence is useful for this discussion. For cultivated plants, discussion may include comments on cultivation practices, evidence to support domesticated status (i.e., a table comparing seed size of specimens from the site to those from other sites), and relevant nutritional information. When discussing wood remains, I consider whether fuel or construction use is most likely and what habitats were used for gathering these materials. It is often useful to present a separate summary table of wood occurrence sitewide, or by component, using percentage frequency of individual species or species grouped by habitat to illustrate patterns of wood use. After discussing each taxon in turn, a general discussion of overall trends in macroremain data serves to pull all components of the analysis together. If a site has multiple components, what changes over time in macroremain occurrence can be demonstrated? Various types of ratios may be used to illustrate trends, as may standard scores, diversity measures, and presence analysis. Do certain species covary over time or space? Statistical techniques may be applied to demonstrate such trends. How do results of the analysis compare to studies at other sites in the same region or cultural tradition? An easy way to compare results to those from other sites is to use the same measures (ratios, presence/absence, etc.). By including basic data in reports, results can be recalculated to allow such direct comparisons.

6. *Conclusions.* The final step in a good report is to review how the data help answer, or at least address, the research questions that guided and structured the study. For example, do the data support a hypothesis of year-

round site occupation? Does an optimal foraging model explain observed patterning in macroremain data? Is there evidence of a shift in site function over time? Only by addressing the relevant archaeological issues is the botanical data used to its fullest potential.

Issues in Interpretation

In closing this chapter on identifying and interpreting macroremains I discuss several general issues about the interpretation of the results of macroremain analysis: sources of remains, the meaning of abundance, sources of bias, and proof and falsification.

Sources of Seeds

There are several excellent articles on the sources of seeds in archaeological sites and the importance of considering what influence this and the context of deposition of seeds has on interpretation of remains (Dennell 1972, 1976; Hally 1981; Hubbard 1976; Lopinot and Brussell 1982; Miller and Smart 1984; Minnis 1981). Leaving aside, for the moment, the related issue of how source and context of deposition of seeds or other macroremains affects the meaning of abundance measures (Pearsall 1988b), let us consider how botanical remains become deposited at a site, and why understanding depositional processes is important for interpreting macroremain assemblages.

Minnis (1981) presents a useful summary of sources of seeds in archaeological deposits. Many seeds recovered by flotation are of modern origin. Seeds may be blown in from afar or may fall from local plants. Seeds accidently kicked into excavation pits end up in flotation samples. Modern seeds are often deeply buried, carried downward by burrowing animals, earthworms, and plowing, or falling downward in soil cracks and root holes. It is usually easy to distinguish modern and ancient seeds, especially at sites where only preservation of charred materials is expected. Lopinot and Brussell (1982), in a study of seed deposition at the Carrier Mills site, found that the assemblage of uncharred seeds recovered from this open-air, mesic site differed significantly from the charred seed assemblage. The uncharred assemblage closely resembled the population of plants growing on the site, whereas the charred assemblage contained taxa not present on the site, such as maygrass and nuts. Comparing seed assemblages to modern surface vegetation can also provide a valuable check on interpretation of assemblages in settings where ancient seeds are dried or waterlogged and not as readily distinguishable from modern seeds.

Minnis (1981) delineates three sources of ancient seeds in archaeological deposits: from direct resource use, from indirect resource use, and from seed rain. Seeds from direct resources are those that become deposited because they are brought into the site to be used. One can further distinguish between seeds that are themselves used (maygrass seeds harvested for food) and seeds that are the waste product of utilized fruits (peach pits, cactus seeds). These may have different pathways of preservation (see below).

Seeds from indirect resources are those that are incorporated as a result of use of the plant, not the seed. An example is grass seed falling from roof thatch. An important source of indirect seed deposition in some regions is animal dung used as fuel (Miller and Smart 1984); this may be the sole source of some taxa commonly found in hearth and ash deposits. A similar situation arises if sod is burned as fuel (Pearsall 1988b). In each case, the type and quantity of seeds deposited may differ by season.

Just as modern seeds may be incorporated into deposits by being blown into a site, so is seed rain a source of deposition both during site occupation and after site abandonment. Through accidental charring, such incidentally introduced seeds become part of ancient seed assemblages. Burning of an abandoned site may result in deposition of considerable quantities of incidentally introduced seeds.

How does one differentiate among these sources of ancient seeds? Mistaking seeds introduced in dung for seeds deliberately collected for consumption can alter interpretation of remains considerably. A number of approaches are useful in tackling this problem. Ethnographic observations are valuable sources of information on potential uses of plants. As Hillman's (1973, 1981, 1984) work has shown, for example, it is possible to determine the types of seeds likely to be recovered from a variety of crop-processing activity areas. Similarly, Miller (1982; Miller and Smart 1984) has demonstrated that study of dung and modern contexts where it is burned can reveal what types of seeds are likely to have this source in the archaeological record. This approach proved invaluable in interpreting seed remains from Panalauca Cave, Peru, where burned camelid dung was found to be a significant source of certain seed taxa (Pearsall 1988b). The ethnographic literature is also a valuable source of general information on plants likely to serve as foods, medicines, fuel sources, construction materials, and so on. Study of floras and personal observation of plants growing near a site provide insight into which taxa are likely to occur as weeds.

Scrutiny of the contexts in which seed assemblages occur is also critical in differentiating among seed sources. Minnis (1981), for example, points out that recovering a dense concentration, or cache, of seeds or observing an association between certain seed types and discrete activity areas or features supports interpretation of direct resource use. Dennell (1972, 1976) also emphasizes the importance of understanding context of deposition for correct interpretation of remains.

Comparing assemblages from different contexts, such as a hearth area and a midden, can lead to mistaken interpretations of differences—mistaking a difference in activity for a change in importance, for example.

As part of the analysis of botanical remains from Panaulauca Cave (Pearsall 1988b), I compared seed occurrence among eight discrete contexts: cave mouth midden, talus slope midden, structures, primary hearth features, in situ ash lenses, secondary ash deposition, pits, and burials (Table 3.14). This analysis revealed similarities between structure and hearth samples, probably related to the fact that seeds were charred in house hearths and then incorporated into floor deposits, and differences between primary burning contexts and secondary or nonburning contexts in occurrence of seeds from dung. Knowledge of variation among contexts then guided comparisons through time. This example is only one of many that could be cited to illustrate that, the more we know about the cultural context of botanical remains, the more informative and accurate our final analysis will be.

The Meaning of Abundance Measures

This discussion of the importance of understanding sources and depositional contexts of botanical remains has implications for interpreting quantitative measures. The other important factor affecting quantities of remains, preservation bias, is considered below. However one measures abundance of remains, a basic question remains after calculations are done: what do the numbers mean? What in fact is being measured?

Quantitative measures have little meaning unless the analyst understands sources of seeds in deposits and cultural contexts of sampling loci (Pearsall 1988b). If one can control for these factors and demonstrate an understanding of factors affecting preservation, then quantitative measures are not only interpretable but a vital part of macroremain analysis. As the examples discussed above illustrate, quantitative approaches often provide insights not gained through qualitative data.

There are contexts, however, in which quantitative assessment of botanical data may yield little additional information over simple determinations of presence or absence. Begler and Keatinge (1979), for example, argue that midden deposits are subject to too many unknown variables of accumulation and distribution of remains for quantitative assessment to be meaningful. Although it is risky to generalize from this case study to all midden contexts, it is important to remember that in complex depositional environments such as middens, where deposition may be episodic and vary considerably in the horizontal dimension, an equally intensive sampling strategy is required to characterize macroremain occurrence adequately. Contexts that are very poor preservational environments may produce inadequate quantities of materials for meaningful abundance measures. Part of the process of making quantitative comparisons of samples is assessing what contexts are appro-

Table 3.14 • Seed Occurrence at Panaulauca Cave (from Pearsall 1988b)

	Cave Mouth Midden	Talus Midden	F 129 Struc.	In situ Hearths	In situ Ash	Secdry Ash	Feats. Lacking Burning
Festuca	2.5	4.0	3.9	2.0		2.8	4.0
Gramineae (sm)	22.9	36.0	28.4	39.4	30.7	16.5	25.0
Gramineae (med)	4.5		5.9	3.4	7.9	4.2	4.0
Gramineae (lge)	8.2	4.0	7.5	9.1	2.6	10.8	3.1
Scirpus	1.0		*				*
Cyperaceae	*						
Cyperaceae, thin			*		*		
Luzula	1.0		1.1	*	*	*	2.7
Sisyrinchium	11.3		7.4	6.2	13.2	9.4	*
Polygonum			*				
Chenopodium	8.7		12.0	14.7	5.3	8.0	6.7
Cf. Chenopodium		8.0			1.8	*	1.8
Amaranthus	*						
Portulaca	*		*				
Calandrinia	8.2	16.0	8.2	4.2	7.0	3.3	7.6
Cf. Cerastium							
Lepidium			*				
Cf. Trifolium	1.5		*	1.1	1.8	*	
Lupinus	*						
Leguminosae	3.7	4.0	3.1	3.1	1.8	9.4	1.8
Cf. Leguminosae							
Geraniaceae							
Malvastrum	1.4		1.2	1.4		1.9	*
Malvaceae	*		*				
Malvaceae, 2	*		*				
Opuntia floccosa	17.5	24.0	8.3	7.1	21.1	26.9	26.3
Echinocactus							
Umbelliferae, R.			*				*
Umbelliferae, P.	*						
Verbena	*						
Labiatae	2.3		2.4	2.5	*	1.4	1.3
Solanum							
Solanaceae							*
Plantago	*		*			*	
Relbunium	1.3	4.0	1.7	*	3.5	*	1.8
Galium							
Compositae	*						*
Compositae, 2	*		*				
Unknowns	3.4		6.7	4.5	*	*	10.7
Seed Count	1644	25	2925	353	114	212	224
No. Samples	17	1	16	5	3	4	2
Seeds/sample	97	25	183	71	38	53	112
% Seeds from dung	46.9	52.0	49.9	53.2	58.5	33.4	36.6

priate for answering what questions. In looking at changes in fuel use over time, for example, comparisons should be made with samples from like contexts (so that measures reflect the temporal dimension, not differences in context) chosen because they were the loci of cooking or other deliberate burning activities.

To illustrate how the same pattern of changing abundance of macroremains may have very different meaning depending on the source of remains, consider Figure 3.49, which illustrates occurrence of a number of seed taxa on five occupation floors at Panaulauca Cave. Note the similarity among the curves for *Sisyrinchium*, small and medium Gramineae, *Calandrinia*, and *Opuntia floccosa*. If all these taxa were considered food plants, one could interpret this as evidence for a pattern of stable use of wild fruit and small seed resources over time. But only *O. floccosa* is a food resource. The same pattern of occurrence for the other four taxa in fact indicates stability in the practice of dung burning.

Sources of Bias in the Paleoethnobotanical Record

The importance of understanding the nature and impact of biases in the archaeological botanical record has been discussed in a number of contexts in this chapter. Because macroremains are usually small and fragile, paleoethnobotanists have traditionally been very concerned about what is "missing" from their samples. There is nothing we can do about utilized plants which do not make it into the archaeological record—trail foods and the like. This is nonrandom data loss for which there is no correction. Such "depositional" bias is part of every analysis; the archaeological record can never be considered complete. In Chapter 2 I detail techniques for recovering representative samples of remains of all size classes which are deposited. Thus, with careful planning and good execution, "recovery" bias can be all but eliminated from a study. This type of bias is important to consider when comparing data from studies using different recovery techniques, especially analyses conducted before the development of flotation techniques.

Biases introduced by differential preservation of botanical macroremains after deposition are of greatest concern for interpretation of macroremain data. Let us consider the most common preservation situation, that in which only material accidentally or deliberately burned is preserved. In the absence of a conflagration, not all botanical material has an equal chance of being charred and preserved. Plants associated with activities involving fire (e.g., fuel plants and foods whose processing requires heating or cooking) are more likely to become charred than are plants not exposed to fire. Equal exposure to fire does not necessarily lead to equal chances of preservation, however. Boiling whole tubers and parching corn kernels are both activities involving fire, but the chances of preservation of recognizable charred remains are quite different for these two food plants. The factors affecting likelihood of preservation are nonrandom. Certain types of remains are always more likely to become accidently charred and preserved than others.

Differential preservation of macroremains has a direct impact on data interpretation. It is difficult to compare occurrences of plants that are subject to very different preservation biases, for example, tubers (rarely preserved) and corn (more often preserved). Application of techniques such as scanning electron microscopy and isotopic analysis, which aid in identifying extremely fragmented remains, may help, but only if some material was charred and preserved to start with. The presence of a small amount of "low-probability" material scattered over a site is probably as strong an indication of importance as large quantities of "high-probability" plants. This is the rationale for using presence, or ubiquity, to indicate abundance of remains.

I find it helpful to categorize plant remains from a site by likelihood of their being charred and preserved in recognizable form, and then to use this "preservation factor" to temper my interpretations. For example, when discussing sitewide plant occurrences, I suggest which taxa are probably underrepresented and how this affects ratios or other measures of abundance. I do not recommend trying to "correct" abundances of underrepresented taxa quantitatively, however. Even if a portion of the site in question was subject to desiccated or waterlogged preservation conditions, relative abundances of remains there may not reflect abundances in other parts of the site.

An interesting illustration of this last point is given by Cohen (1972–1974), who discusses preservation of plant remains at two late prehistoric sites in similar settings in the coastal Peruvian desert. Examining quantities of plant materials recovered from the sites, Cohen found significant differences in amount of vegetable refuse between the two. He argues that differential loss of material, rather than difference in quantities of plants used, explains these results. Although sites in desert settings may have similar preservation conditions, the presence of a wide range of plants in refuse does not mean that all types of refuse are preserved equally. Sampling context and source of remains must always be considered when comparisons are made.

Proof and Falsification

The final issue in macroremain interpretation I discuss is how one demonstrates that interpretations drawn from data are, or are not, correct, and whether it is always possible to do so. Although I have come to this issue last, it is actually a point that should be in the analyst's mind throughout the early stages of compiling and looking for patterning in data.

To begin by restating a basic, but critical point, the seed, wood, and other macroremain data recovered from sites do not directly reflect plant use by the site inhabitants. As was discussed above, paleoethnobotanists interpret recovered remains despite biases introduced by patterns of garbage disposal, differential preservation, post-depositional disturbance, recovery efficiency, limitations imposed by site sampling, and so forth. It should be clear that a list of identified seeds and nuts recovered is not a list of everything that was eaten by a population. Similarly, abundance data cannot be translated directly into the relative contributions of foods in diet.

Given that one gets beyond this level of (mis)interpretation, a more interesting point to consider is the process of interpretation itself. What is the logic used to arrive at a conclusion such as "chenopod seeds were used as food"?

Most paleoethnobotanical interpretation is modeling based on analogy, usually ethnographic. We use ethnographies, histories, and traveler's accounts to glean information on how plants have been used by people. Some researchers do their own detailed observations of people carrying out activities involving plants, as in Hillman's (1984) work on crop processing. Ethnoarchaeology is another approach to obtaining information on what plants were used by a household and the patterns of their disposal. Experimentation is a second source of analogy. For example, one can plant an experimental garden of postulated food plants, try various methods of harvesting, threshing, and storing the produce, and collect and analyze residues from these activities. This is done in archaeology to "observe" activities which are no longer part of living cultures; two well-known examples are knapping flint to study debitage patterns and using reproduction tools to study use wear.

Analogy is used in different ways in paleoethnobotanical interpretation. A common approach is to propose uses for archaeological plant remains by analogy with their use by traditional peoples. For example, I proposed various uses for seeds and other macroremains from Pachamachay Cave in Peru (Pearsall 1980b) by analogy with contempory and contact period plant use by Andean peoples. By using as much data as possible from early contact observations, I avoided most bias from uses introduced by Europeans. If one can demonstrate some degree of cultural continuity between the site population and the population used for analogy, interpretations are more convincing. But did I prove my interpretations of plant use by using such analogies? Does demonstrating that the Inca ate chenopod seeds (quinoa), for example, prove that chenopod recovered at Pachamachay was used as a food?

Proof, or verification, of the interpretive model can come only from independent tests of the model. This final step is rarely accomplished. If, for example, chenopod seeds were found in human coprolites, this would verify their use as food. If chenopod seeds were consistently found associated with grinding tools, encrusted in cook pot residue, or in some other clear association with food preparation, this would be strong support of their use as foods. In the latter example, one could test for a statistically significant correlation. The negative case (not finding seeds in coprolites, not finding clear association patterns) does not falsify the interpretation, however, since preservation or recovery biases may be at work.

Analogy is also used to interpret assemblages of seeds, rather than to propose uses for particular taxa. A certain assemblage of seeds and other refuse may be correlated, either by ethnographic observation or experimentation, with an activity, such as a step in crop processing. If the same pattern is observed archaeologically, a statistical correlation between the archaeological pattern and the ethnographic pattern can be used to demonstrate that that activity occurred. This assumes, however, that no unknown activity existed which could have produced an identical pattern of remains.

In addition to interpretation based on analogy, paleoethnobotanical analysis is commonly focused on demonstrating patterning in data (regardless of what interpretation may be drawn subsequently). As described above, for example, one calculates ratios, frequencies, or percentage presence and documents how these change over time or space. Demonstrating that observed changes in data are actually statistically significant should be the final step in such analyses.

References

Appleyard, H. M., and A. B. Wildman
    1970   Fibres of archaeological interest: Their examination and identification. In *Science in archaeology: A survey of progress and research* (2nd Ed.), edited by Don Brothwell and Eric Higgs, pp. 624–633. New York: Praeger.

Asch, David L., and Nancy B. Asch
    1977   Chenopod as cultigen: A reevaluation of some prehistoric collections from eastern North America. *Mid-Continental Journal of Archaeology* 2:3–45.
    1985   Prehistoric plant cultivation in west-central Illinois. In *Prehistoric Food Production in North America*, edited by Richard I. Ford. *Anthropological Papers, Museum of Anthropology, University of Michigan* 75:149–203.
    1988a   *Archaeological plant remains: Applications to stratigraphic analysis.* In *Current Paleoethnobotany: Analytical Methods and Cultural Interpretations of Archaeological Plant Remains.* Edited by C. Hastorf and V. Popper. Chicago: University of Chicago Press, pp. 86–96.

Asch, Nancy B., and David L. Asch
    1978   The economic potential of *Iva annua* and its prehistoric importance in the lower Illinois Valley. In *The nature and status of ethnobotany*, edited by Richard I. Ford. *Anthropological Papers, Museum of Anthropology, University of Michigan* 67:301–341.

Asch, Nancy B., Richard I. Ford, and David L. Asch
    1972   Paleoethnobotany of the Koster site: The Archaic horizons. *Illinois State Museum, Springfield, Report of Investigations* 24.

Begler, E. B., and R. W. Keatinge
    1979   Theoretical goals and methodological realities: Problems in the reconstruction of prehistoric subsistence economies. *World Archaeology* 2:208–226.

Benson, Lyman
    1959   *Plant classification.* Lexington, Massachusetts: D. C. Heath.

Bertsch, Karl
    1941   *Früchte und Samen: Ein Bestinunungsbuch zur Pflanzenkunde der Vorgeschichtlicken Zeit. Handbücher der Praktischen Vorgeschichtforschung,* edited by Hans Reinerth, Band 1. Stuttgart: Verlag Ferdinand Enke.

Bird, Robert M.
    1980   Maize evolution from 500 B.C. to the present. *Biotropica* 12(1):30–41.
    1984   South American maize in Central America? In *Pre-Columbian plant migration,* edited by Doris Stone. *Papers of the Peabody Museum of Archaeology and Ethnology, Harvard University* 76:39–65.

Bodner, Connie Cox, and Ralph M. Rowlett
    1980   Separation of bone, charcoal, and seeds by chemical flotation. *American Antiquity* 45:110–116.

Bohrer, Vorsila L.
    1986   Guideposts in ethnobotany. *Journal of Ethnobiology* 6(1):27–43.

Bohrer, Vorsila L., and Karen R. Adams
    1977   *Ethnobotanical techniques and approaches at Salmon Ruin, New Mexico. San Juan Valley Archaeological Project Technical Series* 2; *Eastern New Mexico University Contributions in Anthropology* 8(1).

Boulter, D., D. A. Thurman, and B. L. Turner
    1966   The use of disc electrophoresis of plant proteins in systematics. *Taxon* 15:135–143.

Brisson, J. D., and R. L. Peterson
    1976   A critical review of the use of scanning electron microscopy in the study of the seed coat. *Scanning Electron Microscopy/1976/II* (Part VII), *Proceedings: Workshop on Plant Science Applications of SEM,* Chicago: IIT Research Institute, pp. 477–495.

Brouwer, Walter, and Adolf Stählin
    1975   *Handbuch der Samenkunde für Landwirtschaft, Gartenbau, und Forstwirtschaft.* Frankfurt: DLG Verlag.

Burkart, Arturo
    1952   *Las leguminosas Argentinas silvestres y cultivadas* (2nd Ed.) Buenos Aires: Acme Agency.

Callen, E. D.
    1967   The first New World cereal. *American Antiquity* 32:535–538.

Catling, Dorothy, and John Grayson
    1982   *Identification of vegetable fibres.* London: Chapman and Hall.

Cohen, Mark N.
    1972–   Some problems in the quantitative analysis of vegetable refuse illustrated by a
    1974   Late Horizon site on the Peruvian coast. *Nawpa Pacha* 10–12:49–60.

Cook, J. G.
    1968   *Handbook of textile fibres I: Natural fibres.* (4th Ed.). Watford, England: Merrow.

Core, H. A., W. A. Côté, and A. C. Day
    1979   *Wood structure and identification* (2nd Ed.). Syracuse, New York: Syracuse University Press.

Corner, Edred John Henry
    1976   *The seeds of dicotyledons.* Cambridge, England: Cambridge University Press.
Cowan, C. Wesley
    1978   The prehistoric use and distribution of maygrass in eastern North America: Cultural and phytogeographical implications. In *The nature and status of ethnobotany,* edited by Richard I. Ford. *Anthropological Papers, Museum of Anthropology, University of Michigan* 67:263–288.
Cronquist, Arthur
    1968   *The evolution and classification of flowering plants.* Boston: Houghton-Mifflin.
Cruz-Uribe, K.
    1988   The use and meaning of species diversity and richness in archaeological faunas. *Journal of Archaeological Science* 15:179–196.
Cutler, Hugh C.
    1956   Plant remains. In *Higgins Flat Pueblo, western New Mexico,* by Paul S. Martin, John B. Rinaldo, Elaine A. Bluhn, and Hugh C. Cutler. *Fieldiana: Anthropology* 45:174–183.
Cutler, Hugh C., and Leonard W. Blake
    1976   *Plants from archaeological sites east of the Rockies.* American Archaeology Reports, No. 1. University of Missouri-Columbia.
Cutler, Hugh C., and Thomas W. Whitaker
    1961   History and distribution of the cultivated cucurbits in the Americas. *American Antiquity* 26(4):469–485.
Damp, Jonathan E., Deborah M. Pearsall, and Lawrence T. Kaplan
    1981   Beans for Valdivia. *Science* 212:811–812.
Dean, H. L.
    1978   *Laboratory exercises: Biology of plants* (4th Ed.). Dubuque, Iowa: William C. Brown.
Decker, Deena S., and Hugh D. Wilson
    1986   Numerical analysis of seed morphology in *Cucurbita pepo. Systematic Botany* 11(4):595–607.
Delcourt, P. A., Owen K. Davis, and Robert C. Bright
    1979   *Bibliography of taxonomic literature for the identification of fruits, seeds, and vegetative plant fragments.* Oak Ridge, Tennessee: Oak Ridge National Laboratory (available from U.S. Dept. of Commerce, National Technical Information Service, Springfield, Virginia).
Delorit, Richard J.
    1970   *An illustrated taxonomy manual of weed seeds.* River Falls, Wisconsin: Agronomy Publications.
DeNiro, Michael J., and Christine A. Hastorf
    1985   Alteration of $^{15}N/^{14}N$ and $^{13}C/^{14}C$ ratios of plant matter during the initial stages of diagenesis: Studies utilizing archaeological specimens from Peru. *Geochimica et Cosmochimica Acta* 49:97–115.
Dennell, R. W.
    1972   The interpretation of plant remains: Bulgaria. In *Papers in economic prehistory,* edited by E. S. Higgs, pp. 149–159. Cambridge, England: Cambridge University Press.
    1976   The economic importance of plant resources represented on archaeological sites. *Journal of Archaeological Science* 3:229–247.

Derbyshire, E., N. Harris, D. Boulter, and E. M. Jope
    1977   The extraction, composition, and intra-cellular distribution of protein in early maize grains from an archaeological site in N.E. Arizona. *New Phytologist* 78:449–504.

deWet, J. M. J., K. E. Prasada Rao, M. H. Mengesha, and D. E. Brink
    1983   Domestication of Sawa millet (*Echinochloa colona*). *Economic Botany* 37(3): 283–291.

Dilcher, D. L.
    1974   Approaches to the identification of angiosperm leaf remains. *The Botanical Review* 40:1–157.

Dimbleby, Geoffrey
    1978   *Plants and archaeology* (2nd Ed.). Atlantic Highlands, New Jersey: Humanities Press.

Doran, J. E., and F. R. Hodson
    1975   *Mathematics and computers in archaeology.* Cambridge, Massachusetts: Harvard University Press.

Downton, W. J. S.
    1971   Check list of C_4 species. In *Photosynthesis and photorespiration*, edited by M. D. Hatch, C. B. Osmond, and R. O. Slatyer, pp. 554–558. New York: John Wiley and Sons.

Engler, A., and K. Prantl
    1889–
    1915   *Die Natürlichen Pflanzenfamilien*, 22 vols. Leipzig: Wilhelm Engelmann. 2nd Ed, 1924–1960, 25 vols. Leipzig: Wilhelm Engelmann; Berlin: Duncker and Humblot.

Esau, Katherine
    1977   *Anatomy of seed plants* (2nd Ed.). New York: John Wiley and Sons.

Fahn, A.
    1982   *Plant anatomy* (3rd Ed.). Oxford, England: Pergamon.

Flannery, Kent V.
    1986   Food procurement area and preceramic diet at Guilá Naquitz. In *Guilá Naquitz: Archaic foraging and early agriculture in Oaxaca, Mexico*, edited by Kent V. Flannery, pp. 303–317. Orlando, Florida: Academic Press.

Ford, Richard I.
    1979   Paleoethnobotany in American archaeology. In *Advances in archaeological method and theory* (Vol. 2), edited by Michael B. Schiffer, pp. 286–336. New York: Academic Press.

Glauert, A. M.
    1965   The fixation and embedding of biological specimens. In *Techniques for electron microscopy* (2nd Ed.), edited by D. H. Kay, pp. 166–212. Philadelphia: F. A. Davis.

Green, Francis J.
    1979   Collection and interpretation of botanical information from medieval urban excavations in southern England. In *Festschrift Maria Hopf*, edited by U. Korber-Grohne, pp. 39–55. Koln: Rheinland Verlag.

Gumerman, George, IV
    1985   *An optimal foraging approach to subsistence change: The Coso Junction Ranch*

site. M.A. thesis, Department of Anthropology, University of California, Los Angeles.

Gunn, Charles R.
1972 Seed collecting and identification. In *Seed biology* (Vol. 3), edited by T. T. Kozlowski, pp. 55–143. New York: Academic Press.

Gunn, C. R., and M. J. Selden
1977 Preparing seed and fruit characters for a computer with discussion of five useful programs. *Seed Science and Technology* 5:1–40.

Gunn, C. R., J. V. Dennis, and P. J. Paradine
1976 *World guide to tropical drift seeds and fruits.* Quadrangle, New York: New York Times Book Co.

Haberman, Thomas W.
1984 Evidence for aboriginal tobaccos in eastern North America. *American Antiquity* 49(2):269–287.

Hally, David J.
1981 Plant preservation and the content of paleobotanical samples: A case study. *American Antiquity* 46:723–742.

Harlow, William M.
1970 *Inside wood, masterpiece of nature.* Washington, D.C.: American Forestry Association.

Harrar, E. S.
1957 *Hough's encyclopaedia of American woods.* New York: Robert Speller and Sons.

Harrington, H. D.
1957 *How to identify plants.* Denver: Sage.

Harrington, James F.
1972 Seed storage and longevity. In *Seed biology* (Vol. 3), edited by T. T. Kozlowski, pp. 145–245. New York: Academic Press.

Hastorf, Christine Ann
1983 *Prehistoric agricultural intensification and political development in the Jauja region of central Peru.* Ph.D. dissertation, Department of Anthropology, University of California, Los Angeles.
1987 Archaeological evidence of coca (*Erythroxylum coca*, Erythroxylaceae) in the upper Mantaro Valley, Peru. *Economic Botany* 41(2):292–301.

Hastorf, C. A., and M. J. DeNiro
1985 Reconstruction of prehistoric plant production and cooking practices by a new isotopic method. *Nature (London)* 315:489–491.

Hastorf, C. A., and V. S. Popper (editors)
1988 *Current Paleoethnobotany: Analytical methods and cultural interpretations of archaeological plant remains.* Chicago: University of Chicago Press.

Helbaek, Hans
1969 Plant collecting, dry-farming, and irrigation agriculture in prehistoric Deh Luran. In *Prehistory and human ecology of the Deh Luran Plain,* edited by Frank Hole, Kent V. Flannery, and James A. Neely. *Memoirs of the Museum of Anthropology, University of Michigan* 1:383–426.
1970 Palaeo-ethnobotany. In *Science in archaeology: A survey of progress and research* (2nd Ed.), edited by Don Brothwell and Eric Higgs, pp. 206–214. New York: Praeger.

Hicks, Arthur James, and Pearl M. Hicks
    1978   A selected bibliography of plant collection and herbarium curation. *Taxon* 27:63–99.

Hillman, Gordon C.
    1973   Crop husbandry and food production: Modern basis for the interpretation of plant remains. *Anatolian Studies* 23:241–244.
    1981   Reconstructing crop husbandry practices from charred remains of crops. In *Farming practice in British prehistory*, edited by Roger Mercer, pp. 123–162. Edinburgh, Scotland: Edinburgh University Press.
    1984   Interpretation of archaeological plant remains: The application of ethnographic models from Turkey. In *Plants and ancient man: Studies in palaeoethnobotany*, edited by W. van Zeist and W. A. Casparie, pp. 1–41. Rotterdam, Holland: A. A. Balkema.

Hubbard, R. N. L. B.
    1976   On the strength of the evidence for prehistoric crop processing activities. *Journal of Archaeological Science* 3:257–265.
    1980   Development of agriculture in Europe and the Near East: Evidence from quantitative studies. *Economic Botany* 34:51–67.

Hymowitz, T., and N. Kaizuma
    1981   Soybean seed protein electrophoresis profiles from 15 Asian countries or regions: Hypotheses on paths of dissemination of soybeans from China. *Economic Botany* 35:10–23.

Isley, D.
    1947   Investigations in seed classification by family characteristics. *Iowa Agricultural Experiment Station Research Bulletin* 351:317–380.

Jane, F. W.
    1970   *The structure of wood* (2nd Ed.), revised by K. Willson and D. J. B. White. London: Adam and Charles Black.

Johannessen, Sissel
    1984   Paleoethnobotany. In *American Bottom archaeology*, edited by Charles J. Bareis and James W. Porter, pp. 197–214. Urbana: University of Illinois Press.
    1985   The potential of electrophoresis in paleoethnobotany. Ms. on file, Department of Anthropology, University of Minnesota, Minneapolis.

Jones, G. E. M.
    1984   Interpretation of archaeological plant remains: Ethnographic models from Greece. In *Plants and ancient man: Studies in palaeoethnobotany*, edited by W. van Zeist and W. A. Casparie, pp. 43–61. Rotterdam, Holland: A. A. Balkema.
    1987   A statistical approach to the archaeological identification of crop processing. *Journal of Archaeological Science* 14:311–323.

Jones, G. E. M., Kenneth Wardle, Paul Halstead, and Diana Wardle
    1986   Crop storage at Assiros. *Scientific American* 254(3):96–103.

Kadane, Joseph B.
    1988   Possible statistical contributions to paleoethnobotany. In *Current paleoethnobotany: Analytical methods and cultural interpretations of archaeological plant remains*, edited by C. A. Hastorf and V. S. Popper, pp. 206–214. Chicago: University of Chicago Press.

Kaplan, Lawrence
　1965　Archaeology and domestication in American *Phaseolus* (beans). *Economic Botany* 19(4):358–368.

Kenward, H. K., A. R. Hall, and A. K. C. Jones
　1980　A tested set of techniques for the extraction of plant and animal macrofossils from waterlogged archaeological deposits. *Science and Archaeology* 22:3–15.

Kim, Jae-On, and Charles W. Mueller
　1978　Introduction to factor analysis: What it is and how to do it. *Sage University Paper Series on Quantitative Applications in the Social Sciences* 07-013. Beverly Hills, California: Sage.

King, Frances B.
　1985　Early cultivated cucurbits in eastern North America. In *Prehistoric food production in North America*, edited by Richard I. Ford. *Anthropological Papers, Museum of Anthropology, University of Michigan* 75:73–97.

Knudsen, J. W.
　1966　*Biological techniques: Collecting, preserving, and illustrating plants and animals.* New York: Harper and Row.

Kosina, R.
　1984　Morphology of the crease of wheat caryopses and its usability for the identification of some species—a numerical approach. In *Plants and ancient man: Studies in palaeoethnobotany*, edited by W. van Zeist and W. A. Casparie, pp. 177–192. Rotterdam, Holland: A. A. Balkema.

Kribs, David A.
　1968　*Commercial foreign woods on the American market* (2nd Ed.). New York: Dover.

Lawrence, George H. M.
　1951　*Taxonomy of vascular plants.* New York: Macmillan.

Leney, Lawrence, and Richard W. Casteel
　1975　Simplified procedure for examining charcoal specimens for identification. *Journal of Archaeological Science* 2:153–159.

Leon, Jorge
　1964　*Plantas alimenticias Andinas. Instituto Interamericano de Ciencias Agrícolas, Zona Andina,* Boletín Técnico 6.

Levy, J. W.
　1970　The condition of "wood" from archaeological sites. In *Science in archaeology* (2nd Ed.), edited by Don Brothwell and Eric Higgs, pp. 188–190. New York: Praeger.

Lopinot, Neal H.
　1984　*Archaeobotanical formation processes and late Middle Archaic human–plant interrelationships in the midcontinental U.S.A.* Ph.D. dissertation, Department of Anthropology, Southern Illinois University. Ann Arbor: University Microfilms.

Lopinot, Neal H., and David Eric Brussell
　1982　Assessing uncarbonized seeds from open-air sites in mesic environments: An example from southern Illinois. *Journal of Archaeological Science* 9:95–108.

MacNeish, Richard Stockton
   1967   A summary of subsistence. In *The prehistory of the Tehuacan Valley: 1, environment and subsistence,* edited by Douglas S. Byers, pp. 290–309. Austin: University of Texas Press.

Mangelsdorf, Paul C.
   1967   Prehistoric wild and cultivated maize. In *The prehistory of the Tehuacan Valley: 1, environment and subsistence,* edited by D. S. Byers, pp. 178–200. Austin: University of Texas Press.
   1974   *Corn: Its origin, evolution, and improvement.* Cambridge, Massachusetts: Belknap Press of Harvard University Press.

Marquardt, William H.
   1974   A statistical analysis of constituents in human paleofecal specimens from Mammoth Cave. In *Archeology of the Mammoth Cave area,* edited by Patty Jo Watson, pp. 193–202. New York: Academic Press.

Martin, A. C.
   1946   The comparative internal morphology of seeds. *American Midland Naturalist* 36:513–660.

Martin, A. C., and W. D. Barkley
   1961   *Seed identification manual.* Berkeley: University of California Press.

Martin, J. T., and B. E. Juniper
   1970   *The cuticles of plants.* New York: St. Martin's.

Martins, R.
   1976   *New archaeological techniques for the study of ancient root crops in Peru.* Ph.D. thesis, University of Birmingham, Birmingham, England.

Maurer, H. Rainer
   1971   *Disc electrophoresis and related techniques of polyacrylamide gel electrophoresis* (2nd Ed.). Berlin: Walter de Gruyter.

McClure, D. S.
   1957   Seed characters of selected plant families. *Iowa State College Journal of Science* 31:649–682.

McGinnes, E. A., Jr., C. A. Harlow, and F. C. Beall
   1976   Use of scanning electron microscopy and image processing in wood charcoal studies. *Scanning Electron Microscopy/1976* (Part VII), *Proceedings: Workshop on Plant Science Applications of Scanning Electron Microscopy,* pp. 543–548. Chicago: IIT Research Institute.

McGinnes, E. A., Jr., S. A. Kandeel, and P. S. Szopa
   1971   Some structural changes observed in the transformation of wood into charcoal. *Wood and Fiber* 3(2):77–83.

McGinnes, E. A., Jr., P. S. Szopa, and J. E. Phelps
   1974   Use of scanning electron microscopy in studies of wood charcoal formation. *Scanning Electron Microscopy/1974* (Part II), *Proceedings: Workshop on Scanning Electron Microscope and The Plant Sciences, Chicago,* IIT Research Institute pp. 469–476.

Metcalfe, C. R.
   1960   *Anatomy of the monocotyledons: I. Gramineae.* London: Oxford University Press.

Miles, Anne
   1978   *Photomicrographs of world woods.* London: Her Majesty's Stationery Office.

Miller, Naomi Frances
    1982   *Economy and environment of Malyan: A third millennium B.C. urban center in southern Iran.* Ph.D. dissertation, University of Michigan. Ann Arbor: University Microfilms.
    1988   Ratios in paleoethnobotanical analysis. In *Current Paleoethnobotany: Analytical methods and cultural interpretations of archaeological plant remains,* edited by C. A. Hastorf and V. S. Popper, pp. 72–85. Chicago: University of Chicago Press.

Miller, Naomi F., and Tristine Lee Smart
    1984   Intentional burning of dung as fuel: A mechanism for the incorporation of charred seeds into the archaeological record. *Journal of Ethnobiology* 4:15–28.

Minnis, Paul E.
    1981   Seeds in archaeological sites: Sources and some interpretive problems. *American Antiquity* 46:143–152.
    1987   Identification of wood from archaeological sites in the American Southwest. I. Keys for Gymnosperms. *Journal of Archaeological Science* 14:121–131.

Montgomery, F. H.
    1977   *Seeds and fruits of plants of eastern Canada and northeastern United States.* Toronto: University of Toronto Press.

Musil, Albina F.
    1963   Identification of crop and weed seeds. *Agricultural Marketing Service, Agricultural Handbook* 219. Washington, D.C.: U.S. Government Printing Office.

Nabham, Gary, and J. M. J. deWet
    1984   *Panicum sonorum* in Sonoran Desert agriculture. *Economic Botany* 38(1):65–82.

O'Brien, T. P., and M. E. McCully
    1981   *The study of plant structure: Principles and selected methods.* Melbourne, Australia: Termacarphi Pty. Ltd.

Ogden, E. C.
    1974   Anatomical patterns of some aquatic vascular plants of New York. *New York State Museum and Science Service Bulletin* 424.

Palmer, P. G.
    1976   Grass cuticles: A new paleoecological tool for East African lake sediments. *Canadian Journal of Botany* 54:1725–1734.

Panshin, A. J., and Carl de Zeeuw
    1980   *Textbook of wood technology* (4th Ed.). New York: McGraw-Hill.

Pearsall, Deborah M.
    1979   *The application of ethnobotanical techniques to the problem of subsistence in the Ecuadorian Formative.* Ph.D. dissertation, University of Illinois–Urbana. Ann Arbor: University Microfilms.
    1980a   Analysis of an archaeological maize kernel cache from Manabi Province, Ecuador, *Economic Botany* 34:344–351.
    1980b   Pachamachay ethnobotanical report: Plant utilization at a hunting base camp. In *Prehistoric hunters of the High Andes,* by John W. Rick, pp. 191–231. New York: Academic Press.
    1983a   *Plant utilization at the Krum Bay site, St. Thomas, U.S.V.I.* Ms. on file, American Archaeology Division, University of Missouri-Columbia.

1983b Evaluating the stability of subsistence strategies by use of paleoethnobotanical data. *Journal of Ethnobiology* 3:121–137.

1985a The origin of plant cultivation in South America. In *Origins of agriculture in world perspective*, edited by P. J. Watson and C. Wesley Cowan. Ms. on file, American Archaeology Division, University of Missouri–Columbia.

1985b Paleoethnobotanical data and procedures. In *Hunting and herding in the high altitude tropics: Early prehistory of the central Peruvian Andes*, edited by John W. Rick. Ms. on file, American Archaeology Division, University of Missouri–Columbia.

1988a An overview of Formative period subsistence in Ecuador: Palaeoethnobotanical data and perspectives. In *Diet and subsistence*, edited by Brenda V. Kennedy and Genevieve M. LeMoine, pp. 149–158. Proceedings of the Nineteenth Annual Chacmool Conference, Archaeological Association of the University of Calgary.

1988b Interpreting the meaning of macroremain abundance: The impact of source and context. In *Current Paleoethnobotany: Analytical methods and cultural interpretations of archaeological plant remains*, edited by C. A. Hastorf and V. S. Popper, pp. 119–144. Chicago: University of Chicago Press.

Pearsall, Deborah M., and Bernadino Ojeda
　1988 Report of analysis of botanical materials from the El Paraiso site, Peru, with a discussion of Cucurbitaceae by A. Hunter. Ms. on file, American Archaeology Division, University of Missouri–Columbia.

Pearsall, Deborah M., Michael K. Trimble, and Robert A. Benfer
　1985 *A methodology for quantifying phytolith assemblages in comparative and archaeological contexts.* Paper presented at the 2nd Annual Phytolith Conference, Duluth, Minnesota.

Pearsall, Deborah M., Eric E. Voigt, M. F. Cornman, and Alice N. Benfer
　1983 *Conservation and cataloging of botanical materials in the American Archaeology Division systematic collection.* Ms. on file, National Science Foundation. Copies available from the American Archaeology Division, University of Missouri–Columbia.

Pickersgill, Barbara
　1969 The domestication of chili peppers. In *The domestication and exploitation of plants and animals*, edited by P. Ucko and G. W. Dimbleby, pp. 443–450. London: Duckworth.

　1984 Migrations of chili peppers, *Capsicum* spp., in the Americas. In *Pre-Columbian plant migration*, edited by Doris Stone, pp. 105–123. Papers of the Peabody Museum of Archaeology and Ethnology 76. Cambridge, Massachusetts: Harvard University Press.

Pielou, E. C.
　1969 *An introduction to mathematical ecology.* New York: John Wiley and Sons.

Plowman, Timothy
　1984 The origin, evolution, and diffusion of coca, *Erythoxylum* spp., in South and Central America. In *Pre-Columbian plant migration*, edited by Doris Stone, pp. 124–163. Papers of the Peabody Museum of Archaeology and Ethnology 76. Cambridge, Massachusetts: Harvard University Press.

Popper, Virginia
　1988 Selecting quantitative measurements in paleoethnobotany. In *Current paleoethnobotany: Analytical methods and cultural interpretations of archae-*

ological plant remains, edited by C. A. Hastorf and V. S. Popper, pp. 53–71. Chicago: Chicago University Press.

Porter, C. L.
    1967   *Taxonomy of flowering plants* (2nd Ed.). San Francisco: W. H. Freeman.

Pozorski, Shelia
    1983   Changing subsistence priorities and early settlement patterns on the north coast of Peru. *Journal of Ethnobiology* 3:15–38.

Radford, A. E., et al.
    1986   *Fundamentals of plant systematics.* New York: Harper and Row.

Ramenofsky, Ann F., Leon C. Standifer, Ann M. Whitmer, and Marie S. Standifer
    1986   A new technique for separating flotation samples. *American Antiquity* 51:66–72.

Renfrew, Jane M.
    1973   *Palaeoethnobotany: The prehistoric food plants of the Near East and Europe.* New York: Columbia University Press.

Rottlaender, R. C. H., and H. Schlichterle
    1980   Gefässinhalte—eine kurz Kommentierte Bibliographie [contents of vessels—a briefly annotated bibliography]. *Archaeo-Physika* 7:61–70.

Rury, Phillip M., and Timothy Plowman
    1983   Morphological studies of archeological and recent coca leaves. (*Erythroxylum* spp.). *Botanical Museum Leaflets* 29(4):297–341.

Russ, J., and I. Rovner
    1987   Stereological verification of *Zea* phytolith taxonomy. *Phytolitharian Newsletter* 4:10–18.

Sauer, Jonathan, and Lawrence Kaplan
    1969   *Canavalia* beans in American prehistory. *American Antiquity* 34(4):417–424.

Schopmeyer, C. S.
    1974   Seeds of woody plants in the United States. *Agricultural Handbook* 450. U.S. Forest Service, Washington, D.C.: United States Department of Agriculture.

Schweingruber, F. H.
    1978   *Microscopic wood anatomy: Structural variability of stems and twigs in recent and subfossil woods from Central Europe.* Translated by K. Baudais-Lundström Birmensdorf: Swiss Federal Institute of Forestry Research.

Shackley, Myra
    1981   *Environmental archaeology.* London: George Allen and Unwin.

Shannon, C. E., and W. Weaver
    1949   *The mathematical theory of communication.* Urbana: University of Illinois Press.

Shewry, P. R., M. A. Kirkman, S. R. Burgess, G. N. Festenstein, and B. J. Miflin
    1982   A comparison of the protein and amino acid composition of old and recent barley grain. *New Phytologist* 90:455–466.

Smith, B. D.
    1984   Chenopodium as a prehistoric domesticate in eastern North America: Evidence from Russell Cave, Alabama. *Science* 226:165–167.

Smith, C. Earle, Jr.
    1967   Plant remains. In *The prehistory of the Tehuacan Valley: 1. Environment and subsistence,* edited by Douglas S. Byers, pp. 220–260. Austin: University of Texas Press.

1980a   Plant remains from Guitarrero Cave. In *Guitarrero Cave: Early man in the Andes*, edited by Thomas F. Lynch, pp. 87–119. New York: Academic Press.
1980b   Plant remains from the Chiriqui sites and ancient vegetational patterns. In *Adaptive radiations in Prehistoric Panama*, edited by Olga F. Linares and Anthony J. Ranere. *Peabody Museum Monographs, Harvard University* 5:151–174.

Smith, C. Earle, and S. G. Stephens
1971   Critical identification of Mexican archaeological cotton remains. *Economic Botany* 25(2):160–168.

Smith, Frank H., and Brian L. Gannon
1973   Sectioning of charcoals and dry ancient woods. *American Antiquity* 38:468–472.

Smith, J. S., and R. N. Lester
1980   Biochemical systematics and evolution of *Zea, Tripsacum,* and related genera. *Economic Botany* 34:201–218.

Stephens, S. G.
1970   The botanical identification of archaeological cotton. *American Antiquity* 35:367–373.
1975   A reexamination of the cotton remains from Huaca Prieta, north coastal Peru. *American Antiquity* 40(4):406–419.

Stephens, S. G., and M. Edward Moseley
1974   Domesticated cottons from archaeological sites in central coastal Peru. *American Antiquity* 39:109–122.

Stewart, Robert B., and William Robertson III
1971   Moisture and seed carbonization. *Economic Botany* 25:381.

Steyermark, Julian A.
1963   *Flora of Missouri*. Ames, Iowa: Iowa State University Press.

Toll, Mollie S.
1988   Flotation sampling: Problems and some solutions. In *Current Paleoethnobotany: Analytical methods and cultural interpretations of archaeological plant remains*, edited by C. A. Hastorf and V. S. Popper, pp. 36–52. Chicago: University of Chicago Press.

Towle, Margaret A.
1961   *The ethnobotany of pre-Columbian Peru*. Chicago: Aldine.

Ugent, Donald, Shelia Pozorski, and Thomas Pozorski
1981   Prehistoric remains of the sweet potato from the Casma Valley of Peru. *Phytologia* 49(5):401–415.
1982   Archaeological potato tuber remains from the Casma Valley of Peru. *Economic Botany* 36(2):182–192.
1983   Restos arqueológicos de tubérculos de papas y camotes del Valle de Casma en el Perú. *Bol. Lima* 25:1–17.
1984   New evidence for ancient cultivation of *Canna edulis* in Peru. *Economic Botany* 38(4):417–432.

Ugent, Donald, and Michael Verdun
1983   Starch grains of the wild and cultivated Mexican species of *Solanum*, subsection Potatoe. *Phytologia* 53(5):351–363.

United States Department of Agriculture (USDA)
  1948   *Woody-plant seed manual. U.S. Forest Service, Miscellaneous Publications 654.* Washington, D.C.: United States Department of Agriculture.
Van der Veen, M., and N. Fieller
  1982   Sampling seeds. *Journal of Archaeological Science* 9:287–298.
Wagner, Gail E.
  1982   Testing flotation recovery rates. *American Antiquity* 47:127–132.
Western, A. Cecilia
  1970   Wood and charcoal in archaeology. In *Science in archaeology: A survey of progress and research* (2nd Ed.), edited by Don Brothwell and Eric Higgs, pp. 178–187. New York: Praeger.
Willcox, G. H.
  1977   Exotic plants from Roman waterlogged sites in London. *Journal of Archaeological Science* 4:269–282.
Williams, Dorian
  1975   Identification of waterlogged wood by the archaeologist. *Science and Archaeology* 14:3–4.
Wilson, D. G.
  1984   The carbonization of weed seeds and their representation in macrofossil assemblages. In *Plants and ancient man: Studies in palaeoethnobotany*, edited by W. van Zeist and W. A. Casparie, pp. 201–206. Rotterdam, Holland: A. A. Balkema.
Wilson, Hugh D.
  1981   Domesticated *Chenopodium* of the Ozark Bluff Dwellers. *Economic Botany* 35(2):233–239.
Wing, E.
  1975   Hunting and herding in the Peruvian Andes. In *Archaeological studies*, edited by A. T. Clason, pp. 302–308. New York: Elsevier.
Wing, Elizabeth S., and Antoinette B. Brown
  1979   *Paleonutrition.* New York: Academic Press.
Wischnitzer, Saul
  1981   *Introduction to electron microscopy* (3rd Ed.). New York: Pergamon.
Wohlgemuth, Eric
  1984   *Some perspectives on sampling for botanical remains from archaeological sites.* Paper presented at the Annual Meeting of the Society for California Archaeology, Salinas, California.
Yarnell, Richard A.
  1972   *Iva annua* var *macrocarpa*: Extinct American cultigen? *American Anthropologist* 74:335–341.
  1978   Domestication of sunflower and sumpweed in eastern North America. In *The nature and status of ethnobotany*, edited by Richard I. Ford. Anthropological Papers, Museum of Anthropology, University of Michigan 67:289–299.
Yellen, John E.
  1977   *Archaeological approaches to the present: Models for reconstructing the past.* New York: Academic Press.

Chapter 4   Archaeological Palynology

Introduction

Palynology is the study of pollen and spores—the biology of their production and distribution, and also their relevance to reconstructions of past environments and the detection of cultural activities. Reconstructing past vegetation and climate are the most familiar goals of palynology, especially research in the European tradition. Permanently waterlogged sediments (bogs, lake bottoms, ocean floors) are preferred sampling locations, since the lack of oxygen inhibits biological decomposition of pollen grains. Pollen and spores released by plants and mixed in the atmosphere fall onto land and water surfaces. This "pollen rain" is a reflection of the vegetation that produced it; the sequence of its buildup over time in sediments is a record, albeit an imperfect one, of past vegetation.

Archaeologists, first in Europe, then in the New World, were quick to realize the potential of palynology as a tool for examining the relationships between humans and their environments. Application of pollen analysis to the study of human coprolites and soil from archaeological sites and other anthropogenic landscapes has allowed questions of human diet and utilization of plant resources to be addressed directly. Pollen analysis can provide valuable data for paleoethnobotanical research, data that complement and strengthen results of analyses of macroremains and phytoliths. My focus here is on the application of palynological techniques in archaeology—that is, on archaeological palynology. However, I also include discussions on interpreting pollen diagrams, including traditional stratigraphic diagrams. Although few paleoethnobotanists will (or should) analyze a stratigraphic lake column or need to know how to delimit pollen zones or correlate pollen spectra to vegetation formations, most will use such studies in their research if they are available. By understanding how vegetation and climatic reconstructions are made from stratigraphic pollen data, the paleoethnobotanist and archaeologist can make more effective, critical use of such studies.

In his survey of archaeological pollen analysis, Dimbleby (1985) contrasts soil pollen analysis, the analysis of soil from archaeological sites and old land surfaces, with traditional stratigraphic palynology, which focuses on analysis of stratified waterlogged deposits. Because pollen is deposited and distributed in these contrasting situations in different ways, understanding the distinctive mechanisms of pollen deposition, movement, and destruction is vital to interpretation of fossil assemblages. Dimbleby discusses these issues in detail and surveys types of depositional environments encountered at archaeological sites.

Dimbleby's *The Palynology of Archaeological Sites* is not the only recent work focusing on application of pollen analysis in archaeology. Bryant and Holloway's (1983; Holloway and Bryant 1986) overviews of the role of pollen analysis in archaeology include basic information on the nature of pollen as well as information on sampling techniques, data analysis, and types of questions which can be approached through pollen analysis. Shackley (1981) includes discussion of pollen and also spores from ferns, mosses, and microorganisms in her *Environmental Archaeology*.

There is, in addition to the works cited above, a wealth of literature on all aspects of traditional stratigraphic palynology, as well as many readily available archaeological pollen studies. However, as was the case for macroremain analysis, recent cultural resource management archaeological palynological studies are difficult to obtain.

I begin this chapter with reviews of the nature and production of pollen, the history of archaeological palynology, and field and laboratory techniques. In each of these sections I cite basic references from which further detail can be obtained. The remainder of the chapter is devoted to the presentation and interpretation of pollen data. Using the results of several recent palynological studies, I illustrate how pollen data are interpreted and applied to reconstructions of vegetation, climate, and subsistence systems. These discussions lead to a final comment on current directions in archaeological palynology and possibilities of better application of pollen data in the future.

Nature and Production of Pollen

Pollen grains are formed in the anther, the male portion of a flower (see Fig. 3.17). They are produced by sporogeneous tissue called pollen mother cells and represent the asexual generation of the flowering plant. On maturity, the wall of the anther breaks, releasing pollen for transfer to the female portion of the flower. The mechanism for this transfer differs among plants and influences both the abundance of pollen produced and its form (Faegri and Iversen 1975).

In general, each pollen mother cell produces four pollen grains. In the tetrad stage, grains are connected; most separate at maturity. Most pollen grains can be described as regular rotation ellipsoids; that is, they are symmetrical around an axis (Fig. 4.1). By definition, the polar axis is the line running from the proximal (part nearest the center of the tetrad) to the distal (part farthest from the center) pole of the grain. The polar axis is usually the axis of symmetry. As Figure 4.1 illustrates, grain shape can be described by the ratio of polar and equatorial axes. Shape can be an important characteristic for distinguishing among taxa. Most grains are ellipses of various types, but triangular forms and irregular types, often with attachments, also occur.

A pollen grain consists of three concentric layers: living cell, intine, and exine (Fig. 4.2). As the name implies, the living cell, making up the center of the grain, is the portion that germinates, effecting fertilization of the female portion of the flower. The intine, the layer immediately surrounding the living cell, is composed of cellulose and other elements, including protein. It is believed that proteins of the intine may be the source of the allergic reactions of humans to pollen. Neither living cell nor intine is preserved in fossil pollen.

The outer wall of the pollen grain, the exine, is composed of sporopollenin, one

Figure 4.1 The major shape classes of pollen grains and a diagram showing orientation of the polar (P) and equatorial (E) axes (from Kapp 1969:4).

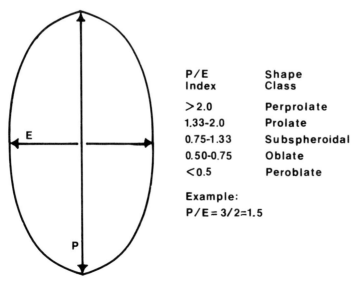

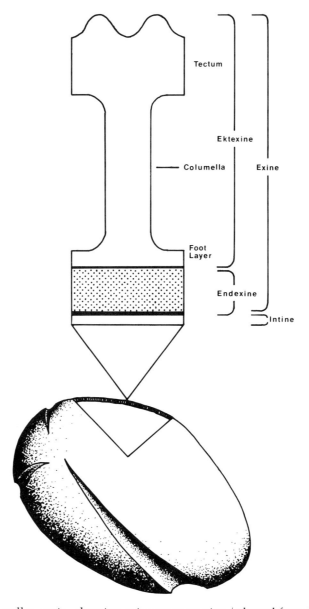

Figure 4.2 A pollen grain, showing exine cross section (adapted from Kapp 1969:7).

Nature and Production of Pollen • 249

of the most resistant natural organic substances. Sporopollenin is susceptible to oxidation, however, and to destruction by mechanical degradation and biological agents. Fossil pollen is identified largely by the structure and sculpturing of the exine and by the form of the openings, or apertures, in it.

Apertures open into the interior of the pollen grain, providing an avenue for emergence of the pollen tube. Apertures are generally divided into two types, pores and furrows. Pores are usually round but can be somewhat elongated with rounded ends. Furrows are boat-shaped in appearance—elongated with pointed ends (Fig. 4.3). The location of pores or furrows on the grain and number of apertures present are characteristics useful for separating pollen types. Some keys, such as Kapp (1969), are arranged by aperturation. Number of apertures varies from none to 40 or so. Many dicotyledon pollen grains have three apertures arranged equidistant from each other. Many monocotyledons, including grasses, are characterized by one aperture at the distal end of the polar axis.

Faegri and Iversen (1975:22–44) present a detailed discussion of terminology used to describe structure, sculpturing, and apertures of the exine. Similar discussions are found in Erdtman (1969:21–50) and Moore and Webb (1978:30–45). Familiarity with terms such as "tectate," "echinate," "rugulate," and "colpate" are necessary to use pollen keys, such as Kapp (1969) and Moore and Webb (1978), but such terminology may be confusing to the novice paleoethnobotanist, or to the archaeologist who just wants to know how to interpret pollen diagrams. As Shackley (1981) points out, the actual number of pollen types one is likely to encounter in an archaeological study is often fairly small. Recognition of types can often be accomplished without keying. I here summarize basic parameters for describing the exine and leave it to the interested reader to grapple with the intricacies of terminology.

The exine is composed of two layers, ektexine and endexine (see Fig. 4.2). The

Figure 4.3 Some pore and furrow patterns (adapted from Kapp 1969:22–25, 27).

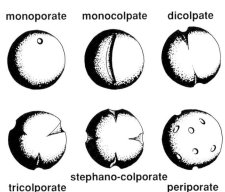

250 • Chapter 4 Archaeological Palynology

outer layer, the ektexine, is easily studied microscopically by viewing its surface and by focusing on its internal components. It has a number of features used to distinguish pollen of different taxa. A basic distinction is whether the outermost surface, the tectum (Latin "roof"), is fused into a continuous surface, as depicted in Figure 4.2; is discontinuous, that is, a "roof" with holes; or is absent altogether. Figure 4.4 illustrates a number of variations of exine structure, and surface patterning. If the tectum is discontinuous, the columellae ("pillars" holding up the "roof") and foot layer may be visible through the openings. If the tectum is absent, the columellae will also be lacking. Patterning produced by surface visibility of these elements, as well as their form, contributes to the distinctive character of the exine as a whole.

In some pollen, the tectum is covered with projections of varying heights, which may be connected to form surface patterns (Fig. 4.4, tectate forms). Such

Figure 4.4 Variations of exine structure and sculpturing. Endexine, black; ektexine, dotted; col, columellae (simpl = simple, dig = digitate); tec (perf), perforate tectum; Psi, psilate. Bottom row: scabrate, verrucate, gemmate, clavate, baculate, and echinate sculpturing (from Faegri and Iversen 1975:30).

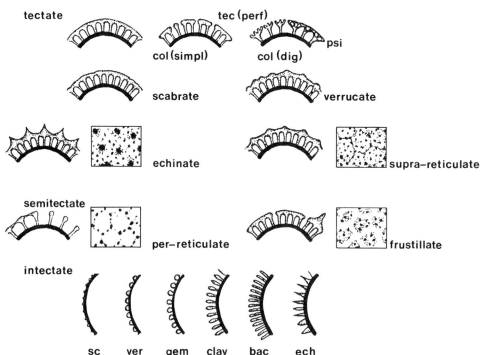

external features, described without reference to internal elements (e.g., beneath the tectum), are known as sculpturing. It is easy to confuse terminologies of structure (the form and arrangement of elements beneath the tectum) and sculpturing (features on the tectum). Texts such as Faegri and Iversen (1975) can be consulted for details of these terminologies.

In addition to characters of the exine and overall form of the grain, size can be a useful criterion for distinguishing pollen. Pollen grains have been observed from 5 microns to more than 200 microns in size. Although the size of living pollen grains of any one taxon can be quite variable, once the living cell and intine have decayed, size stabilizes and is rather constant for a species. Faegri and Iversen (1975) caution, however, that processing and embedding procedures alter the size of fossil pollen grains. Measurements made using different mounting media may vary considerably; this should be taken into account when using size as an identification criterion.

Over- and underrepresentation of pollen in the archaeological record is an important consideration in paleoecological reconstructions. Pollen representation can be discussed in two aspects: (1) differential production and dispersal, and (2) differential destruction (Faegri and Iversen 1975).

There are four major mechanisms for pollen dispersal, the action of transporting pollen from the anthers to the styles of flowers. The quantity of pollen produced and the form of the grains are dependent in large part on dispersal mechanism, which in turn influences the quantity of pollen deposited over the landscape.

Anemophilous plants are those pollinated by wind. Wind-pollinated taxa produce the greatest quantity of pollen of any group of plants, generally 10,000–70,000 grains per anther. Many wind-pollinated taxa have smooth pollen grains with little sculpturing. These grains also tend to be dry. These characteristics make the pollen very aerodynamic, aiding in free dispersal. Some wind-pollinated types, such as pine, have "bladders" to increase buoyancy and are capable of being transported thousands of kilometers (Fig. 4.5). The mix of pollen transported by air currents, eventually being deposited over the land surface, is referred to as pollen rain.

Wind-pollinated taxa contribute differentially to the pollen rain of a region. In a forested setting, for example, the uppermost plant layer (upper canopy) contributes a higher percentage of pollen to pollen rain than do understory plants. Small, buoyant grains are transported farther than large, heavy grains. The latter, although wind-pollinated, may be transported only a few hundred meters from the source plant. Maize is a good example of a wind-pollinated plant with a short dispersal distance. Other factors influencing the quantity of anemophilous pollen which becomes part of pollen rain are production of pollen per anther, number of anthers per flower, timing of pollination in relation to leaf production, and air turbulence.

Given the complexity of the interaction of factors affecting production and dispersal of pollen from wind-pollinated taxa, how does one determine which taxa

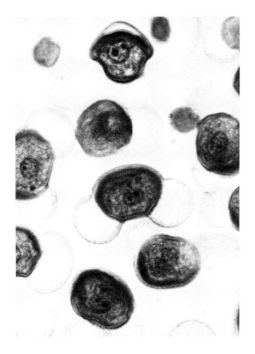

Figure 4.5 Pine pollen, with flotation bladders (unstained).

will be underrepresented and which overrepresented in soil or sediment samples? This problem is approached by sampling contemporary pollen rain and soils from known vegetation formations. These approaches are discussed in "Presenting and Interpreting Results," below.

Another common dispersal mechanism for pollen is by animal vector. Such plants are referred to as zoophilous. Insects, birds, or bats visiting flowers to feed on pollen or nectar brush against anthers and carry pollen from one flower to another. Such pollination does not occur by pollen being blown from one flower to another. In some cases, flower and animal vector are very precisely matched, with one animal species affecting pollination of one plant species. This is the case with orchids, for example.

Zoophilous taxa tend to produce far fewer pollen grains than anemophilous plants, in the range of 1000 or fewer per anther. This is an effective reproductive strategy, since pollination does not depend on mass movement of pollen. There are exceptions to this generalization, however, such as plants that produce large quantities of pollen as an attraction for pollen-eating animals. A few zoophilous taxa, such as *Tilia*, produce quantities of pollen equal to those of wind-pollinated plants.

Pollen grains of zoophilous taxa are often covered with sticky oils or are highly sculpted, so they adhere readily to the pollinating animal. There is no advantage to buoyancy or small size. The result of the lower pollen production and adherence to pollinators is that zoophilous pollen rarely becomes part of the pollen rain, and so zoophilous plants tend to be underrepresented in pollen assemblages from soil or sediment. Exceptions occur when a pollinating animal is itself incorporated into deposits, when soil samples are taken from where blossoms have fallen, or when a high-producing taxon like *Tilia* is present. Underrepresentation can create analytical problems if animal-pollinated taxa form an important component of local or regional vegetation. Such taxa may be completely absent from pollen rain, only occurring in situations where entire flowers or pollinators have become incorporated into deposits.

From the archaeologist's perspective, finds of pollen concentrations from incorporation of flowers into soil can give valuable insight into subsistence or ritual practices. Perhaps the most famous example of this is the Neandertal burial Shanidar IV (Leroi-Gourhan 1975; Solecki 1971), where concentrations of pollen from at least seven species of flowers were identified in grave soil. Flowers were apparently incorporated into boughs used to line the grave. Similarly, presence of significant quantities of pollen from zoophilous plants in human coprolites can be used to argue for ingestion of flowers, honey, or other plant parts to which pollen adheres (Bryant 1974a; Bryant and Holloway 1983).

The other major pollination mechanisms, water-pollination (hydrophilous or hydrogamous) and self-pollination (autogamous), also lead to poor representation of taxa in pollen rain. In the case of water-pollinated taxa, although one would think that pollen from aquatics would be well represented in lake bottom deposits, pollen of most aquatic plants have thin exines and do not preserve well (Faegri and Iversen 1975). Self-pollinating plants also contribute little to the pollen record, since few grains are produced and flowers usually do not open until after pollen has germinated. In fact, some autogamous taxa have flowers that never open. Since several important economic plants, including wheat and a number of legumes, are self-pollinated, underrepresentation of pollen of these taxa in soil or sediment can become an important interpretive issue.

Another aspect of bias in representation of pollen in archaeological soils or sediments is differential destruction of pollen once it is deposited. A number of factors determine whether pollen grains are preserved over time. Forces affecting preservation do not act on all pollen types equally. These forces fall into three related categories: mechanical degradation, chemical destruction, and action of biological agents (Bryant 1978; Bryant and Holloway 1983).

Destruction of pollen by mechanical degradation begins almost immediately upon deposition of grains. Exine surfaces can become abraded by soil particles; this eventually leads to the breakup of grains. Abraded pollen is also more susceptible to destruction by fungi and bacteria. Loss of sculpturing detail can render damaged grains more difficult to identify. Alternating episodes of soil wetting and drying, such as occurs in soils subject to periodic flooding or seasonal rainfall, can also weaken grain exines, leading to their breakup.

Although sporopollenin, the major constituent of pollen exines, is very resistant to chemical destruction, pH and Eh (oxidation potential) of sediments do affect preservation. In general, pollen preservation is enhanced in acid soils (pH <7). Holloway (1981, cited in Bryant and Holloway 1983) reported that eight of nine compounds found in laboratory tests to cause destruction of exine walls were of basic pH. Dimbleby (1957) demonstrated that quantity of pollen preserved dropped in sediments with a pH greater than 6.0. Other researchers have shown that pollen can be preserved in base-rich (alkaline) soils, although grains may be deteriorated and difficult to identify (Bryant and Holloway 1983). In a recent discussion of pollen-poor soils, Dimbleby (1985) points out that, if pollen does occur in base-rich soils, it does so in low concentrations, making counting laborious. It might be necessary to scan twenty or thirty slides, for example, to obtain a count of 200 grains. And, "assuming one can reach an adequate pollen total (200) for each sample, how meaningful are the results" (Dimbleby 1985:19)? There is considerably more biological activity in alkaline than in acidic soils, and assemblages of surviving pollen would be highly selected for types most resistant to decay. This factor operates in addition to the direct effects of basic chemical compounds on the exine.

The oxidation potential of soil is closely related to pH as a factor affecting pollen preservation (Bryant and Holloway 1983). Pollen grains are not highly resistant to oxidation; a reducing environment is much more conducive to their preservation. Such environments can be produced by anaerobic bacteria that release hydrogen as a component of by-products of respiration. This produces lower pH and low Eh potential, enhancing preservation of pollen.

Biological agents such as fungi, bacteria, earthworms, and millipedes can greatly alter assemblages of pollen in soil. Humus-feeding arthropods mix the soil layer in which they occur, causing homogenization of pollen assemblages. This is an important consideration in interpretation to which I return in the last section of this chapter. Fungi and bacteria cause extensive destruction of pollen unless their activities are checked (e.g., in waterlogged or very dry sediments, where activity is inhibited). As mentioned above, anaerobic bacteria can also contribute to pollen preservation. In acidic soil, where earthworms are not present, there tends to be a characteristic distribution of pollen with depth (Fig. 4.6). Absolute quantity of pollen (APF, absolute pollen frequency) is highest near the surface, where recent pollen rain falls. Absolute quantity declines deeper in the profile. Not only is there less recent pollen, but the longer grains lie in soil, the greater the number destroyed by biological agents. Fungi and bacteria attack the grains, weakening the exine. Some types of pollen are more susceptible to this type of attack than others (Bryant and

Holloway 1983). The few grains present deep in a soil profile should be predominately ancient (see Fig. 4.6, percentage frequency).

In their overview of pollen analysis in archaeology, Bryant and Holloway close their discussion of preservation of pollen by discussing whether there is any reliable way to assess the "pollen-preservation potential" of soils prior to analysis. This would certainly save archaeologist and palynologist alike time and effort. There is unfortunately no easy way this can be done, although Bryant (1978) has suggested a rule of thumb that uses the percentage of organic matter in soil: soils showing less than 1% organic matter are not likely to contain pollen. The worst environments for pollen preservation seem to be (1) highly alkaline (base-rich) environments, whether water-deposited sediments or soil, (2) soils with a high percentage of leaf mold, and (3) clay river soil. Acidic soils, acid sediment deposits (waterlogged), presence of metallic salts (e.g., cupric salts from decaying copper artifacts), and high aridity (desert areas, dry caves) are more likely to result in pollen preservation. Each of these factors inhibits microbiological decay, which is the key to preservation.

Before leaving this topic, I list some types of sites and depositional environments in which pollen has been recovered. Archaeological pollen analysis has been carried out successfully in tropical as well as temperate latitudes, in moist as well as dry environments, and in habitation sites and anthropogenic landscapes such as raised agricultural fields (Wiseman 1983). Among types of sites successfully studied using pollen analysis are the following, summarized from Dimbleby (1985) and Bryant and Holloway (1983):

1. sites at edges of water courses, lakes, or sea shores (waterlogged deposits)

Figure 4.6 Curves showing the theoretical distribution in soil of pollen of different ages (from Dimbleby 1985:7).

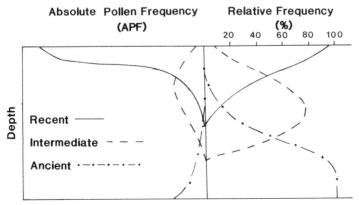

2. sites located in peat deposits or covered by peat formation after abandonment
3. sites in which lower levels or the entire occupation area became saturated by a rising water table
4. features in otherwise dry sites which are cut into the water table, either deliberately or accidentally: wells, sumps, storage pits
5. sites buried by volcanic ash
6. sites buried by blowing sand
7. sites buried by soil washed over occupation areas, or sites (including portions of sites) deliberately buried: old surfaces under earth mounds, flagstone pavements, monolithic stones
8. sites located in arid or semiarid environments
9. cave or rockshelter sites in which arid conditions prevail
10. artifacts such as metates (grinding stones), mortars, baskets, or intact ceramic vessels
11. dried human feces (coprolites)
12. agricultural features: buried field surfaces, sunken gardens, sedimented irrigation or drainage ditches, raised fields.

This is an incomplete inventory of types of sites and contexts at which pollen has been recovered archaeologically. Even in sites on base-rich soils or clay alluvial sediments, contexts may be present where pollen preservation is locally enhanced. It is best to assume that any sampling context may produce at least some pollen grains. Whether to take a sample from a particular context depends on how interpretable the data will be—if some locus has pollen, what does this mean? In the final analysis, it is a relatively simple task to take pollen samples in the course of excavation. By consulting the palynologist before and during excavation, one can make sampling decisions that maximize chances for good recovery and productive interpretation. If processing of a test set of diverse samples reveals poor preservation, little time or money has been lost and rewards can be considerable.

History of Pollen Analysis

A review of the history of pollen analysis, especially its application in archaeology, can be found in Bryant and Holloway (1983). The technique has its roots in investigation of vegetation preserved in peat bogs. Studies of seeds and other small materials preserved in peat led to the discovery that pollen grains were also present in such deposits. A number of studies of fossil pollen were conducted in the nineteenth century, but the real potential of the method was not realized until Lennart von Post presented the first modern percentage pollen analysis in 1916. Von Post's study is considered a landmark in the field because use of percentage calculations permitted direct quantitative description of past vegetation patterning. The history of changes in vegetation and climate during the Quaternary period was an issue of considerable interest in Scandinavia at that time, and pollen analysis was quickly applied to the topic.

Archaeologists were also quick to see the potential of the new technique. One of the earliest applications of pollen analysis in archaeology was as a tool to determine relative dating within a site, or among sites in a region. In regions where the sequence of vegetation changes during late glacial and Holocene periods had been established from studies of peat bog and lake deposits, pollen assemblages from archaeological sites could be compared to the sequence and used to date site occupation. Another early direction was investigation of human impact on the environment. One of the earliest studies of this kind was Iversen's (1949) use of pollen data to date the beginning of the Neolithic in Denmark.

By the 1960s, application of pollen analysis in archaeology had become fairly widespread. Dimbleby is among those active in the area of archaeological palynology in the Old World (Dimbleby 1954, 1957, 1961, 1970, 1978). Other active researchers include Van Zeist (1967), Wright (1984), and Bryant and Murry (1982). Dimbleby advocates use of absolute pollen frequencies rather than relative abundances (percentages based on a pollen sum) to express patterning in palynological data. Although use of absolute pollen frequencies is still less commonly used than relative measures, Dimbleby presents convincing arguments for its utility in soil (as contrasted to sedimentary) pollen analysis. The relative merits of these and other quantitative approaches in pollen analysis are discussed below under data interpretation.

In the New World, application of palynology in archaeology lagged behind its use in Old World sites. Sears, who worked actively in the American Southwest, was one of the first researchers in the United States to apply pollen analysis to issues of archaeological importance (Sears 1937, 1952, 1982; Sears and Clisby 1952). Research remained focused on the arid Southwest for some time, with the work of Martin and associates (Martin 1963; Martin and Byers 1965; Martin and Sharrock 1964), as well as Bryant (1974b, 1978; Bryant and Williams-Dean 1975), Hill and Hevly (1968), and Schoenwetter (1962), among others. But pollen analysis was soon incorporated into archaeological investigations in many regions of the United States, as the work of Bryant (1974c), King and associates (King and Allen 1977; King and Lindsay 1976; King *et al.* 1975), McMillan and Klippel (1981), and Schoenwetter (1974) illustrates.

There is an long tradition of palynological research in Latin America focused on vegetation and climate reconstruction. Much of this work is of importance for archaeology; for example, the work of van der Hammen and associates in northern South America (van der Hammen 1963, 1966, 1981; van der Hammen *et al.* 1973;

van der Hammen and Gonzalez 1960, 1964, 1965; van Geel and van der Hammen 1973; Wijmstra and van der Hammen 1966; Wijmstra 1967), Wright and associates in Peru (Hansen *et al.* 1984; Wright 1983), and Bartlett and Barghoorn (1973) in Panama. There is also active research in archaeological pollen analysis. Among recent studies are those by Bryant (1975), Clary (1980), Kautz (1980), Schoenwetter and Smith (1986), Weir and associates (Weir and Bonavia 1985; Weir and Eling 1986), and Wiseman (1983).

Field Sampling

As is the case for all paleoethnobotanical techniques, pollen sampling has two aspects, the activity of taking samples and the planning of a sampling strategy that permits research questions of interest to be addressed adequately. Before discussing the mechanics of taking samples, a task familiar to many archaeologists, I review types of research questions that can be approached through archaeological pollen analysis and give an example of planning sampling strategies. The goals of the research structure choice of sampling locations and final selection of samples for analysis.

There are similarities between the following discussion on field sampling for pollen analysis and the Chapter 5 discussion of phytolith sampling. Part of this similarity is due to a borrowing of palynological sampling techniques by phytolith analysts. Parallels in mechanics as well as strategies of sampling design between the two approaches also have less trivial roots, namely, that both pollen and phytolith analysis are concerned with recovery and interpretation of microscopic bodies whose incorporation into and extraction from soil, identification, and quantification are similar in a number of ways. Areas of dissimilarity of analysis and interpretation, due in large part to differences of origin, preservation, and precision of identification of these microremains, become apparent in this chapter and the next.

Sampling Strategies

Bryant and Holloway (1983) provide a summary of archaeological research questions that can be investigated through pollen analysis. They emphasize that archaeological deposits are not ideal for attempting vegetation reconstructions. Archaeological sites represent the results of human activities. Although wind-borne pollen from local and regional vegetation is continuously deposited on sites, as it is over the entire landscape, there are many other sources of pollen in archaeological deposits. Use of plants as building materials, food, medicines, and tools and creation of disturbed habitats encouraging weedy plants all provide opportunities for pollen deposition. It can be very difficult to distinguish natural pollen rain, the record of past vegetation, from pollen introduced by human use of plants. And natural pollen rain on sites may itself be biased. As an example, consider pollen samples taken from a floor in the interior of a structure. Loci far from the entrance may lack wind-deposited pollen, while samples in the entrance area may have a fairly complete representation of pollen rain. Thus, although archaeological pollen samples may allow description of past vegetation in general terms, analysis of noncultural sediments is essential for detailed paleoenvironmental reconstruction. Additionally, sampling of agricultural field areas may provide insight into past vegetation as well as subsistence (e.g., Weir and Eling 1984; Wiseman 1983). If plans for such sampling are included in overall research design of a project, comparison between cultural and noncultural depositional contexts can give valuable insight into interpretation of all paleoethnobotanical data.

Archaeological pollen analysis, like analysis of macroremains and phytoliths, is a source of data on prehistoric diet. One of the most direct sources of such data are human feces or coprolites, as research by Bryant and associates (Bryant 1974a, 1974b, 1974c, 1975; Bryant and Holloway 1983; Bryant and Williams-Dean 1975; Holloway and Bryant 1986; Trevor-Deutsh and Bryant 1978; Williams-Dean and Bryant 1975), Callen and associates (Callen 1965, 1967, 1970; Callen and Cameron 1960; Callen and Martin 1969), Hantzchel and associates (Häntzchel et al. 1968), Heizer and Napton (1969), Martin and Sharrock (1964), Schoenwetter (1974), and Weir and Bonavia (1985) demonstrates. Especially useful are groups of contemporary coprolites which represent more than one meal or individual. Pollen in human coprolites may document economic plants used by the population. Large quantities of pollen, especially from zoophilous taxa, can demonstrate eating of flowers, honey, or other parts of plants to which pollen has adhered. In addition to economic pollen, background pollen from natural pollen rain occurs in coprolite samples. Pollen may fall on coprolites before they are buried, or wind-borne pollen may be present on foodstuffs, in water, or on utensils. Wind-borne pollen may give insight into vegetation patterning, but it should be used cautiously, as discussed above. In addition to information on foods ingested, pollen from coprolites can also help document season of occupation of sites. If the quantity of pollen present suggests that flowers were eaten, the meal probably took place during the flowering period of that species. (I discuss extraction and analysis of pollen and other remains in coprolites in "Laboratory Analysis," below).

Unfortunately, coprolites are usually preserved only in sites where aridity leads to preservation of uncharred botanical remains. In other situations, pollen originally in coprolites becomes part of general midden, floor, or pit fill once the coprolite disintegrates. Soil from such contexts may contain pollen from a variety of sources, from disintegrated coprolites and decayed vegetable material with adhering pollen to ancient pollen rain. Analysis of such sediments can give insight into plants in the diet, for example by recovery of pollen from cultivars or wild edible plants, and also into function of the sampled context, for example by recovery of pollen spectra indicative of a food-processing or storage area.

Ritual and medicinal practices can also be investigated through soil pollen analysis. Sampling in burial chambers or graves may reveal what foods, herbs, flowers, or other items were left as offerings or used in grave construction (e.g., Leroi-Gourhan 1975). Similarly, pollen assemblages from structures thought to be ceremonial in character may be contrasted to assemblages from domestic contexts.

Pollen data, like data from macroremain and phytolith analyses, may be useful in determining the functions of artifacts and features as well as structures. Analysis of pollen adhering to ground stone artifacts allows testing of hypotheses of tool function (e.g., Bryant and Morris 1986; Bryant and Murry 1982). Sampling from the lining, primary fill, and secondary fill of pit features may reveal sequences of use and reuse of such features. Analysis of adobe brick has been found a good source of information on vegetation as well as on season of construction (O'Rowrke 1983). In situations where pollen preservation is good, data may be used to distinguish among activity areas in a structure, among a group of domestic buildings, or within a site precinct (e.g., Berlin *et al.* 1978; Hill and Hevly 1968).

As an illustration of how research questions guide development of a systematic sampling strategy, I discuss the research design described by Bohrer and Adams (1977:49–54) for pollen and flotation sampling at Salmon Ruin, New Mexico. This strategy was used to select samples for analysis from a large body of systematically collected samples.

Bohrer and Adams's goal was to reconstruct lifeways of people occupying Salmon Ruin pueblo. Encompassed within this broad goal were questions concerning subsistence, medicinal and ceremonial uses of plants, reason for site abandonment, seasonality of plant procurement, and habitats used for plant procurement. In addition, a specific goal of the project, to obtain information on prehistoric social organization, was to be approached through an analysis of spacial organization, specifically room-function analysis. Botanical data would form one data set for this analysis.

The primary area of sampling was the room—a bounded space, usually delimited by four walls. Successive floors identified within rooms were primary units for interpretation. Samples were obtained from features, artifacts, and open space in floors. Specifically, paired pollen and flotation samples were chosen for analysis from both a metate and a firepit, if present, and from each room grid square. For ceremonial rooms (kivas), additional samples were analyzed from wall plaster, niches, and pits.

To test hypotheses of room function, predictions of seed and pollen occurrence given assumptions of feature or room functions were developed from ethnographic sources. Conceptually, this approach is comparable to Hillman's (1984) more detailed study of crop-processing stages and associated plant residues and seeds (see Chapter 3). Actual patterns of seed and pollen occurrence were compared to predicted patterns, and an evaluation of function was made. Bohrer and Adams also researched differential use of kiva floors, pits, niches, and vegetable components of pigments and plasters used in decoration.

In addition to floors and their associated features and artifacts, rooms also contained two types of fill: occupation fill and post-occupation fill (fill deposited after room abandonment). In general, distinguishing between these types of fill is important for interpreting pollen spectra or macrobotanical remains recovered from them. Occupation fill may be a valuable source of economic information if it represents debris dropped during use of the room. It may be impossible to determine the association of post-occupation fill, however, making it less useful for analysis. During excavations at Salmon Ruin, pollen samples were taken from both types of fill.

Another common sampling locus at Salmon Ruin was the open plaza area. Sampling and analysis were focused on living surfaces and activity areas. In such areas, samples were taken from each excavation grid square. All other areas were sampled less frequently.

The final sampling locus discussed by Bohrer and Adams is the burial. Research into burial customs among pueblo peoples revealed an array of uses of plants in burial contexts. Through pollen and flotation sampling in burials, the authors hoped to gain insight into traditional locations for food offerings, as well as what foods or other items were commonly used as offerings. Association of certain foods with special religious responsibilities or other indications of status would be investigated.

Bohrer and Adams argue convincingly that seed and pollen data can be interpreted most effectively if flotation and pollen samples are paired, that is, taken from the same locations over the site. Presence of seeds may sometimes explain presence of pollen, since pollen often adheres to seeds or fruits. Plants used before seeds set may be identified only by pollen. The two data sets are usually complementary. One exception to this general rule is sampling in hearths or other areas of high charcoal concentration. These contexts should be avoided in pollen sampling, since charcoal is almost impossible to remove from samples.

Bohrer and Adams also advocate use of control samples as an aid to interpretation of archaeological botanical data. Sampling for pollen in an area undisturbed by human activity helps determine background pollen spectrum. Taking flotation samples from noncultural fill aids in identifying contaminants and assessing preservation biases.

Taking Soil Samples

Turning now to procedures for taking pollen samples, I begin with three procedures for sampling archaeological contexts. This should give a fair idea of how to take samples and how to adapt sampling to fit the needs of particular field situations. I then briefly discuss sampling in sedimentary bog or lake deposits and sampling for comparative purposes.

262 • Chapter 4 Archaeological Palynology

Bryant and Holloway (1983) Procedure

Bryant and Holloway (1983:199) give these guidelines for sampling profiles to minimize contamination problems often associated with pollen sampling:

1. Clean the outer surface of the excavation pit profile prior to sampling.
2. Always use a clean trowel or other type of digging implement.
3. Collect between ½ and 1 liter of material. This ensures sufficient sample size for a second analysis, if necessary.
4. Carefully clean the sampling tool before collecting each new sample.
5. Always use a sterile, uncontaminated, leak-proof container for each soil sample. If the sample is damp, add a few drops of fungicide or 100% ETOH (absolute alcohol, or ethanol) to prevent microbial activity, which may destroy pollen in the sample.
6. Correctly label each sample with a permanent ink pen.
7. When collecting samples from a profile, sample within one stratum (whenever possible) rather than mixing strata in a single sample.
8. Sample a profile starting at the bottom and work toward the top; this ensures that material falling from the upper samples does not contaminate the lower samples.
9. Movement of the trowel should be lateral, following the plane of the stratum; again, this prevents any contamination between strata.
10. Avoid taking samples from hearths or any archaeological features that appear to have been burned or that contain large amounts of ash and charcoal. Avoiding these charcoal-rich areas is important, since charcoal is often impossible to remove from samples during laboratory processing and pollen is often destroyed in areas that have been subjected to intense heating.

Bryant and Holloway also recommend collecting samples from adjacent areas not on the site to serve as control samples. Such samples aid in distinguishing between pollen spectra from cultural activities and those resulting from natural processes. Obtaining surface samples from identified plant communities is also recommended by these authors. Such samples serve as analogues between vegetation and the soil pollen spectrum.

Dimbleby (1985) Procedure

The following points are summarized from Dimbleby's (1985:20–25) discussion of sampling in archaeological deposits:

1. Take serial samples in most situations. In other words, take a sequence of samples from a deposit, so that a pollen curve can be built up to serve as the basis for interpretation. Spot sampling is generally valueless,

since any one sample is likely to contain pollen of various ages. The exception is sampling an old land surface or other situations for which all pollen in the sample is contemporaneous.

2. Whenever possible, take contiguous samples, that is, samples that touch one another along a profile. In very long profiles, samples can be spaced at 5- or 10-cm intervals but should be more closely spaced above and below features. Contiguous sampling allows buried surfaces to be identified more readily. Samples should not be taken by soil horizon, since horizon formation may be recent, postdating site occupation.

3. Sample size varies according to the physical nature of the deposit and the concentration of pollen. In soils with abundant pollen per gram of soil, about 30 g per sample is adequate.

4. Take samples from a freshly exposed face to avoid contamination by pollen in air, dust, or water. Take samples from the bottom of a profile upward.

5. Sampling tools and techniques vary according to the nature of the deposit. For highly organic deposits, or soils free of stones or sand, samples can be cut out in small blocks or sampled with a tubular borer. Cut blocks are preferred for contiguous sampling. In stony soils, fine material can be separated from stones at the time of sampling by using a metal soil sampler. A plastic bag is fit over the end and fine material dislodged from around stones and shaken into the bag.

6. Label each sample clearly with sampling depth and provenience information. Samples should be cross-referenced to a profile description in which soil color and texture, occurrence of humus-rich layers, iron pans, root channels, land snails, evidence of earthworm activity, artifact layers, and charcoal are recorded.

7. If samples cannot be processed within 48 hours, they should be treated to retard microbiological attack. Dry samples can be treated with a vaporizing disinfectant such as toluol. Such compounds do not permeate all soil aggregates in moist samples, however. Moist samples can be stored at 5° C or lower or dried. Oven drying at 90–100° C is the safest practice, although sun drying can be used if precautions are taken to avoid contamination. High temperatures should be avoided, since distortion of grains can occur. Chemical methods of preventing decay are not very reliable.

Bohrer and Adams (1977) Procedure

The following procedures are summarized from Bohrer and Adams's (1977:20–28) detailed discussion of sampling procedures developed for research at Salmon Ruin pueblo:

Taking soil samples (features, floors, fill, vessel contents).
1. With a clean trowel, free of dust and soil, scrape a clean surface for sampling.
2. Use the trowel to take pinches of dirt, and put the dirt into a previously unopened bag.
3. Collect 60 ml (about ⅓ cup) of soil per sample; this is twice the quantity needed for extraction. Because sandy soils generally contain less pollen than silty soils, samples from sandy deposits should be 50% larger.
4. Put a clean bag with the pollen sample into a second clean bag, close, and tag. This procedure protects the sample from contamination with airborne pollen.

Pollen wash. A pollen wash is a laboratory procedure applied to whole vessels or lithic artifacts to remove pollen adhering to use surfaces. At Salmon Ruin, whole vessels and ground stone artifacts were fairly common, so only a sample of each type was designated for pollen washes. The remainder were sampled for pollen as described above.

Objects to be washed need special handling in the field. Cover the artifact after exposure to avoid contamination with modern pollen. Leave as much of the dirt matrix as possible in or on the artifact to protect the use surface. Once the object is in the lab, remove loose soil to expose the use surface. Test the surface with a few drops of dilute HCl to see if calcium carbonate deposits are present. If the acid reacts, use dilute HCl rather than distilled water for the wash, since carbonates may inhibit loosening of pollen from the surface. Wash the surface with a stream of liquid, catching it in a storage container. If water is used, acidify it to inhibit fungal activity during storage.

The dry conditions at Salmon Ruin pueblo, and the excellent pollen preservation that resulted, made possible the detailed sampling strategy described earlier. There are a number of general points in Bohrer and Adams's discussion which are valuable for any archaeological pollen study. Among these are the value of sampling in and on whole artifacts recovered in good contexts, the importance of distinguishing between sediments associated with cultural activities and those representing fill, and of choosing appropriately between scatter samples (a sample made up of a series of subsamples from a defined area) and point samples, and the important role control samples can play in interpretation.

Sampling in Bog or Lake Deposits

In excavations of a habitation site in a lake edge setting or other waterlogged situation (e.g., Dimbleby 1985:31–43), pollen samples should be taken as described above for archaeological sampling. Detailed sampling such as described by Bohrer and Adams (1977) may be highly productive, since pollen preservation is often enhanced in waterlogged settings. What I review briefly here are traditional sampling techniques for noncultural lake or bog deposits. If such contexts are included in an archaeological research design, direct data on paleovegetation can be recovered. This can in turn be used to model past climate and to identify modifications in landscape caused by human manipulation of the environment. Ideally, coring sedimentary deposits should be carried out by a palynologist, who then also conducts subsequent analysis. If the archaeologist must take cores, procedures should be reviewed with a professional in advance of fieldwork. In regions where appropriate deposits are present and paleoenvironmental data are lacking or scarce, such efforts may yield a considerable return.

There are a number of descriptions of equipment and field technique for coring lake or bog deposits (e.g., Deevey 1965; Faegri and Iversen 1965, 1975:85–100; Moore and Webb 1978:16–22; Shackley 1981:72–74; Wright et al. 1965). The following discussion is taken primarily from Faegri and Iversen (1975).

The aim of sampling sedimentary deposits is to obtain samples that are uncontaminated by modern pollen or pollen from adjacent strata, to identify types of sediments present, and to explain the stratigraphy. This is most easily done by sampling profiles cut into peat or filled lakes. In these situations, the profile is cleaned and samples are taken with a sampling tube, forceps, and spatula, or as a continuous column, much as described above for sampling soil profiles.

Usually, however, there is no easy way to cut a deep profile into a bog or lake deposit, and sampling must be done from the surface downward. This involves coring. Peat or filled lakes are still easiest to sample, since a firm surface is available on which to work. Coring in open water adds a number of difficulties: establishing a floating coring platform, keeping the sampler vertical, finding the bore hole between samples, dealing with loose, unconsolidated surface sediments, and so on. Techniques and equipment have been developed to deal with these difficulties; these are described in the sources cited above.

The first problem encountered when sampling for noncultural pollen spectra is choice of sampling location. If the object of research is to obtain "an integrated picture of the whole of the vegetational development" (Faegri and Iversen 1975:86), a location is needed where regional pollen rain can be easily distinguished from local pollen. Faegri and Iversen recommend sampling in lake deposits rather than peat and suggest that "too small" and "too large" lakes be avoided (the ideal size being around 5000 m^2). Cores should be taken from protected places where there has been little wind or current. Places where streams enter or leave a lake should be avoided to lower risk of contamination, movement of sediments by current, and loss of pollen by oxidation. In small bogs or lakes one sampling location may be sufficient, but it should be located near the sedimentation center of the deposit. It is usually desirable to sample several loci, however, so that the nature of the deposits can be clearly understood. The more complex the depositional history of the sediments, the shorter the distance between sampling locations.

Once a location is chosen, the appropriate coring equipment is determined. There are two basic corers used for palynological sampling, side-filling and bottom-filling types. Both have closed chambers that are pushed down to the desired depth, opened, rotated (side-filling) or pushed (bottom-filling) to fill the chamber, closed, and extracted. Auger-type samplers, which are screwed into deposits, are not acceptable for palynological sampling, since turning motion disturbs deposits. Similarly, open-chamber samplers are unacceptable because of the high risk of contamination as the sampler is lowered and raised.

Length of the corer chamber is limited because of the danger of breakage and friction between sampler and sediment. For a deep deposit, the corer is pushed into the bore hole many times, using rigid rods to extend length, and a sequence of samples is removed. Pushing a wide corer into sediments is not easy; some sort of mechanical device is needed to pull the corer up as well as push it into the deposit.

As samples are removed from the bore hole, they may be retained in their entirety for subsampling in the laboratory or sampled on the spot. It is easier to extract the sample in its entirely from a bottom-filling sampler, since the whole core is simply pushed out. Using expendable chamber tubes is an efficient way of retaining whole segments for later subsampling; the sample is in a plastic or metal tube within the corer chamber, which when removed and closed serves as the storage container for the sample. For side-filling samplers, the chamber is opened and small samples taken out at regular intervals. Whether sampling is done at the coring site or in the lab, it is important to cut off the outermost portion of the core (the part in contact with the corer wall), since contamination between samples may occur.

Samples from lake or bog deposits must be kept moist until analyzed. Faegri and Iversen note that some humic colloids may become difficult to disperse if sediments are allowed to dry, and pollen may be damaged by soil shrinkage. It is of course essential that all samples be carefully labeled so that depth and deposit type (as well as which end of the core segment is up) are known.

Comparative Sampling

The final aspects of pollen sampling I discuss here are sampling modern surface soils and collecting samples of airborne pollen. Collecting modern soil samples may fall to the archaeologist if a pollen analyst is unable to visit the study area. Modern surface soils are studied to compare pollen deposition to known vegetation cover. Quantitative comparisons of plant cover and pollen deposition are used to establish correspondences that can be used to reconstruct ancient patterns of plant cover from fossil pollen assemblages.

In Chapter 5 I present a detailed discussion of surface sampling and its role in interpreting phytolith assemblages. Lack of keys and descriptions of phytolith assemblages for many vegetation formations makes such analyses vital for archaeological phytolith research. In the case of pollen analysis, there are many more published sources of information on specific vegetation formations and pollen produced by them. Webb and McAndrews (1976) is an example of such a study for central North America. It is still very useful for archaeologist or archaeological palynologist to collect soil samples from modern surfaces, especially if regional pollen studies have not been done for the research area.

Surface soil samples are taken as "scatter" or "pinch" samples. A delimited area (e.g., a 10- × 10-m square) is marked off and small portions of soil collected from around the area. Clean plastic bags or vials should be used to hold soil. The tip of a trowel is a useful collecting tool. All pinches are combined in one container to give a composite sample for the square. Care should be taken to collect only surface soil and to avoid the area around any pollinating plants. Containers should be labeled immediately with sample location and referenced to a detailed description of vegetation covering the sampled area. If samples are not to be processed immediately, appropriate measures to preserve pollen should be taken.

Collect surface samples from distinctive vegetation formations in the study area. How finely one subdivides area vegetation depends on the complexity of vegetation, how disturbed it is, existence of prior studies, and how much time and funding can be expended on analysis. I recommend sampling whatever vegetation formation covers the site and formations representing basic floristic divisions (e.g., if the site is near a prairie–forest boundary, collect good prairie and good forest samples). If investigation of agricultural systems is part of the research design, sampling in actively cropped, fallow, and "pristine" vegetation is important. Any opportunity to collect samples from contexts that might parallel prehistoric depositional episodes should be pursued (e.g., food-processing or storage areas, abandoned houses).

Another useful source of information on pollen assemblages produced by known vegetation is airborne pollen. Trapping pollen rain as it falls to the ground, in addition to extracting it from surface soil, can give a good approximation of pollen assemblages before components are lost by mechanical degradation, chemical destruction, or action of biological agents in the soil. Knowing what pollen is in the air during fieldwork can also alert the analyst to possible contaminants in archaeological samples and to long-distance pollen sources.

Methods of sampling airborne pollen are described in Kapp (1969:13–16) and

Lewis and Ogden (1965). The simplest method is to expose a microscope slide or petri dish smeared with a sticky substance (glycerin or silicone oil) or covered with double-sided adhesive tape at ground level for a period of time. Pollen falling onto the adhesive sticks and can be examined with no processing beyond staining for easier viewing. A simple method of trapping a cumulative sample of pollen rain is to use a rain gauge or widemouth jar set at ground level. The opening should be covered with screen to keep out insects and windblown debris. After the sampling period, water is removed from the sampler and centrifuged to concentrate pollen grains. There are also samplers available which trap pollen from a known volume of air, rotating into the prevailing wind. Such devices are useful in obtaining pollen accumulation data.

Sampling Modern Vegetation

As I discussed in detail in Chapter 3, study of modern vegetation in conjunction with an archaeological project serves two purposes: to obtain comparative materials and to understand composition and distribution of vegetation. Modern vegetation patterning in areas highly altered by humans may give little insight into ancient patterns, but study of even limited relic stands can alert the analyst about species that may occur archaeologically.

When collecting plants for comparative pollen samples, follow the procedures discussed in Chapter 3, with these additions and cautions. Since pollen is produced in the flower, extra flowers must be collected for processing (take voucher and extras from the same plant if possible). Well-developed but unopened flower buds are an excellent source of pollen. Dry samples to prevent mold growth, but do not use excessive heat, which may damage pollen.

If the palynologist cannot visit the study area, the vegetation setting of the site should be described. The analyst will need to know what is growing immediately around the site and what plants are in flower during excavation. Information about natural and human-altered vegetation formations near the site, with names of dominant plants, is also useful. Before going to the field, meet with the palynologist to discuss strategies for comparative as well as archaeological sampling.

The beginning archaeological palynologist working in the temperate latitudes may be able to identify much of the pollen in archaeological samples without building a comparative collection. Keys to pollen and spore types are available for the temperate floras of both the Old and New Worlds. Although keying is difficult for the beginner, some time spent studying terminology, examining photographs, and practicing with known types helps facilitate use of keys. Among general pollen keys or descriptions of types are the following: Erdtman (1952, 1957, 1965, 1969), Erdtman and Sorsa (1971), Faegri and Iversen (1975), Hueng (1972), Hyde and Adams

(1958), Kapp (1969), Kremp (1965), McAndrews *et al.* (1973), Moore and Webb (1978), and Wodehouse (1959). Consult these sources for additional references to specific taxonomic groups.

The beginning archaeological palynologist must also be cautious—and must be very critical of level of identification of pollen types. If a key is used, the taxonomic level at which a type is distinctive (i.e., family, genus, species) should be determined. If a comparative collection is used, pollen from related taxa should be examined before identifications are finalized. If species-level identification is being attempted, floras for the study area should be consulted and all species examined. If a pollen type is only distinctive at the genus or family level, this should be stated clearly. This is especially important when dealing with economic plants—not every grass pollen grain is corn.

Laboratory Analysis

In Chapter 5 I discuss equipment and supplies needed to set up processing and microscopy laboratories for phytolith analysis. These laboratories are patterned after facilities used for pollen extraction and identification. Although procedures for extracting pollen and phytoliths from soil are distinct, basic equipment needs are similar: fume hood with acid drain, centrifuge, hot water bath, and so on. I do not repeat the discussion of laboratory facilities here; review the Chapter 5 discussion, and note in the description of processing procedures in this chapter differences in small equipment, glassware, and chemical needs. For detailed descriptions of pollen laboratory setup and equipment, consult Gray (1965a).

Another similarity between phytolith and pollen processing is the use of toxic chemicals—and the need for careful handling of these substances. Extraction should never be attempted without a fully functioning fume hood. Always wear protective coat, gloves, and goggles. A processing lab should have an emergency eyewash station, first aid kit, and fire extinguisher. Although pollen processing is not technically difficult, the beginner needs basic chemistry laboratory experience and should process the first few times under the supervision of someone experienced in handling corrosive chemicals. Training by a pollen analyst is the best way to learn. It is important to emphasize here that I am not advocating "cookbook" palynology. The procedures described in this section are not appropriate in every situation. Consult an expert before attempting extractions and invite expert criticism of interpretations. Inexperience can lead easily to regrettable results that could damage the reputation of the field and its practioners.

In the remainder of this section, I discuss basics of soil processing for extracting pollen and give two examples of processing procedures. I then briefly discuss processing coprolites, preparing floral material for comparative specimens, and mounting slides. Finally, I review counting and identification procedures, including relative and absolute counting techniques.

Soil Extraction Techniques

The basic aim of soil pollen extraction is to concentrate pollen and render grains as visible as possible. Concentration is the process of removing everything else from a sample with a series of chemicals, leaving only pollen; rendering the grains visible involves removing clays, mounting residue in a suitable medium, and staining. Faegri and Iversen (1975:101–122) present a general discussion of processing; their guidelines have been widely adapted and used. Brown (1960) is a compilation of many procedures, organized by sediment type. Gray (1965b) summarizes many techniques by four steps of recovering pollen: cleaning, disaggregation and dispersal, chemical extraction, and density separation. The last two references are sources of information on specific problem matrices or processing steps. The specific procedure followed in a study depends on the soil matrix and pollen preservation. Experience is the best guide to selecting a procedure, so I advise the beginner to consult an experienced researcher who has handled sediments of the type to be analyzed (e.g., peaty, with high clay content) or to experiment with a procedure used for such sediments before beginning full analysis.

Soil contains inorganic and organic components which, from the viewpoint of the palynologist, obscure pollen and make counting difficult and time consuming. Carbonates, humic compounds, silica (phytoliths, diatoms, etc.), and cellulose are among substances routinely removed. Carbon (residue from burning) is very difficult to remove, which can be a problem in samples taken from hearths, ash lenses, and similar contexts. The following summary of removal of each of these common soil components, summarized from Bryant and Holloway (1983), serves to illustrate the basic approach of pollen extraction:

Carbonates found in archaeological sediments can usually be removed with concentrated hydrochloric acid (35%). In some cases glacial acetic acid must be used.

Silicates fall into two classes for purposes of removal from soil: large, coarse-grained silicates and smaller, fine-grained silicates. The former can be removed by heavy liquid separation or by swirling. In heavy liquid such as solutions of $ZnCl_2$ or $ZnBr_2$ set at specific gravity 1.95–2.0, pollen (with specific gravity between 1.45 and 1.52) floats while heavier silica particles sink. As discussed in Chapter 5, fine-grained silicates (from silt and fine sand fractions), including phytoliths and diatoms, are quite light and float in heavy liquids. Swirling is a technique used to separate particles of different specific gravity in water by means of their differential settling rates. Swirling may be used to remove heavier particles, but does not separate particles of similar specific gravities (i.e., pollen and phytoliths).

Concentrated hydrofluoric acid (HF) is used to remove fine-grained silicates from sediments. Bryant and Holloway recommend using 70% HF, rather than the more readily available 44–45% concentration, to remove silicates rapidly. They report good results removing both classes of silicates by using sonication (see below) and heavy liquid separation, followed by HCl and HF treatment.

Faegri and Iversen (1975:104) recommend that removal of siliceous material take place prior to acetolysis (see below) and after deflocculation and removal of humic compounds.

Clays. Presence of clays in sediments obscures visibility of pollen, making counting and identification difficult. The process of disaggregating clays so that they can be removed is referred to as deflocculation. A variety of substances can be used to deflocculate sediments; a number of them were discussed in Chapter 2 as aids in flotation of clayey soil. Bryant and Holloway have found dimethyl sulfoxide (DMSO) to be very effective. Each sample is placed in a solution of 99% DMSO, which is then heated in a hot water bath for 1 to 1½ hours. Although DMSO quickly deflocculates sediments, it has one serious drawback: it can quickly enter the bloodstream through skin contact. Extreme caution must be used in its handling, and it should be completely removed from samples before other reagents are used (it can act as a carrier for other hazardous chemicals).

Sonication is a technique for removing clays widely used by palynologists in the oil industry. According to Bryant and Holloway, a 10-minute sonication with a Delta D-5 sonicator does not damage pollen grains, although they advise that tests for damage be conducted prior to use. Faegri and Iversen (1975:102) caution, however, that even short ultrasonic treatment may damage some grains. The basic procedure is to dissolve samples in a commercial dispersing solution, such as Darvan, and place them in the sonicator. The dispersing solution aids in freeing pollen from the matrix of the sample.

Samples can sometimes be deflocculated simply by soaking in detergent, as described in Chapter 5 for phytolith extraction.

Organic material. Soil samples often contain a variety of organic compounds, residues of incompletely decayed plant and animal material. Bryant and Holloway recommend the acetolysis method for removing organic compounds from samples. Acetolysis involves rinsing samples in glacial acetic acid (to dehydrate), then adding the acetolysis mixture (9 parts acetic anhydride, 1 part H_2SO_4) and placing samples in a boiling water bath for 3 to 5 minutes. A second glacial acetic acid rinse (to prevent cellulose acetate from precipitating in KOH or water steps) is carried out, followed by a KOH treatment (10% KOH, 2–3 minutes in boiling water bath) to stop further activity by the acetolysis mixture. Acetolysis is necessary to remove cellulose from samples and is useful in general for processing highly organic samples.

If little organic material is present, heating samples in a boiling water bath with 10% NaOH or 10% KOH is usually effective for removing humic compounds (unsaturated organic soil colloids). Treatment with 10% KOH can be repeated if samples contain much humic acid; higher KOH concentrations should not be used, however. Treatment with KOH also helps deflocculate samples.

Bryant and Holloway advise against using nitric acid (HNO_3), bleach (NaOCl), or ammonium hydroxide (NH_4OH) for removing organic material from samples, except as a last resort, since these chemicals can damage pollen exines. The exine is organic and so is susceptible to damage by strong oxidizing agents. If oxidation is necessary, acid oxidants are preferable to alkaline solutions, since the pollen exine is more easily damaged in an alkaline environment; Faegri and Iversen recommend saturated $NaCLO_3$ and concentrated HCl.

Charcoal. In traditional palynology, one avoids taking samples from deposits with heavy charcoal inclusions, since this is often a sign of human-altered sediments. In archaeological palynology, such samples are often of particular interest. Charcoal is inert and not easily destroyed by chemicals used in pollen extraction. Too much charcoal in a sample renders slides difficult to scan and can hinder concentration of pollen. Bryant and Holloway recommend removing charcoal debris by mechanical means. Their procedure is as follows:

1. Screen material through a 150-micron brass screen; this removes the larger charcoal fragments.
2. Sonicate the sample in a saturated solution of Darvan several times; this frees the remaining small charcoal fragments from other debris in the sample.
3. Using a heavy liquid set at 1.9 specific gravity, float off small charcoal fragments. Repeat flotation using a liquid set at 1.65 specific gravity. Pollen grains are lighter than 1.65 and will sink and remain in the sample.
4. Place samples in centrifuge tubes and add a 1.15 specific gravity solution, centrifuge, and decant the supernatant. The remaining light charcoal will float, while the heavier pollen grains sink. Repeat several times.

To illustrate how the extraction procedures outlined above are combined into a complete processing procedure for pollen extraction, I present two procedures currently in use. The first, from Dean (1988), is a procedure designed for arid sediments from the American Southwest. The second, from Shackley (1981:75–76), incorporates elements from Jones and Cundhill (1978). Before using these or any other procedures, the beginner should consult an experienced palynologist, should be familiar with basic chemistry laboratory techniques (mixing solutions, decanting, application of heat), and should understand completely what each step in the procedure accomplishes and how it affects pollen. All extractions are carried out under a fume hood.

Dean (1988) Procedure

1. Screen each sample through a tea strainer (mesh openings of about 2 mm) into a beaker to a total screened weight of 25 g. Sediments should be dry. Add three tablets of pressed *Lycopodium* (clubmoss) spores to each sample. Pay attention to the batch number; numbers of spores per tablet vary from batch to batch and are specified in the literature sent with the tablets.
2. Add concentrated HCl (38%) to remove carbonates, cover, and allow the samples to sit overnight.
3. Add distilled water to the samples, and wash out the acid and dissolved carbonates by repeated centrifugation at 2000 RPM in tapered 50-ml tubes. Transfer the concentrated residues back into numbered beakers and add more distilled water. Swirl the water–sediment mixture, allow to sit 10 seconds, and decant the fine fraction off of the settled heavy residue through a 195 micron mesh over another beaker. This process, essentially similar to bulk soil flotation, differentially floats off light materials, including pollen grains, from heavier, nonpalynological matter. Concentrate the fine "floated" fractions by centrifugation at 2000 RPM; discard the heavy fraction remaining in the beaker after checking a drop under the microscope for any pollen that may have settled out..
4. Transfer the fine fractions back into numbered plastic beakers and mix with 49% HFl to remove smaller silicates. Stir occasionally, cover, and allow to sit overnight.
5. Add distilled water to dilute the acid–residue mixture, and transfer it to the 50-ml centrifuge tubes again. Repeat centrifugation and washing of the compacted residue with distilled water at least three times as above to remove acid and dissolved siliceous compounds.
6. Mix trisodium phosphate (5% solution), a wetting agent, with the residue and centrifuge. This mixture is very basic and potentially harmful to degraded pollen; expose the sample to this risk as briefly as possible. Repeat the centrifuge-assisted rinses with distilled water to remove fine charcoal and small organic matter (minimum of five rinses). Wash resi-

due with glacial acetic acid to remove remaining water in preparation for acetolysis.

7. Add freshly mixed acetolysis mixture (9 parts acetic acid anhydride to 1 part concentrated sulfuric acid) to the residues in the plastic centrifuge tubes to destroy small organic particles. Heat the tubes in a boiling water bath for 5 minutes, followed by cooling in another water bath for about 5 minutes. Centrifuge to compact the residues and decant the acetolysis mixture. Mix residues with glacial acetic acid, centrifuge and decant, and then wash at least three times with distilled water to remove remaining traces of acid and dissolved organic compounds. Total exposure of the residue to acetolysis mixture should be about 15 minutes.

8. Wash samples in warm dilute methanol, made by adding about 100 ml methanol to 400 ml of distilled water in a wash bottle (the reaction is exothermic). Float off small silicates, organic material, and charcoal from the palyniferous residues by centrifugation at 2000 RPM for periods varying from 60 to 90 seconds (decant carefully). Stain remaining residues with safranin O mixed with liquid glycerol and store in 3-dram stoppered vials with a small amount of glycerol. Use one or two drops of the glycerol-mounted residue to prepare microscope slides in which the grains can be turned over during observation to facilitate identification.

Shackley (1981) Procedure

Initial treatment.
1. Examine the sample and remove pebbles or plant parts by wet sieving.
2. Concentrate the sample by centrifuging at 3000 RPM for 5 minutes for 50-ml tubes or 2 minutes for 15-ml tubes. (Note: rinsing and concentration by centrifuging is carried out between each of the following stages of processing.)

Removal of carbonates and alkali-soluable humic compounds.
1. Place about 0.5 cm^3 (2 g) of sample in a 50-ml polypropylene boiling tube. Add a little 10% HCl, then fill the tube to $\frac{2}{3}$ full, unless a violent reaction occurs.
2. When the reaction is complete, and effervescence stops, centrifuge and decant.
3. Add a few drops of 10% NaOH to the residue, mix, add a further 20 ml of NaOH, and place in a boiling water bath for 20 to 60 minutes.
4. Stir well to break any remaining lumps or use a sample mixer.
5. Filter the sample through a sieve into a polypropylene centrifuge tube.

Wash the sieve residue with distilled water and add this water to the centrifuge tube.

6. Centrifuge and decant. If the supernatant liquid is still dark, humic material is still present; the sample must be washed, centrifuged, and decanted until the liquid is clear.
7. Wash the residue on the sieve toward the center using a water jet. Invert the sieve onto a petri dish, washing out the debris.

Removal of silica by digestion with hydrofluoric acid.

1. Add a few drops of deionized water to the sample, which should be in a polypropylene centrifuge tube, and mix thoroughly.
2. Add 20 ml of 40–60% HF to the sample and place it in a boiling water bath for 20 to 60 minutes. Stir with a polypropylene rod to determine when siliceous material is no longer present.
3. Centrifuge and decant; make sure that water is running in the fume hood sink while decanting HF.
4. Add a few drops of 10% HCl and mix. Add an additional 20 ml of 10% HCl and place the sample in a boiling water bath for 15 minutes. This step removes colloidal silicon dioxide and silicofluorides.
5. Centrifuge and decant.
6. Rinse the sample in water and a few drops of 10% NaOH. Centrifuge and decant.
7. Wash, centrifuge, and decant.

Acetolysis.

1. Add 10 ml of glacial acetic acid to the sample residue, which should be in a polypropylene test tube. Mix, centrifuge, and decant in a fume hood sink with running water. Repeat.
2. Prepare an acetolysis mixture (1 ml concentrated H_2SO_4, 9 ml acetic anhydride).
3. Add a few drops of the acetolysis mixture to the sample, mix, then add 20 ml of mixture. Put in a boiling water bath for 3 minutes, stirring carefully. Do not leave the stirring rods in the tubes or let water into them, as this will result in a violent reaction.
4. Centrifuge and decant the supernatant into a large beaker of water in the fume hood.
5. Add glacial acetic acid to the sample, mix, centrifuge, and decant.
6. Wash the sample with water containing a few drops of 10% NaOH. Mix, centrifuge, and decant. Repeat using water only.

As indicated in this procedure, it is advisable to decant all noxious liquids into a fume hood acid-resistant drain, with water running continuously.

Processing Coprolites

As mentioned earlier, archaeologists have found coprolites a valuable source of information on human diet (Bryant 1974; Bryant and Holloway 1983). Not only are remains of vegetable foods (fragmented or whole seeds, fiber, fruit skins) often present in human coprolites but also fragments of bone, hair, or feathers which reflect the meat component of meals. Pollen also survives the digestive process and may be preserved in coprolites if soil conditions are favorable.

Although coprolites are sometimes analyzed for either macroremains or pollen, processing remains to recover both types of data increases their interpretive potential. Phytoliths have also been observed in coprolites and are another source of data on diet. Phytolith extraction procedures are described in Chapter 5; be sure to reserve a portion of specimen for phytolith extraction, since silica is destroyed during processing for pollen.

The following procedure for processing coprolites for macroremains and pollen is summarized from Bryant (1974a):

1. Measure, weigh, photograph, and describe the general appearance of each specimen. Remove material obviously of nonhuman origin: rodent pellets, herbivore droppings (identifiable by shape, size, and high grass and fiber content), and carnivore feces (identified by a hard outer coating of dried intestinal lubricant).

2. Clean the surface of each specimen thoroughly to remove pollen deposited after the coprolite was excreted.

3. Place each specimen in an airtight container; divide it into portions for analysis. For small specimens, the entire specimen is used for both pollen and macroremain analysis. For larger specimens, a portion can be reserved for each analysis or for replicate studies.

4. Once a portion is selected for analysis, place the specimen in an airtight container and add a 0.5% solution of trisodium phosphate to cover the specimen. Seal. Do not use a stronger solution; destruction of plant tissues may result. Soak specimens a minimum of 72 hours, or until softened (several weeks may be required).

5. Working under a fume hood (to vent odors), open each container and note the color and smell of the specimen and whether a thin scum appears on the surface of the solution. Color and odor are indicators of origin; carnivore coprolites usually turn trisodium phosphate white, pale brown, or yellow brown and have a musty odor; herbivore coprolites turn the solution pale yellow to light brown and are musty; human coprolites turn the solution dark brown or black, change it from translucent to opaque, and

have the original intense odor. Presence of surface scum indicates that meat was part of the diet.

6. Wash each specimen with distilled water through a 20 mesh geological sieve, catching and reserving the water. Break up debris on the screen to liberate pollen grains. Set solid residue from the sieve aside.

7. Pass liquid and small solid remains through a 100 mesh geological sieve, reserving the water. Break up debris on the screen to liberate pollen grains. Set the solid sieve residue aside.

8. Dry both size fractions of solid residue and store in closed containers for analysis of macroremains (these are sorted and identified as described in Chapter 3 for analysis of flotation samples).

9. Transfer liquid to centrifuge tubes; centrifuge and decant, combining residues until all material from the specimen is in one tube. Reserve for pollen analysis: hydrofluoric acid to remove silicates, acetolysis to remove organic material, alcohol washes to dehydrate and stain, benzene wash, and storage in silicon oil (Williams-Dean and Bryant 1975).

Processing Floral Specimens

Identifying fossil pollen grains after soil processing is perhaps the most difficult aspect of pollen analysis for the beginner. One way to learn to recognize pollen types is by beginning a "pollen herbarium." Study of comparative plant specimens processed to render modern pollen similar in appearance to fossil pollen (i.e., lacking cell contents, oils, and the like), is useful for learning types; photographs can be difficult to use for identification.

The basic method used to prepare modern specimens is the acetolysis technique, introduced and popularized by Erdtman (1960; Traverse 1965). The acetolysis procedure destroys floral material containing pollen and all organic components of the grain except for the resistant exine. The exine is changed in some respects, however, and grains tend to swell. For this reason, some pollen analysts prefer alkali maceration over acetolysis. This procedure involves boiling floral material in 5–10% KOH (see Traverse 1965 for details of this procedure). Although alkali maceration is gentler on delicate pollen grains, acetolysis is still the more common procedure. As Traverse notes, pollen that cannot survive acetolysis is unlikely to survive deposition in sediment or chemical extraction.

Processing floral specimens for comparative samples requires no additional equipment beyond that used in soil processing. It is important, however, that archaeological samples not be contaminated with comparative pollen material. Different glassware should be used for comparative and fossil samples. Centrifuge, fume hood, and work areas should be thoroughly cleaned before switching from one processing procedure to the other. And while floral processing is going on, soil samples should be tightly closed and glassware and other materials used for soil processing removed from the processing area. For details on acetolysis, see Traverse (1965) and Faegri and Iversen (1975:117–122).

Mounting Slides

Slide mounting is an important part of the process of pollen analysis, since all data for subsequent interpretation are derived from counts of mounted samples. A few commonsense rules can help the beginner make usable, consistent mounts:

1. Use the same mounting medium for all slides to be compared during an analysis—soil samples as well as comparative samples. Mounting media vary in refractive index, reaction with stains, impact on grain size, and longevity. Using one mounting medium throughout minimizes extraneous variability in samples.
2. Mount a consistent amount of material per slide. Do not change technique in the middle of an analysis.
3. Include "marker" pollen or spores in samples; this facilitates absolute pollen counting (see next section).
4. Staining is a matter of personal preference, but it can help bring out structural details and enhance recognition of broken grains. These are advantages to the beginner, especially when scanning archaeological samples where grains may be battered and scarce. To be most effective, stain should give the maximum possible contrast with shortwave light.

Shackley (1981:76–77) summarizes procedures for mounting samples in glycerol (temporary mounts), and silicone oil (permanent mounts). These are probably the most common fluid mounting media in use. Andersen (1965) gives a detailed review of mounting media, discussing the advantages of media of different viscosities. Fluid media, including glycerol and silicone oil, allow rotation of grains, but they must be carefully sealed and handled so that grains do not move and skew counts. Solid media permit more easy handling of slides but do not allow rotation to view grains. Although careful handling is needed, fluid mounts are preferred. Shackley's procedures are as follows:

Temporary mounts.
1. Add a few drops of distilled water to pollen extract (still in the centrifuge tube following extraction), then add one or two drops of safranin or fuchsin stain (if staining is desired). Mix, centrifuge, and decant.
2. Invert the tube over filter paper and carefully allow any remaining water to drain out.

3. Add 3–6 drops of glycerol to the pollen extract and mix thoroughly. The precise amount needed depends on the concentration of pollen present and the amount of residue. Start with the lesser amount.
4. Using a small spatula, transfer a small drop of material to a clean glass slide. Cover with a cover slip. Label with sample number and other provenience information. Prepare three or four slides of each sample.
5. Seal edges of the cover slips with clear nailpolish.
6. Transfer any remaining extract to a small glass vial, close tightly, and label with number and provenience information.

Permanent mounts.
1. Add distilled water to pollen extract, still in the centifuge tube, centrifuge, and decant. Repeat.
2. Add 1 ml of distilled water and 5 ml of 100% ethanol. Centifuge and decant. Add one or two drops of safranin (if staining is desired).
3. Add 100% ethanol, mix, centifuge, and decant.
4. Add 1 ml of toluene and pour the mixture from the centifuge tube into a labeled glass vial. Close tightly. This step should be carried out under the fume hood.
5. Carefully lower the vials into the centrifuge and centrifuge at 750 RPM (no faster) for 10 minutes.
6. Decant the toluene into a beaker under the fume hood.
7. Add 2–6 drops of silicone fluid (200/200 cSt viscosity) and stir well (otherwise grains will clump together).
8. Allow excess toluene to evaporate for 24 hours under the fume hood.
9. Prepare slides, using one small drop of material per slide. Cover slips should be fixed with a dab of nailpolish at each corner. There is no need to completely seal the mounts.

Dimbleby (1985:155) recommends using a calibrated dropper (such as a hypodermic syringe or disposable pipette) for placing one drop of extract on the slide. He also recommends making up a uniform volume of sample (2 to 5 ml) with a 1:1 glycerol–water mixture rather than just adding a few drops to samples.

Counting and Identifying Pollen

Slide scanning, the process of counting and classifying pollen grains and spores, is the final laboratory step in pollen analysis. It is a critical step, since validity of final results depends not only on correct identifications but on counting procedures that yield statistically significant samples of pollen present in sediments. Understanding how pollen counts are made will help the general archaeologist utilize these valuable data critically.

There are two types of counts in common use in palynology: counts that yield relative or percentage occurrence data and counts that yield absolute frequency data. Analysts disagree over which method is preferable, but relative counts are much more commonly used.

In a relative pollen count, occurrence of each type is expressed as a proportion of some specified pollen sum. Although common sense might dictate that a pollen sum should consist of all pollen types encountered in a sample, this is not always the case. In the early days of pollen analysis, interest focused on arboreal pollen, since reconstruction of vegetation in northern Europe relied heavily on what tree taxa were present. It was not uncommon for the pollen sum to consist only of arboreal pollen (AP). Occurrence of each AP taxon was expressed as a proportion of total AP. Counts of 150 to 200 AP grains were commonly used (Dimbleby 1985:26–29; Shackley 1981:77–83). Nonarboral pollen (NAP) encountered during counting was sometimes also tallied and its occurrence expressed as a proportion of the AP sum.

A common approach today is to count all pollen types encountered and use total pollen (TP) as the basis for the pollen sum. This is done either by counting all pollen types as encountered, until a predetermined total is reached, or by counting all grains until a certain number of arboral types are encountered. The latter technique gives a variable total pollen sum, ideally over 500 total grains. Birks and Gordon (1985) suggest that 300–500 grains is usually an adequate pollen sum.

The pollen sum should be a good approximation of abundance of pollen types of greatest interest to the analyst. Moore and Webb (1978) recommend tailoring the sum to fit the goals of research by making a series of counts, calculating relative occurrence data, and then assessing the results. Some pollen types may be clearly overrepresented (e.g., local water-edge plants). These can then be eliminated from the sum, permitting higher counts of rarer, more ecologically significant taxa. If local vegetation is of greatest interest, types transported long distances can be eliminated, or vice versa.

How does one determine how many pollen grains should be included in the sum? Number needed for a reliable count depends, in part, on diversity of types present (Moore and Webb 1978:78–83). The more types present, and the more uneven their distribution (i.e., some types are common, other types very rare), the higher counts must be to give a good approximation of relative abundance of all types on the slide. Dimbleby (1985:26–29) gives as a rule of thumb that for twenty taxa counts of 150 to 200 grains are adequate, while for more complex assemblages (e.g., one hundred or more taxa), counts of 1000 or more may be necessary. Pollen counts of 200 grains are common. The "200-grain count" is derived from the work of Barkley (1934), Dimbleby (1957:13–15), and Martin (1963:30–31) and gives about 75–85% accuracy for common taxa (G. Dean, personal communication, from examination of Dimbleby's data on some twenty taxa). Counting more than 200 grains allows rarer taxa to be represented more accurately, but at a cost of increased scanning time. If very rare taxa are of interest, counts may have to be even higher. Eliminating some taxa from the sum can simplify the assemblage, allowing lower counts.

What can one do if, because of poor preservation, few grains are present on a slide? One solution is to scan several slides for each sediment sample and to combine counts. If a low count is being used (e.g., 200 grains), scanning two or three slides is not overly time consuming. However, very low pollen concentrations may not accurately represent original deposition.

The major drawback of relative counting is that, when one pollen type increases in occurrence, one or more types must decrease, because all proportions must total 1. There is no way of knowing which taxa are actually changing and which are merely responding to change. For example, localized drought conditions might cause a decline in local plant growth, reducing deposition of those pollen types; this would lead to higher proportions in the sample of pollen transported from long distance, such as pine, giving the false impression that pine forests were expanding. Similarly, a dramatic change in pollen contribution can be masked if the type in question is already a heavy contributor to the record. For example, suppose that slide 1 has 90 grains of A and 10 grains of B, giving relative counts of 90% A and 10% B (TP = 100), and that slide 2 has 180 grains of A and 10 grains of B, giving counts of 95% A and 5% B (TP = 190). Type A has doubled its contribution to pollen rain in slide 2, but this is masked in percentage occurrence calculations, which change only from 90 to 95%. In this case, the absolute count gives a better indication of magnitude of change.

The technique of calculating absolute pollen frequencies was developed as an alternative to relative or percentage occurrence calculations. Absolute pollen frequency estimates the annual deposition of each pollen type independently of all other types (pollen concentration) and independently of changing rates of sedimentation (pollen accumulation) (Birks and Gordon 1985:10–18). Pollen concentration is the number of grains deposited per unit volume or unit weight of soil. Pollen accumulation (or pollen deposition rate, or pollen influx) is the number of grains deposited per unit area of sediment surface per unit of time.

There are several methods of arriving at absolute pollen frequencies. Each is a twofold process: determining absolute density of grains in a sample of known volume or weight, and determining how the amount of pollen present varies with depth (i.e., the sedimentation rate). Arriving at absolute grain density figures can be achieved using volumetric, weighing, or exotic marker methods. The most common, the exotic marker method, is described here. A known concentration of pollen-size spheres (polystyrene or plastic), exotic pollen, or spores is added to a sample.

Some commonly used exotics are *Nyssa, Ailanthus,* and *Eucalyptus* pollen or *Lycopodium* spores. The exotic type should be one that is unlikely to occur naturally in samples and should be of average size, easy to recognize, and nonclumping in the mounting medium. The exotic marker can be added either before or after processing. Final pollen extract (or unprocessed sample) is first weighed, and then a weighed quantity of exotic marker material in suspension is added. Ideally, the exotic should constitute 20–40% of the total pollen in the sample. After mounting, the entire slide does not have to be scanned, since the concentration of each pollen type encountered can be calculated relative to occurrence of the exotic, as follows:

$$\frac{E_n}{E_t} = \frac{X_n}{X_t} \text{, so}$$

$$X_t = \frac{E_t X_n}{E_n}$$

where E_n is the number of exotic grains counted in one scan, E_t is the total number of exotics in the sample, X_n is the number of grains of one archaeological pollen type counted in the scan, and X_t is the calculated number of archaeological type grains in the sample. For example, if 30,000 exotic *Lycopodium* spores are added to a sample, and 40 are counted during a scan in which 10 grass grains are also counted, then

$$\text{Grass}_t = \frac{(30,000)(10)}{40} = 7500 \text{ grains}$$

If, in the same scan, 30 chenopod-amaranth grains were also counted, then chenopod-amaranth$_t$ = 22,500 per unit of soil. To determine the total number of fossil grains in a sample, simply multiply the number of fossil grains of all types in one scan by the ratio E_t/E_n.

Once pollen concentration figures for each sample are determined by the exotic marker method, deposition rates must be calculated. Without deposition rates, variation in samples resulting from change in vegetation cannot be distinguished from that due to change in sedimentation rate. To estimate sedimentation rates, a number of strata along the sequence must be radiocarbon dated. Length of time of deposition for each stratum can then be extrapolated.

Birks and Gordon argue that calculating absolute pollen frequencies introduces larger errors to an analysis than those arising from use of relative counts. There are many possibilities for error in estimating pollen concentrations, for example. Among a few obvious ones are losing extract during processing, uneven distribution of grains in liquid suspension, and differential destruction of exotics and naturally occurring pollen during processing. Error can also be introduced in estimating sedimentation rates. If few radiocarbon dates are available, accurate extrapolation may be difficult. As archaeologists are well aware, there is always error associated with the dates themselves.

Dimbleby, by contrast, argues strongly for use of absolute pollen frequencies, particularly in soil (as opposed to sediment) pollen analysis. In conventional analysis of sediments (i.e., lake and peat deposits), strata are steadily accumulating, and all pollen in a stratum can be assumed to be of approximately the same age. In soils, however, there is continuous downward movement of water which brings about downward movement of pollen. At any given depth, pollen of various ages may be found (see Fig. 4.6). If only relative counts are calculated, this distribution is obscured, whereas it is revealed by calculating absolute frequencies. Dimbleby advocates calculating both types of occurrence measurements.

Another reason to calculate the total number of grains present in a unit of soil is to determine if the sample meets the "1000-grain per gram" rule (Hall 1981). Encountering fewer than 1000 grains in a gram of soil is an indication that pollen degradation was high or that deposition was restricted in some way.

Although I have so far only discussed counting techniques, counting does in fact take place hand-in-hand with identification of grains. Earlier in this chapter, I described some characteristics of grains useful in identification. Among these are grain shape and size, structure of the exine, sculpturing patterns, and aperture number, shape, and patterning. The beginner will have difficulty using the precise terminology developed for these characteristics or in following a key such as Kapp (1969); identifying pollen comes with practice and guidance from published sources and an experienced researcher.

To begin the process of identification, scan a number of slides quickly to get an idea of the variation in pollen types present. No formal counts need be done at this preliminary stage, although a rough estimate of pollen abundance can be made. Make sketches of entire and fragmented pollen types. Although photographs are very useful for confirming identifications and for including in publications, hand sketching types helps one learn to recognize important grain characters.

This preliminary assessment of pollen in samples may lead to quick identification of especially distinctive pollen or to determination of the major group to which grains belong (see, e.g., Kapp 1969:21–27, picture key to major spore and pollen groups). Within these major groups it may be necessary to key types, however.

Learning to key pollen grains, like learning to use any kind of dichotomous botanical key, requires familiarization with terminology. Long lists of definitions of terms can be overwhelming; an easy way to start is by keying out a diverse group of known pollen types. Since an important part of the whole process of pollen analysis is preparation of comparative collections, use those specimens for keying practice. It is also important that the key used contains all the common pollen types likely to occur in a sample. Check this by keying out known floral dominants and important economic taxa.

For archaeological grains, try keying unknown types. Identifications arrived at through keying should be considered only preliminary. An archaeological type should be carefully compared to comparative materials and published photographs of the suggested type. Take care not to confuse taxa that produce very similar pollen, or to propose too precise an identification for damaged grains. Unidentifiable grains should be described and given letter or number designations as unknowns.

After the test-sample pollen has been identified, types to include in the pollen sum are chosen and the systematic count is begun.

Presenting and Interpreting Results

The discussions which follow have a dual purpose: to help the nonspecialist read traditional sedimentary pollen diagrams and understand how stratigraphic pollen data are interpreted, and to show the novice archaeological palynologist and nonspecialist how archaeological pollen data may be presented and interpreted. As discussed earlier, many paleoethnobotanists and archaeologists will have occasion to use results of analyses of lake or bog cores even if they never do such work themselves. Complex diagrams can be difficult to read and evaluate. I review the conventions used in such diagrams and discuss three important levels of interpretation of these data: zonation, vegetation reconstruction, and climatic reconstruction. Simplified diagrams and tabular presentation of archaeological pollen data are discussed, and two examples of interpreting these data, for reconstructing subsistence and for documenting ecological change, are considered. I conclude this chapter by discussing briefly current directions and issues in archaeological palynology.

Presenting Data

Traditional Pollen Diagrams

The construction of diagrams to present stratigraphic pollen data is described in most palynology texts. The following description is taken largely from Moore and Webb (1978:83–84) and Faegri and Iversen (1975:128–139). Figure 4.7, a stratigraphic diagram of sediments from Lake Junín, Peru, illustrates most of the following points about traditional diagrams:

   1. The vertical axis of the diagram depicts depth; the horizontal axis depicts relative (and/or absolute) pollen abundance. Figure 4.7 illustrates relative abundance; pollen sums are shown in a column in the right-hand

area of the diagram, setting off pollen counted outside the sum (cryptogams and aquatics).

2. The stratigraphic sequence is shown in a column on the left vertical axis. Sediment types may be depicted with symbols or described. Horizontal lines may be used to separate sediment types to increase clarity in complex diagrams. Radiocarbon dates, accumulation rates, or pollen concentration data may also be shown on the vertical axis. In Figure 4.7, pollen concentration is shown on the right axis, dates on the left axis.

3. For each deposit sampled, abundance of pollen types is depicted by a bar histogram or by a point on a continuous curve (as in Fig. 4.7). Major types or types of particular interest are given individual curves. Curves may be superimposed using standardized symbols for major taxa (see Faegri and Iversen 1975:129, for pollen and spore symbols for northern Europe), but this reduces clarity in a complex diagram. The pollen sum for each sample should be included on the diagram. Ruled vertical lines between pollen types may enhance clarity.

4. The scale used for relative or absolute frequency data should be consistent throughout the diagram and clearly indicated. In Figure 4.7, the scale is marked for the first curve, Cyperaceae (bottom of the diagram), and is then repeated under each subsequent curve. If it is necessary to use an exaggerated scale for minor types, this should be clearly marked.

5. The conventional order for types along the top horizonal axis is arboreal taxa, shrubs, herbs, and spores. Alternative orderings include ecological groupings (e.g., puna, sub-puna shrubland, Andean forest, cryptogams, aquatics, as in Fig. 4.7) or distance groupings (local, regional, long distance).

6. Summary or composite curves indicating how the sum is divided into arboreal, shrub, and herbaceous components, or into other subunits, are often included on stratigraphic diagrams. In Figure 4.7, four composite curves are depicted on the left axis: Gramineae, other puna, sub-puna, and Andean forest types. These are the components of the pollen sum (and sum to 100%).

Diagrams vary considerably in complexity, depending on the nature of the stratigraphic sequence, the complexity of the pollen assemblage, and what the analyst is trying to illustrate. Figure 4.7, for example, is a fairly complex diagram with two sections. In the right three-quarters of the diagram, 51 pollen types are individually depicted. This is a resolved diagram, one that presents a separate curve or histogram for each taxon. Such diagrams can be very large if a pollen assemblage is complex. The left quarter of Figure 4.7 is a composite diagram that helps summarize data in

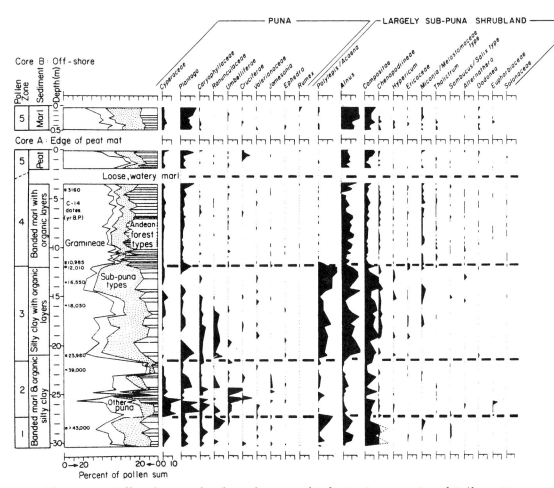

Figure 4.7 Pollen diagram for the sediments of Lake Junín, cores A and B (from Hansen *et al* 1984:1460–1461). (Figure continues.)

the complex assemblage. If pollen assemblages are fairly simple, a composite diagram may not be needed. Note that pollen zones, shown on the far left vertical axis of Figure 4.7, have been defined. I return to definition of pollen zones below.

Although resolved and composite diagrams are the most common form of stratigraphic data presentation, other types of diagrams are sometimes used to focus on particular aspects of data or to present summary or regional data. In a survey diagram, for example, pollen types are grouped by ecological or phytogeographic categories to illustrate paleoecological or plant geographic changes. If one is interested in spacial variation in plant occurrence over a region, circular diagrams or

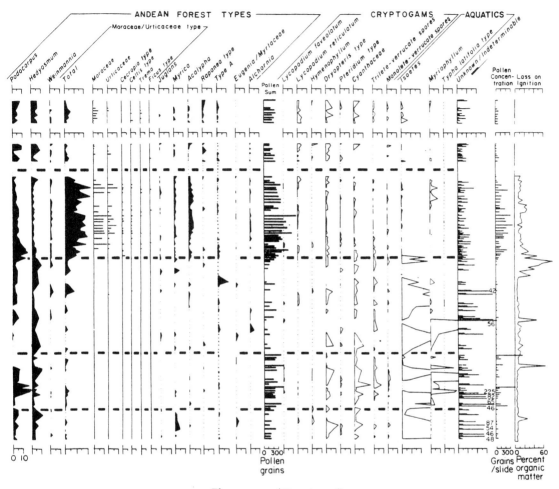

Figure 4.7 (Continued).

resolved maps can be drawn for a given period. In these types of diagrams, a map forms the basis for the diagram. In a resolved map, abundance of one species during the time interval of interest is indicated by sizes of circles or other symbols (see Faegri and Iversen 1975:140–143). A circular diagram is similar, except that for each location relative abundances of taxa are depicted by sections in a pie chart. Again, the map presents data for one time interval, focusing on the geographic dimension.

Another approach to presenting summary stratigraphic data in regional perspective is construction of isopollen maps. Abundance values for a pollen taxon are depicted as contour lines on a map. A series of maps can be used to illustrate changes over time. Isopollen maps are useful for summarizing large amounts of

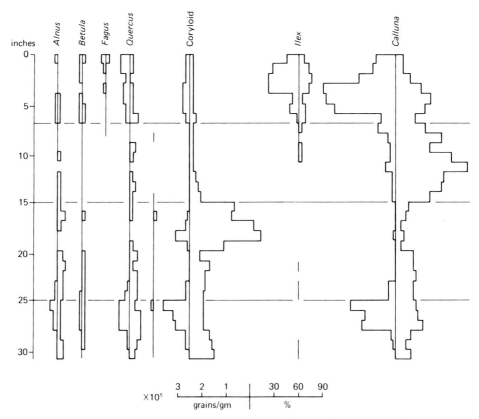

Figure 4.8 Pollen diagram of a section through a small round barrow at Burley, New Forest, Hampshire (from Dimbleby 1985:82–83).

stratigraphic data, and for depicting correlations of key pollen types to topography or soils, but they must be carefully contructed using only well-dated sequences.

Archaeological Pollen Data

Figure 4.8 is an example of a simplified diagram style used by Dimbleby (1985) for presenting results of stratigraphic archaeological soil analysis, in this case a small earthwork. Both relative and absolute (grains/ gram) abundances are depicted by using double histograms (relative abundance to the right, absolute to the left). This is a simple resolved diagram with 13 pollen types. Note that the absolute count data are given as pollen concentration rather than pollen accumulation (i.e., grains/gram, not grains/area/unit of time). Pollen sums have been omitted, which weakens the presentation.

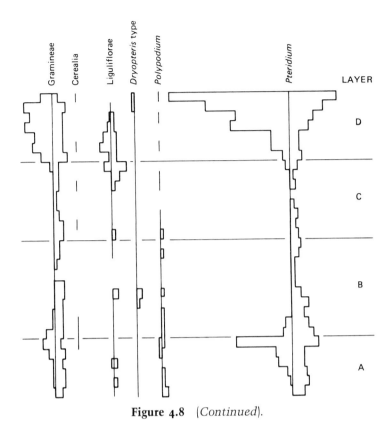

Figure 4.8 (*Continued*).

Figure 4.9, another example of a stratigraphic archaeological soil diagram, presents pollen data from Guilá Naquitz cave, Oaxaca (Schoenwetter and Smith 1986: 203). This is a simple resolved diagram with 27 pollen types depicted, including unknowns and cultivated types. The lower portion of the diagram shows individual samples; the upper section sums these by archaeological zone. This illustration would be improved by the addition of depths and radiocarbon dates.

It is rarely possible to publish primary pollen data (i.e., raw counts of all types for all samples analyzed) and sometimes impossible to include complete relative or absolute frequency data on diagrams. However, it is important that enough data be included on a diagram (or in tabular form) to illustrate the basis for major conclusions and to allow the reader to judge validity of interpretations. By the same token, it is generally undesirable to present data only as smoothed curves, since the original structure of the data is lost. Presenting smoothed curves in addition to primary relative or absolute frequency data can help illustrate trends, however.

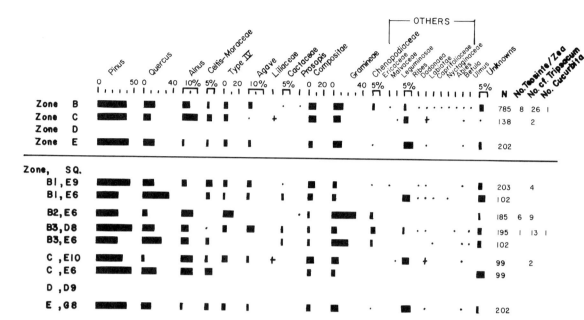

Figure 4.9 Pollen spectra from Guilá Naquitz: upper, population spectra of the stratigraphic zones; lower, pollen spectra of individual samples (from Schoenwetter and Smith 1986:203).

Results of analyses of surface samples or nonstratified archaeological samples can also be presented in a simple diagram. If surface samples are collected from a transect through different vegetation zones, results can be presented on a transect diagram, which is merely a resolved diagram where distance along the transect, rather than depth, is depicted on the vertical scale. Pollen taxa (horizontal scale) are generally grouped by vegetation type. Such diagrams can also be used for depicting results of discontinuous surface sampling. In such a case, the vertical scale does not indicate distance but is merely a sequence of samples representing discrete vegetation types or sampling locations. Figure 4.10, which depicts pollen assemblages from upland valley plots in the Oaxaca area, is an example of this type. Relative abundance is depicted by histograms. Note that each vegetation zone is represented by several sampling loci, and that cultivated and uncultivated areas are separated; this reflects how Schoenwetter and Smith will use the data to help interpret the archaeological sequence depicted in Figure 4.9.

Diagrams can also be used to present data from discrete archaeological samples, such as a series of pits, floor samples, or coprolites. It is also easy to present such data clearly in tabular form, however. Table 4.1, for example, shows the results of pollen analysis of coprolites from Upper Salts Cave, Kentucky (Schoenwetter 1974). In this presentation, pollen types are listed on the vertical axis, sample numbers on the horizontal. Specimen proveniences are given in the text, rather than on the

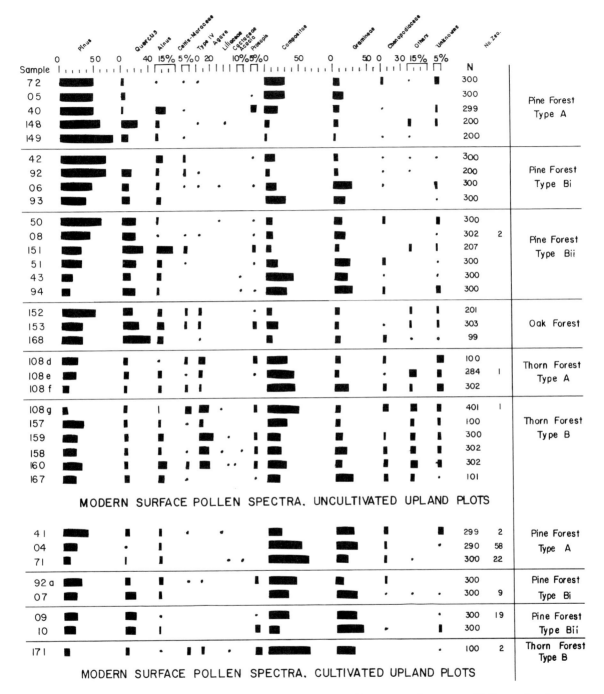

Figure 4.10 Spectra (cultigens excluded) of modern pollen rains of Oaxacan Valley upland plots (from Schoenwetter and Smith 1986:188).

Table 4.1 • Grain Counts of Pollen Observed in Coprolites from Upper Salts Cave

Pollen type	\multicolumn{8}{c}{Fecal specimen number}							
	1	2	3	4	6	7	8	9
Wind-pollinated (anemogamous)								
Pinus (pine)	1						1	
Quercus (oak)	6	8		1			12	2
Gramineae (grass)	5	1		5	2	2	14	24
Chenopodiineae (chenopod–amaranth group)	4	4		95	130	28	8	2
Ambrosieae (sumpweed group)	5		2	86	66	165	6	11
Insect-pollinated (zoogamous)								
Other Compositae (e.g., sunflower)	4		2	13	2	5	2	159
Opuntia (cactus)	1							
Caryophyllaceae								1
Polygonum—tricolporate type	2							
Polygonum—periporate type	7							
Rosaceae				1				
cf. Liliaceae				8			157	
cf. Iridaceae—Amaryllidaceae			187					
cf. *Acorus* (sweet flag)	164	187						
Unknowns	1							1
	200	200	200	200	200	200	200	200

Source: Schoenwetter (1974:52).

diagram, which makes it less complete. Note that this table presents raw data (counts) rather than percentages.

Interpreting Sedimentary Data

To help the nonspecialist evaluate results of sedimentary pollen analyses, such as the core data presented in Figure 4.7, I next discuss three important levels of interpreting such data: zonation, vegetation reconstruction, and climatic reconstruction. Delimiting pollen zones in a stratigraphic pollen sequence is one of the first steps in interpreting any stratigraphic pollen data set (Fig. 4.11). Pollen deposited in a lake or bog is a direct function of vegetation. Interpreting fossil pollen assemblages in terms of past vegetation is a primary goal of sedimentary pollen analysis. The ultimate goal of pollen analysis may be much broader than identifying past vegetation, however. Reconstructing past climate is a topic of considerable interest in archaeology which can be approached through interpretation of sedimentary pollen data.

Delimiting Pollen Zones

Figure 4.7 is subdivided on the far left on the horizontal dimension into pollen zones. These zones are not determined by stratigraphy or radiocarbon dates but by the pollen data. Because it is difficult to interpret a long stratigraphic sequence,

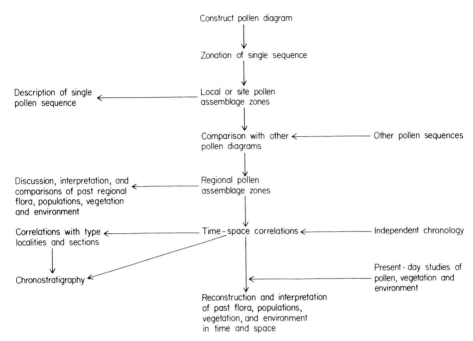

Figure 4.11 Stages of paleoecological reconstruction and interpretation (from Birks and Gordon 1985:49).

simplifying that sequence by dividing it into smaller units, pollen zones, is a necessary first step. Most analysts follow E. J. Cushing's (cited in Birks and Gordon 1985:48) concept of a pollen zone as "a body of sediment distinguished from adjacent sediment bodies by differences in kind and amount of its contained fossil pollen grains and spores, which were derived from plants existing at the time of deposition of the sediment." The key is that a pollen zone is a biozone, or assemblage zone, defined solely on the basis of the contained fossil pollen and spores. Pollen zones should not be viewed as time periods or climatic periods but as local schemes based on pollen assemblages (Moore and Webb 1978:89–96). Zonation by cultural levels, such as illustrated in Figure 4.9, should be clearly distinguished from pollen assemblage zonation (biozones).

As Figure 4.11 suggests, delimiting appropriate pollen zones in a stratigraphic analysis affects subsequent data interpretation. It is therefore essential that decisions about dividing up a sequence into zones be made carefully. There are two approaches to this process: the "subjective" approach, and the "objective," or numerical, approach (Moore and Webb 1978:89–96).

In the subjective approach, zone lines are placed in the sequence where the analyst sees several concurrent changes in pollen type. Hansen et al.'s (1984) study of cores from Lake Junín relied on zonation by nonnumerical methods. In Figure 4.7, five pollen zones are depicted: zone 5, characterized by high occurrence of puna taxa; zone 4 by high, steady values of Gramineae and high proportion of Andean forest taxa; zone 3 by virtual absence of forest taxa and high percentage of Compositae and *Polylepis–Acaena* (shrubland); zone 2 by low forest and shrubland occurrence and high occurrence of puna types; and zone 1, which resembles zone 3. I discuss this study further on pp. 298–299.

It is still very common for analysts to use the "subjective" approach for delimiting pollen zones. Birks and Gordon (1985:47–90), in their detailed discussion of numerical methods of delimiting zones, emphasize that the numerical approach should not really be considered more objective than the nonnumerical approach. Assumptions and decisions made in the process of applying numerical methods can greatly affect outcomes.

The reader interested in numerical approaches in palynology should consult Birks and Gordon's *Numerical Methods in Quaternary Pollen Analysis* (1985). Four kinds of approaches are used for numerically delimiting pollen zones: (1) comparison of pairs of neighboring levels to detect points where assemblages change, (2) application of standard numerical classification procedures without consideration of stratigraphic position, (3) application of numerical classification with stratigraphic constraints, and (4) use of standard scaling procedures with no stratigraphic input in order to check zonations suggested by other numerical procedures. Birks and Gordon feel that numerical methods allow the analyst to deal with masses of data more effectively, to apply the same criteria for delimiting zones in every analysis, avoiding bias from prior results, and to formulate precise criteria for comparison of results to those of other analysts.

One of the limitations of the numerical approach is the difficulty of translating intuitive feelings about importance of certain species as ecological indicators into mathematically weighted factors in an analysis. It is not uncommon, when reanalyzing "subjective" pollen zonations by numerical methods, to miss zones defined in the original analysis by presence of indicator species or those defined by one aspect of the overall assemblage (e.g., zones defined on the basis of arboreal taxa only, rather than on arboreal and herbaceous taxa). Another common occurrence is creation of "extra" zones by numerical analysis. These are usually zones recognized but ignored by the analyst as being of little interpretive importance. In general, "objective" and "subjective" approaches to dividing stratigraphic pollen sequences into pollen assemblage zones produce very similar results (Webb and Moore 1978).

Vegetation Reconstruction

The process of interpreting sedimentary pollen data is the process of interpreting changes in pollen assemblages in terms of vegetation. Although the ultimate goal of analysis may be reconstruction of past climate or paleoenvironment, such goals are one step further removed; pollen deposited at a site is a direct function of vegetation, not of climate.

There are three common methods of correlating vegetation pattern and pollen assemblages: the comparative, or analogue, approach, the representation factor (R-factor) approach, and the indicator species approach. Each of the three relies on analyses of modern pollen rain from known vegetation formations, which model the complex relationship between number of pollen grains of a taxon deposited in sediment and number of individual plants of that taxon growing in the area around the sampling site. This relationship is affected by (1) frequency of occurrence of the taxon in the area, (2) absolute pollen production, which is influenced by flowering frequency, pollen production per flowering episode, and growth conditions (open or closed stand), and (3) pollen dispersal mechanism. Sedimentation and preservation conditions also influence pollen deposition at a site. Rather than try to model each of these processes, one studies their end results—pollen assemblages produced by known vegetation stands—and attempts to determine the corresponding pattern for the vegetation and pollen assemblage under study.

In the comparative, or analogue, approach, past vegetation around a sampling site is reconstructed from fossil pollen assemblages by direct reference to corresponding patterns between present-day vegetation and contemporary pollen assemblages. Figure 4.10 is an example of qualitative classification of vegetation types from the Valley of Oaxaca, with corresponding surface pollen assemblages. If several samples from each vegetation type are collected, the range of variation in the pollen assemblage produced can be assessed.

The next step in the comparative approach is to determine whether there are unique corresponding patterns between modern vegetation and surface pollen assemblages. Although clear differences in vegetation may be evident, differential production, deposition, and preservation of pollen may result in crucial taxa being poorly represented or even invisible in the pollen record. Invisibility may be a problem in formations dominated by zoophilous taxa, for example.

It is often possible to identify corresponding patterns by inspection. By calculating mean values of pollen occurrence for each vegetation type, or by observing range of variation in occurrence among samples from the same formation, some correspondences may be readily recognized. For example, based on data illustrated in Figure 4.10, Schoenwetter and Smith (1986) determined that the pine forest zone (Pine Forest Type A, Type Bi, and Type Bii) could be distinguished from other upland zones in Oaxaca by occurrence of pine pollen greater than or equal to 40% *and* oak pollen less than or equal to 13%. This approach has its limitations, however. In the Oaxaca example, the authors note that "there are ecological conditions we can identify in the Valley of Oaxaca—and we desire to reconstruct for the preceramic period—that are not evidenced as characteristic variations in pollen frequency values" (1986:194). A multivariate statistical technique (discriminant function analysis) was then employed to distinguish among pollen assemblages of different vegetation types (see pp. 301–302, see Birks and Gordon 1985:146–182, for details on quantitative methods of characterizing pollen assemblages).

Once modern pollen assemblage groups have been defined, their variability determined, and their correspondence to known vegetation formations assessed, the pollen assemblage of each fossil sample (or average values for a fossil pollen zone) is compared to each modern group.

A second approach to reconstructing vegetation patterns from fossil pollen assemblages is the pollen-representation factor (R-value) approach (Birks and Birks 1980:195–206; Birks and Gordon 1985:182–204, 225–236). This method was formalized by Davis (1963). In this approach, an R-value, the ratio between abundance of a pollen type in recent sediment and abundance of the source plant species in vegetation, is calculated for each species as follows (from Birks and Birks 1980:196). For a given plant species,

$$R_m = \frac{P_m}{V_m}$$

$$V_f = \frac{P_f}{R_m}$$

where R is the representation factor, P is the pollen percentage, V is the vegetation percentage, m indicates the modern value, and f indicates the fossil value. By assuming that representation factors are invariant in space and time (i.e., that species are consistently over- or underrepresented), the process can be applied in reverse to fossil pollen assemblages; for each species, pollen percentage is converted to vegetation percentage (Birks and Gordon 1985:182). A difference between the pollen-representation factor and the comparative or analogue approaches is that, in the former, abundances of *individual taxa* in past vegetation are estimated from abundances of fossil pollen, whereas, in the latter, the nature of the *entire past vegetation formations* is characterized by comparing fossil pollen assemblages to assemblages from known vegetation.

R-values have been calculated for many common taxa (see, e.g., Faegri and Iversen 1975:155, Table 8). However, it has been found that, although some plant genera are always less well represented in the pollen record than others, the precise representation factor differs among species and among data taken from different vegetation formations. The representation factor approach is used less often than the analogue approach, which is considered the soundest method currently available for reconstructing past plant communities (Birks and Birks 1980:231–261).

Still, pollen-representation factors can be useful interpretive tools if carefully applied. For details of the pollen-representation factor approach, refer to Birks and Gordon's discussion.

A third method commonly used to correlate fossil pollen assemblages and past vegetation is by indicator species. There is an excellent discussion of this approach in Birks and Birks (1980:231–237). In the indicator species approach, presence of several pollen types (indicator species) is used to infer occurrence of a certain plant community in the past. Observation of modern plant communities shows that species tend to grow together in recurring combinations; this is due to factors of the physical or nonliving environment (such as soil, aspect, and moisture) as well as the living environment (sociological factors such as competition). Although many species do occur as components in more than one plant community, others are more sensitive to local conditions and are more restricted in distribution. It is usually necessary to use several taxa of similar ecological habitat to define a plant community, however.

Like the comparative approach, the indicator species approach relies on analogy between present and past ecological relationships; in other words, one assumes that plants observed in association in modern plant communities were also associated in past communities. There are a number of problems with this assumption. Obviously, neither the comparative nor the indicator species approach (nor the R-factor method, if extinct species are involved) can be applied if the past community has no modern analogue. This may be the case for vegetation during glacial periods, for example, when individual plant distributions changed in response to changing environmental factors (Colinvaux 1987). In such a case, one is limited to defining recurrent groups of pollen (types observed to co-occur systematically) and assuming that these represent plant communities. A related limitation is the necessity of relying on modern analogues that may have been considerably influenced by humans. Janssen (1970) notes, for example, that a number of modern plant communities may only have originated in the late Holocene, arising as a result of human impact on the environment. For example, humans are responsible for introducing new species, which changes competitive relationships among plants. In addition, soils have not remained static; in the late glacial period soils were less leached, and more open conditions prevailed than today. These problems suggest that the indicator species approach is best applied in fairly recent situations (i.e., late Holocene).

On the practical side, using indicator species often means identifying pollen at the species level; species in the same genus may have very different ecological requirements. Such precision is impossible for some important groups (e.g., within the grass family). Indicator species are often minor components of vegetation—plants that are more sensitive to local conditions. As such, they may not show up in quantity in regional pollen rain (Janssen 1970).

298 • Chapter 4 Archaeological Palynology

Climatic Reconstruction

As discussed earlier in this section, pollen deposited in a lake or bog is a direct function of vegetation. The ultimate goal of analysis may be much broader than identification of vegetation patterning, however. I illustrate application of sedimentary pollen data to reconstructing past climate using Hansen et al.'s (1984) study of pollen cores from Lake Junín, Peru. Many studies illustrate the use of pollen data to reconstruct past climate. Hansen et. al.'s work at Lake Junín is especially interesting to me because of its application to interpretation of archaeological sequences at puna cave sites such as Pachamachay and Panaulauca (Pearsall 1980, 1985; Pearsall and Moore 1985; Rick 1980, 1985).

As Figure 4.7 illustrates, the Lake Junín pollen sequence can be divided into five pollen zones. As discussed earlier, these are biozones, defined by the nature of deposited pollen. Vegetation contributing to pollen rain changed over the time period represented in the core (modern to 43,000 B.P.). Stratification of vegetation zones by altitude in the Andes (highest to lowest) is super-puna, puna, sub-puna shrubland, dry montane forest or savanna, humid mountain forest, and subtropical humid forest.

Hansen et al. interprete the upper 12 m of the core (zones 4 and 5), dated to 12,000 B.P. and after, as deposited after the final Andean glaciation (Punrun glacier). It is during this period, the Holocene, that the first evidence of human occupation of the puna zone occurs (Rick 1980, 1985). Zone 4 is particularly interesting from the archaeologist's perspective. Dating from about 3000 to 12,000 B.P., zone 4 is characterized by steady values of grass pollen, ranging between 45 and 55%, and a high proportion of Andean forest taxa. Presence of forest taxa is attributed to long-distance transport from forests below the puna on the eastern Andean slopes. These forests were restricted to lower elevations during glacial maxima and contributed less to pollen rain on the puna then. The 9000-year interval represented by zone 4 appears to be a period of stability in vegetation and, by inference, in basic climatic parameters such as rainfall and temperature. This stability is reflected in the archaeological record from sites in the center of the Lake Junin puna, where subsistence practices of small hunting and gathering groups show little change over many millenia.

Pollen records from older segments of the Lake Junín core reflect vegetation changes in response to glacial conditions in the Junín basin. Zone 3, for example, is interpreted as the last phase of the Punrun glaciation, when ice advanced into the Junín basin. Silt content is very high, resulting from outwash over sparsely vegetated terrain. Most pollen was blown into the lake from the east. However, in comparison to zone 4, Andean forest taxa occur at low frequency, while there are high levels of sub-puna shrubland types. This reflects altitudinal depression of vegetation zones (i.e., the forest was lower and farther away) and perhaps drier conditions on the eastern slopes (expansion of the dry shrub zone at the expense of the humid forest). Pollen concentration levels are very low, again reflecting low influx of local pollen.

An abrupt change in the pollen record between zone 3 and zone 2 probably represents an unconformity. Radiocarbon dates, showing a gap of almost 15,000 years, support a hiatus in the record, which may encompass the entire interglacial. Lake Junín was probably dry during this period. The lower sections of the core are interpreted as deposited during an earlier glacial phase (Rio Blanco), more than 39,000 years ago. Both the younger (24,000–12,000 B.P.) and older (>39,000 B.P.) glacial phases correlate to glaciations in North America.

This brief summary of Hansen *et al.*'s study of pollen cores from Lake Junín should serve to illustrate, not only how pollen analysts argue from vegetation patterning to climatic reconstruction, but also how such information may be useful in archaeological interpretation. One issue not discussed in this analysis is possible degradation and loss of pollen from the record. Low pollen concentrations in some samples could be the result of high loss rather than low deposition.

Interpreting Soil Data

There are several differences between pollen analysis of sediments and of soils. Deposition of pollen in lake bottom sediments is a process of continuous accumulation; pollen rain falls on the water surface, sinks to the bottom, and is incorporated into a steadily building deposit. Although deposits may become mixed through storm turbulence or winter freeze, in theory all pollen in a given level is of approximately the same age.

When pollen falls on an active soil surface, it may be subject both to erosion and accumulation processes. Accumulation of soil may be by both cultural and natural means in an archaeological site. Rainwater percolates through most soil deposits and can have an effect on pollen deposition. As Dimbleby (1985) discusses, pollen in soil is mostly bound up in slowly decaying surface humus layers and is therefore not free-moving. Some pollen may, however, be carried downward by water movement. This produces the profile of soil pollen illustrated in Figure 4.6. Both water percolation and erosion, which can remove pollen-bearing deposits, are inhibited in sheltered sites such as caves or rockshelters, within structures, or in other protected situations. Steady soil accumulation also helps protect deposited pollen. Since it is often possible to identify areas of erosion or deflation of sediments, archaeological pollen sampling can be limited to areas of accumulating soil. Earthworm activity serves to accentuate pollen movement, resulting in a zone of "homogenized" deposition. In analyzing pollen profiles from open soil settings, it is important to (1) identify the earthworm activity zone, if earthworms are present, (2) analyze modern surface samples to identify the pollen assemblage of the active surface, and

(3) calculate absolute pollen concentrations, as well as proportional counts, to determine distribution of pollen with depth. This permits better understanding of processes of deposition and preservation of pollen in the site environment and helps one separate modern and ancient components of the pollen spectrum.

Subsistence Reconstruction

A number of studies illustrate use of soil pollen data in reconstructing subsistence. I have chosen Schoenwetter's (1974) analysis of pollen in Salts Cave coprolites because it is one of the early examples of this application of pollen analysis and because it utilizes both pollen and macroremain data for interpretation. Another early coprolite study is that of Martin and Sharrock (1964).

Schoenwetter details procedures used to analyze fecal specimens and discusses the security of the identifications summarized in Table 4.1. The class Chenopodiineae (chenopod–amaranth group) comprises grains that may belong to either family but that cannot be distinguished without electron microscopy. Schoenwetter feels that most are probably *Chenopodium* in his study material; chenopod seeds in fact occur in most specimens. Schoenwetter also supplies an informative discussion of sources of pollen in coprolites, with four primary means of introducing pollen to coprolites detailed: eaten as part of food or beverage, transferred to another stored food and eaten, part of pollen rain on food, and inhaled.

Schoenwetter argues that most pollen contained in Salts Cave coprolites represents foods eaten or stored. Many samples, for example, contain high percentages of pollen from insect-pollinated taxa (Table 4.1). It is unlikely that such pollen was deposited on coprolites or inhaled as part of pollen rain. Also, wind-pollinated taxa common in pollen rain are lacking in coprolites. This supports the interpretation that pollen entered coprolites through eating of flower buds, fresh foliage with adhering pollen, or seeds that were commonly gathered with pollen adhering to them (e.g., various composites, chenopod). Foods that are rarely gathered with adhering pollen, such as maygrass and hickory, would not be expected to contribute pollen to coprolites, even if macroremains were present.

Schoenwetter draws several conclusions about plant use by the inhabitants of Salts Cave based on the pollen spectra in Table 4.1. First, the pollen data suggest very early spring use of the cave. Early spring-flowering taxa (sweet flag, lily, iris) are represented; late spring and summer-flowering taxa are mostly lacking. Second, the pollen data support macroremain evidence for ingestion of stored foods. In particular, association of early spring pollen such as sweet flag with late summer foods like hickory nuts and sunflower in a number of coprolites is suggestive. Finally, Schoenwetter feels that some, if not all, of the ingestion of fresh spring flowers buds and foliage may have been for medicinal purposes.

Documenting Ecological Change

Schoenwetter and Smith's (1986) pollen analysis of archaeological and surface samples from the Valley of Oaxaca, Mexico, was part of a multidisciplinary investigation of the Oaxaca Archaic by Flannery and associates (Flannery 1986). I use this detailed and thorough study to illustrate how pollen data may be used to document ecological change.

Schoenwetter and Smith begin with a general discussion of the analogue method and the nature of pollen rain. This section provides the reader with considerable insight into the approaches used and introduces a major focus of the research, investigation of surface pollen records. In the authors' words, "the objective of our study is to identify those aspects of the pollen rain that vary directly with variations in ecological conditions" (1986:183). In other words, by studying surface pollen from modern vegetation, for which parameters of annual temperature, annual precipitation, effective moisture, and canopy were known, direct relationships between pollen assemblages and ecological conditions in the valley would be developed.

To achieve this goal, several surface pollen samples were collected from plant associations in the valley. Associations for pollen sampling were established through detailed study of plant ecology of the valley by project personnel and by analysis of surface pollen samples from two transects in the least disturbed arm of the valley. This initial study showed that certain plant associations, including ones considerably influenced by humans, could be grouped; even though vegetation differed, they produced similar pollen spectra because underlying ecological variables, such as relative effective moisture level, were similar. An example of data resulting from the final study is given in Figure 4.10, which shows pollen recovered from upland valley plots.

As a result of these analyses, Schoenwetter and Smith found that certain pollen frequencies characterized specific ecological parameters and could therefore be used to identify the occurrence of those parameters in the past. There were, however, other ecological conditions not readily recognizable by direct observation of pollen spectra. Discriminant function analysis was chosen as a means of further differentiating among ecological conditions of interest.

In the discriminant function analysis, pollen frequency data were used as dependent variables; units of the ecological classification were used as groups to which pollen data were assigned. For each category of the ecological classification (effective moisture groups, annual rainfall groups, annual temperature groups), pollen spectra assigned to groups were mathematically evaluated to generate models for maximum separation of groups. Overall classification strength of each model was then assessed. For example, using standard pollen sums (cultivars not excluded), discriminant function analysis of samples in the effective moisture category revealed that pollen spectra could be used to separate effective moisture groups in the highland valley area quite accurately (91.43% of cases were correctly reclassified to group using the discriminant function model).

Schoenwetter and Smith went on to use the discriminant function models generated from modern pollen data to reconstruct paleoenvironment from fossil pollen assemblages from sites such as Guilá Naquitz (Fig. 4.9). For each archaeological horizon, pollen spectra were used to reconstruct effective moisture, annual temperature, and annual rainfall; these ecological parameters were then used to model past vegetation. For the period from 9500 to 8000 B.P. (Zones B and C at Guilá Naquitz cave), for example, fossil pollen spectra suggested low-moderate effective moisture, moderate annual temperature, and low to high-moderate annual rainfall. This represented a drop in effective moisture and annual rainfall levels from previous periods, leading Schoenwetter and Smith to suggest local establishment of thorn forest and marked vegetation change in upper alluvium and lower piedmont slope areas. Using this approach, pollen data from four preceramic sites, Martinez, Guilá Naquitz, Cueva Blanca, and Gheo-Shih, were used to reconstruct vegetation patterns in Oaxaca during the period from before 12,000 to 4000 B.P. Vegetation and ecological data were used to assess availability of wild plant resources and agricultural potential of various parts of the valley.

This is only the briefest overview of Schoenwetter and Smith's analysis of modern and archaeological pollen data from the Valley of Oaxaca; there are many aspects that deserve fuller discussion. One of the most striking points in this analysis, from a paleoethnobotanist's perspective, is that, although most modern pollen data used in this study came from human-impacted locations (where "natural" vegetation may have occurred only in hedgerows, grazed areas, or relic stands), pollen rain was still a reflection of underlying ecological parameters. This is encouraging for the application of the analogue method in a world where archaeological sites are rarely located in pristine settings.

Directions in Archaeological Palynology

Throughout this chapter I have contrasted archaeological palynology, the study of pollen deposited in the soil of archaeological sites and other anthropogenic landscapes such as agricultural fields, to traditional, stratigraphic palynology, which is concerned with study of accumulating sediments (bogs, lakes, ocean floor). Palynologists who work with archaeological soils face a number of challenges either not shared by their nonarchaeological colleagues or of lesser impact in analysis and interpretation of sedimentary data. I now turn to a review of some of these concerns, which help structure current directions in archaeological palynology.

I have already discussed some of the interpretive problems caused by differential pollen preservation. These problems can be acute in soil analysis. Although pollen degradation and destruction may be of lessened impact in arid or waterlogged settings, where microbial activity is reduced, soils of most archaeological sites do not provide the best setting for pollen preservation. An issue of ongoing concern is how, and whether, severely reduced pollen assemblages can be interpreted. Given that pollen grains may be recoverable from most archaeological contexts if enough soil is processed and enough slides are read, further research is needed to establish the limits of interpretation in samples with low absolute pollen counts.

As in all paleoethnobotanical analysis, the interpretive power of pollen data is also linked closely with sampling context. I have discussed several approaches to sampling archaeological deposits for pollen. There are other considerations beyond locations for sampling, however. For example, there is little known about how pollen is transported around a habitation site after its initial deposition. Does pollen recovered from a feature reflect that feature's initial use, or have post-depositional processes introduced new pollen? Since pollen is destroyed by burning, "hearth" pollen, for example, probably does not reflect the period of active use of the hearth but rather the period after its use ceased or changed. Another important issue with regard to sampling archaeological contexts is the number of samples that should be taken to obtain a representative sample of pollen in a context. This is closely related to the issue of how pollen is initially deposited on a site, and to that of the nature of post-depositional movement. "Enough" samples must be taken to characterize variability within and between contexts. A further contextual issue is the difference between deposition of wind-borne pollen within enclosed areas, such as structures, and that in open-site areas, such as plazas. Although manner of transport (wind vs. animal vector) and nature of depositional situation must have an impact on interpretation of data, detailed modeling of such variables to aid interpretation is lacking.

Other, more technical concerns in archaeological palynology deal with issues of processing, counting, and identification. For example, although much more work has been done on standardizing procedures than is the case for phytolith analysis, there are still disagreements, such as whether heavy liquid flotation or chemical extraction gives the best results for removing pollen from soil matrix. Identification criteria for plants of special interest to archaeologists, domesticates and their related weedy, wild, and semidomesticated forms, are scattered in the literature or incompletely published.

In general, the type of ethnographic modeling of deposition and recovery of remains discussed for macroremain interpretation might be very useful for addressing some of the issues discussed above. Alternatively, computer simulation of depositional and preservation situations might aid in addressing these issues. Another valuable direction in archaeological palynology would be the regular consideration of pollen data in conjunction with other biological data. For example, Schoenwetter's (1974) analysis of coprolites, incorporating both pollen and seed data, illustrates how these data augment each other. Analysis of wood charcoal created by burning may complement analyses of pollen assemblages that lack taxa utilized in burning contexts. Phytoliths, inorganic remains which are preserved where pollen or macroremains may not be, can also add missing components to analysis.

References

Anderson, S. T.
    1965   Mounting media and mounting techniques. In *Handbook of paleontological techniques*, edited by Bernhard Kummel and David Raup. pp. 587–598. San Francisco: W. H. Freeman.

Barkley, Fred A.
    1934   The statistical theory of pollen analysis. *Ecology* 15(3):283–289.

Bartlett, A.S., and E.S. Barghoorn
    1973   Phytogeographic history of the Isthmus of Panama during the past 12,000 years (a history of vegetation, climate, and sea-level change). In *Vegetation and vegetational history of northern Latin America*, edited by A. Graham, pp. 203–299. Elsevier.

Berlin, G. L., J. R. Ambler, R. Hevly, and G. Schaber
    1978   Identification of a Sinaqua agricultural field by aerial thermography, soil chemistry, pollen/plant analysis, and archaeology. *American Antiquity* 42:588–600.

Birks, H. J. B., and H. H. Birks
    1980   *Quaternary palaeoecology*. London: Edward Arnold.

Birks, H. J. B., and A. D. Gordon
    1985   *Numerical methods in Quaternary pollen analysis*. London: Academic Press.

Birks, H. J. B., and B. Huntley
    1978   Program POLLDATA.MK5 Documentation relating to FORTRAN IV program of 26 June 1978. Ms. on file, Subdepartment of Quaternary Research, University of Cambridge. Cited in Birks and Gordon 1985.

Bohrer, Vorsila L., and Karen R. Adams
    1977   *Ethnobotanical techniques and approaches at Salmon Ruin, New Mexico. San Juan Valley Archaeological Project Technical Series 2; Eastern New Mexico University Contributions in Anthropology* 8:(1).

Brown, Clair Alan
    1960   *Palynological techniques*. Baton Rouge, Louisiana.

Bryant, Vaughn M., Jr.
    1974a   The role of coprolite analysis in archaeology. *Bulletin of the Texas Archeological Society* 45:1–28.
    1974b   Prehistoric diet in Southwest Texas: The coprolite evidence. *American Antiquity* 39:407–420.
    1974c   Pollen analysis of prehistoric human feces from Mammoth Cave. In *Archeology of the Mammoth Cave area*, edited by P. J. Watson, pp. 203–249. New York: Academic Press.
    1975   Pollen as an indicator of prehistoric diets in Coahuila, Mexico. *Bulletin of the Texas Archaeological Society* 46:86–106.

1978 Palynology: A useful method for determining Paleo-environment. *The Texas Journal of Science* 30:25–42.

Bryant, Vaughn M., Jr., and Richard G. Holloway
1983 The role of palynology in archaeology. In *Advances in archaeological method and theory* (Vol. 6), edited by M. S. Schiffer, pp. 191–223. New York: Academic Press.

Bryant, Vaughn M., Jr., and Don P. Morris
1986 Uses of ceramic vessels and grinding implements: The pollen evidence. In *Archaeological investigations at Antelope House*, edited by Don P. Morris, pp. 489–500. National Park Service, U.S. Department of the Interior, Washington, D.C.

Bryant, Vaughn M., Jr., and R. E. Murry, Jr.
1982 Preliminary analysis of amphora contents. In *Yassi Ada* (Vol. 1), edited by G. F. Bass and F. H. van Doorninck, pp. 327–330. College Station: Texas A & M University Press.

Bryant, Vaughn M., Jr., and Glenna Williams-Dean
1975 The coprolites of man. *Scientific American* 232:100–109.

Callen, Eric O.
1965 Food habits of some pre-Columbian Indians. *Economic Botany* 19:335–343.
1967 Analysis of the Tehuacan coprolites. In *The prehistory of the Tehuacan Valley*, volume 1: *Environment and subsistence*, edited by D. S. Byers, pp. 261–289. Austin: University of Texas Press.
1970 Diet as revealed by coprolites. In *Science in archaeology: A survey of progress and research* (2nd Ed.), edited by Don Brothwell and Eric Higgs, pp. 235–243. New York: Praeger.

Callen, Eric O., and T. W. M. Cameron
1960 A prehistoric diet revealed in coprolites. *New Scientist* 8:35–40.

Callen, Eric O., and P. S. Martin
1969 Plant remains in some coprolites from Utah. *American Antiquity* 34:329–331.

Clary, Karen H.
1980 The identification of selected pollen grains from a core at Isla Palenque. In *Adaptive radiations in prehistoric Panama*, edited by O. F. Linares and A. J. Ranere, pp. 488–491. *Peabody Museum Monograph 5*, Harvard University.

Colinvaux, Paul A.
1987 The changing forests: Ephemeral communities, climate, and refugia. *Quarterly Review of Archaeology* 8:1, 4–7.

Davis, M. B.
1963 On the theory of pollen analysis. *American Journal of Science* 261:897–912.

Dean, G.
1988 A processing procedure. In *Archaeological Investigatons at Sites 030-3895 and 030-3900, Dona Ana County Fairgrounds, New Mexico*, by T. J. Seaman, P. Gerow, and G. Dean, 77 pp., Office of Contract Archaeology, Universtiy of New Mexico, Albuquerque.

Deevey, Edward S., Jr.
1965 Sampling lake sediments by use of the Livingstone sampler. In *Handbook of paleontological techniques*, edited by Bernhard Kummel and David Raup, pp. 521–529. San Francisco: W. H. Freeman.

Dimbleby, G. W.
- 1954 Pollen analysis as an aid to the dating of prehistoric monuments. *Proceedings of the Prehistoric Society* 20:231–236.
- 1957 Pollen analysis of terrestrial soils. *New Phytologist* 56:12–28.
- 1961 Soil pollen analysis. *Journal of Soil Science* 12:1–11.
- 1970 Pollen analysis. In *Science in archaeology: A survey of progress and research* (2nd Ed.), edited by Don Brothwell and Eric Higgs, pp. 167–177. New York: Praeger.
- 1978 Changes in ecosystems through forest clearance. In *Conservation and agriculture*, edited by J. G. Hawkes, pp. 3–16. London: Duckworth.
- 1985 *The palynology of archaeological sites.* London: Academic Press.

Erdtman, G.
- 1952 *Pollen morphology and plant taxonomy. Angiosperms. (An introduction to palynology, I)* Stockholm: Almqvist and Wiksell.
- 1957 *Pollen and spore morphology/plant taxonomy. Gymnospermae, Pteridophyta, Bryophyta. (An introduction to palynology, II)* Stockholm: Almqvist and Wiksell.
- 1960 The acetolysis method—a revised description. *Svensk Botanisk Tidskrift* 54:561–564.
- 1965 *Pollen and spore morphology/plant taxonomy. Gymnospermae, Bryophyta. (An introduction to palynology, III)* Stockholm: Almqvist and Wiksell.
- 1969 *Handbook of Palynology.* New York: Hafner.

Erdtman, G., and P. Sorsa
- 1971 *Pollen and spore morphology/plant taxonomy. Pteridophyta. (An introduction to palynology, IV)* Stockholm: Almqvist and Wiksell.

Faegri, K., and J. Iversen
- 1965 Field techniques. In *Handbook of paleontological techniques,* edited by Bernhard Kummel and David Raup, pp. 482–494. San Francisco: W. H. Freeman.
- 1975 *Textbook of pollen analysis* (3rd Ed.), revised by Knut Faegri. New York: Hafner.

Flannery, Kent V.
- 1986 Food procurement area and preceramic diet in Guilá Naquitz. In *Guilá Naquitz: Archaic foraging and early agriculture in Oaxaca, Mexico,* edited by Kent Flannery, pp. 303–317. Orlando, Florida: Academic Press.

Gray, Jane
- 1965a Palynological techniques. In *Handbook of paleontological techniques,* edited by Bernhard Kummel and David Raup, pp. 471–481. San Francisco: W. H. Freeman.
- 1965b Extraction techniques. In *Handbook of paleontological techniques,* edited by Bernhard Kummel and David Raup, pp. 530–587. San Francisco: W. H. Freeman.

Hall, Stephen A.
- 1981 Deteriorated pollen grains and the interpretation of Quaternary pollen diagrams. *Review of Palaeobotany and Palynology* 32:193–206.

Hansen, Barbara C. S., H. E. Wright, Jr., and J. P. Bradbury
- 1984 Pollen studies in the Junín area, central Peruvian Andes. *Geological Society of America Bulletin* 95:1454–1465.

Häntzschel, W., F. El-Baz, and G. C. Amstutz
- 1968 Coprolites: An annotated bibliography. *Memoirs of the Geological Society of America* 108.

Heizer, Robert F., and L. K. Napton
    1969   Biological and cultural evidence from prehistoric human Coprolites. *Science* 165:563–568.

Hill, J. N., and R. H. Hevly
    1968   Pollen at Broken K. Pueblo: Some new interpretations. *American Antiquity* 33:200–210.

Hillman, Gordon
    1984   Interpretation of archaeological plant remains: The application of ethnographic models from Turkey. In *Plants and ancient man*, edited by W. van Zeist and W. A. Casparie, pp. 1–41. Rotterdam, Holland: A. A. Balkema.

Holloway, R. G.
    1981   *Preservation and experimental diagenesis of pollen exine*. Ph.D. dissertation, Department of Biology, Texas A & M University, College Station.

Holloway, Richard G., and Vaughn M. Bryant, Jr.
    1986   New directions of palynology in ethnobiology. *Journal of Ethnobiology* 6:47–65.

Hueng, Tseng-Chieng
    1972   *Pollen flora of Taiwan*. Taipei: National Taiwan University, Botany Department Press.

Hyde, H. A., and K. F. Adams
    1958   *An atlas of airborne pollen grains*. London: Macmillan.

Iverson, Johannes
    1949   The influence of prehistoric man on vegetation. *Danmarks Geologiske Undersoegelse* 4(3),6. 25 pp. Cited in Faegri and Iversen 1975.

Janssen, C. R.
    1967   Stevens Pond: A post-glacial pollen diagram from a small typha swamp in northwestern Minnesota, interpreted from pollen indicators and surface samples. *Ecological Monographs* 37:145–172.
    1970   Problems in the recognition of plant communities in pollen diagrams. *Vegetatio* 20:187–198.

Jones, R. L., and P. Cundhill
    1978   Introduction to pollen analysis. *British Geomorphological Research Group Technical Bulletin* 22. Cited in Shackley 1981.

Kapp, Ronald O.
    1969   *How to know pollen and spores*. Dubuque, Iowa: William C. Brown.

Kautz, Robert R.
    1980   Pollen analysis and paleoethnobotany. In *Guitarrero Cave: Early man in the Andes*, edited by T. S. Lynch, pp. 45–63. New York: Academic Press.

King, J. E., and William H. Allen, Jr.
    1977   A Holocene vegetation record from the Mississippi River Valley, southeastern Missouri. *Quaternary Research* 8:307–323.

King, J. E., W. E. Klippel, and R. Duffield
    1975   Pollen preservation and archaeology in eastern North America. *American Antiquity* 40:180–190.

King, J. E., and E. H. Lindsay
    1976   Late Quaternary biotic records from spring deposits in western Missouri. In *Prehistoric man and his environments: A case study in the Ozark highland*, edited W. R. Wood and R. B. McMillan, pp. 63–81. New York: Academic Press.

Kremp, G. O. W.
    1965   *Morphologic encyclopedia of palynology.* Tucson: University of Arizona Press.

Leroi-Gourhan, A.
    1975   The flowers found with Shanidar IV: A Neanderthal burial in Iraq. *Science* 190:562–564.

Lewis, Donald M., and Eugene C. Ogden
    1965   Trapping methods for modern pollen rain studies. In *Handbook of paleontological techniques,* edited by Bernhard Kummel and David Raup, pp. 613–626. San Francisco: W. H. Freeman.

Martin, P. S.
    1963   *The last 10,000 years: A fossil pollen record of the American Southwest.* Tucson: University of Arizona Press.

Martin, P. S., and William Byers
    1965   Pollen and archaeology at Wetherill Mesa. In *Contributions of the Wetherill Mesa archaeological project,* edited by D. Osborne, pp. 122–135. Society of American Archaeology Memoirs 19.

Martin, P. S., and Floyd W. Sharrock
    1964   Pollen analysis of prehistoric human feces: A new approach to ethnobotany. *American Antiquity* 30:168–180.

McAndrews, J. H., A. A. Berti, and G. Norris
    1973   Key to the Quaternary pollen and spores of the Great Lakes region. *Royal Ontario Museum of Life Science Miscellaneous Publications* 61 pp. Cited in Faegri and Iversen 1975.

McMillan, R. Bruce, and Walter E. Klippel
    1981   Post-glacial environmental change and hunting–gathering societies of the Southern Prairie Penninsula. *Journal of Archaeological Science* 8:215–245.

Moore, P. B., and J. A. Webb
    1978   *An illustrated guide to pollen analysis.* New York: John Wiley and Sons.

O'Rourke, Mary Kay
    1983   Pollen from adobe brick. *Journal of Ethnobiology* 3:39–48.

Pearsall, Deborah M.
    1980   Pachamachay ethnobotanical report: Plant utilization at a hunting base camp. In *Prehistoric hunters of the high Andes,* by John W. Rick, pp. 191–231. New York: Academic Press.
    1985   Paleoethnobotanical data and procedures. In *Hunting and herding in the high altitude tropics: Early prehistory of the central Peruvian Andes,* edited by John W. Rick. Ms. on file, American Archaeology Division, University of Missouri–Columbia.

Pearsall, Deborah M., and Katherine Moore
    1985   Prehistoric economy and subsistence. In *Hunting and herding in the high altitude tropics: Early prehistory of the central Peruvian Andes,* edited by John W. Rick. Ms. on file, American Archaeology Division, University of Missouri–Columbia.

Rick, John W.
    1980   *Prehistoric hunters of the high Andes.* New York: Academic Press.
    1985   *Hunting and herding in the high altitude tropics: Early prehistory of the central Peruvian Andes.* Ms. on file, Stanford University, Stanford, California.

Schoenwetter, James
- 1962 The pollen analysis of eighteen archaeological sites in Arizona and New Mexico. In *Chapters in the prehistory of eastern Arizona, I.* by Paul S. Martin et al. *Fieldiana: Anthropology* 53:168–209.
- 1974 Pollen analysis of human paleofeces from Upper Salts Cave. In *Archaeology of the Mammoth Cave area,* edited by P. J. Watson, pp. 49–58. New York: Academic Press.

Schoenwetter, James, and Landon Douglas Smith
- 1986 Pollen analysis of the Oaxaca Archaic. In *Guilá Naquitz: Archaic foraging and early agriculture in Oaxaca, Mexico,* edited by Kent V. Flannery, pp. 179–237. Orlando, Florida: Academic Press.

Sears, P. B.
- 1937 Pollen analysis as an aid in dating cultural deposits in the United States. In *Early man,* edited by G. G. MacCurdy, pp. 61–66. Philadelphia: Lippincott.
- 1952 Palynology in southern North America: I. Archaeological horizons in the basins of Mexico. *Geological Society of America Bulletin* 63:241–254.
- 1982 Fossil maize pollen in Mexico. *Science* 216:934–935.

Sears, P. B., and K. H. Clisby
- 1952 Two long climatic records. *Science* 116:176–178.

Shackley, Myra
- 1981 *Environmental archaeology.* London: George Allen and Unwin.

Solecki, R. S.
- 1971 *Shanidar: The first flower people.* New York: Alfred A. Knopf.

Traverse, Alfred
- 1965 Preparation of modern pollen and spores for palynological reference collections. In *Handbook of paleontological techniques,* edited by Bernhard Kummel and David Raup, pp. 598–613. San Francisco: W. H. Freeman.

Trevor-Deutsch, B., and V. M. Bryant, Jr.
- 1978 Analysis of suspected human coprolites from Terra Amata, Nice, France. *Journal of Archaeological Science* 5:387–390.

Van der Hammen, T.
- 1963 A palynological study on the Quaternary of British Guiana. *Leidse Geologische Mededelingen* 29:125–180.
- 1966 The Pliocene and Quaternary of the Sabana de Bogotá (the Tilatá and Sabana formations). *Geologie en Mijnbouw* 45:102–109.
- 1981 The Pleistocene changes of vegetation and climate in the northern Andes. In *The glaciation of the Ecuadorian Andes* (Appendix IV), by S. Hastenrath, pp. 125–145. Rotterdam, Holland: A. A. Balkema.

Van der Hammen, T., and E. Gonzalez
- 1960 Upper Pleistocene and Holocene climate and vegetation of the Sabana de Bogotá. *Leidse Geologische Mededelingen* 25:261–315.
- 1964 A pollen diagram from the Quaternary of the Sabana de Bogotá (Colombia) and its significance for the geology of the northern Andes. *Geologie en Mijnbouw* 43:113–117.
- 1965 A pollen diagram for Laguna de la Herrera (Sabana de Bogotá). *Leidse Geologische Mededelingen* 32:183–191.

Van der Hammen, T., J. H. Werner, and H. Van Dommelen
  1973 Palynological record of the upheaval of the northern Andes: A study of the Pliocene and Lower Quaternary of the Colombian Eastern Cordillera and the early evolution of its high-Andean biota. *Review of Palaeobotany and Palynology* 16:1–122.
Van Geel, B., and T. Van der Hammen
  1973 Upper Quaternary vegetational and climatic sequence of the Fuquene area (E. Cordillera, Colombia). *Palaeogeography, Palaeoclimatology, Palaeoecology* 14:9–92.
Van Zeist, W.
  1967 Archaeology and palynology in the Netherlands. *Review of Palaeobotany and Palynology* 4:45–65.
Webb, T., III, and J. H. McAndrews
  1976 Corresponding patterns of contemporary pollen and vegetation in central North America. *Geological Society of America Memoirs* 145:267–299.
Weir, Glendon H., and Duccio Bonavia
  1985 Coprolitos y dieta del Precerámico tardio de la costa Peruana. *Bulletin Institut français d'études andines* 14:85–140.
Weir, Glendon H., and Herbert H. Eling
  1986 Pollen evidence for economic plant utilization in ridged fields of the Jequetepeque Valley, northern Peru. In *Andean archaeology*, edited by Ramiro Matos and Solveig Turpin. Institutions of Archaeology, University of California—Los Angeles, Monograph XXVII.
Wijmstra, T. A.
  1967 A pollen diagram from the Upper Holocene of the Lower Magdelena Valley. *Leidse Geologische Mededelingen* 39:261–267.
Wijmstra, T. A., and T. Van der Hammen
  1966 Palynological data on the history of tropical savannas in northern South America. *Leidse Geologische Mededelingen* 38:71–90.
Williams-Dean, Glenna, and Vaughn M. Bryant, Jr.
  1975 Pollen analysis of human coprolites from Antelope House. *Kiva* 41:97–111.
Wiseman, Frederick M.
  1983 Analysis of pollen from the fields at Pulltrouser Swamp. In *Pulltrouser Swamp*, edited by B. L. Turner II and Peter D. Harrison, pp. 105–119. Austin: University of Texas Press.
Wodehouse, R. P.
  1959 *Pollen grains: Their Structure, Identification, and Significance in Science and Medicine.* New York: Hafner.
Wright, H. E., Jr.
  1983 Late-Pleistocene glaciation and climate around the Junín Plain, central Peruvian highlands. *Geografiska Annaler* 65A(1–2):35–43.
  1984 Palaeoecology, climatic change, and Aegean prehistory. In *Contributions to Aegean archaeology*, edited by Nancy C. Wilkie and William D. E. Coulson, pp. 183–195. Dubuque, Iowa: Kendall/Hunt.
Wright, H. E., E. J. Cushing, and D. A. Livingstone
  1965 Coring devices for lake sediments. In *Handbook of paleontological techniques*, edited by Bernhard Kummel and David Raup, pp. 494–520. San Francisco: W. H. Freeman.

Chapter 5 Phytolith Analysis

Introduction

Increasing acceptance and application of phytolith analysis, the identification and interpretation of plant opal silica bodies, in archaeology has enhanced recovery of data pertaining to human and plant relationships from sites with poor preservation of macroremains or pollen (Pearsall 1982a; Rovner 1983a). Analysis of phytoliths also complements and strengthens interpretations drawn through study of other archaeological plant materials. Because soil phytolith analysis is a relatively new paleoethnobotanical technique, analytical procedures are still being developed, and a variety of extraction and counting procedures exist. There are still gaps in our knowledge of phytolith production in many plant families, and a paucity of keys and published regional studies makes applying the technique difficult for the beginner. I hope to demonstrate in this chapter, however, that phytolith analysis has made great strides since Rovner brought the technique to the attention of archaeologists in his 1971 publication "Potential of Opal Phytoliths for Use in Paleoecological Reconstruction."

Nothing is more indicative of this progress than the 1988 publication by Piperno of *Phytolith Analysis: An Archaeological and Geological Perspective*. This excellent overview of the technique and Piperno's research in the New World tropics, published as this chapter reached final revision, will give the interested reader additional detail on all aspects of phytolith research as well as photographs and keys to phytoliths produced by many tropical plant taxa.

I begin this discussion of phytolith analysis with a brief overview of the nature and occurrence of phytoliths in plants, a history of the study of opaline silica bodies, and an assessment of the current status of the technique as it is applied in archaeology. I then discuss field sampling, laboratory procedures, and interpretation and presentation of results of phytolith analyses. Two case studies, applying the phytolith technique for identifying maize and for reconstructing paleoenvironment, are used to illustrate data interpretation.

Although I focus on analysis of soil silica, this is not the only application of phytolith analysis in archaeology. It has been demonstrated that carbon contained in some opaline silica bodies can be radiocarbon dated (Wilding 1967). Preliminary research into dating opaline silica directly by thermoluminescence has given encouraging results (Rowlett and Pearsall 1980; 1988). Siliceous residue on stone tools can sometimes aid in determining tool function (Shafer and Holloway 1979). Silica can also cause wear on teeth of grazing animals (Walker *et al.* 1978). Study of wear patterns and identification of phytoliths in tooth calculus (Armitage 1975) can give insight into mammalian diet. Phytoliths contained in burned cooking residue in ceramic vessels can aid in interpreting vessel function as well as diet (Pearsall 1987c; Thompson 1988).

Nature and Occurrence of Phytoliths

Phytoliths occur in stems, leaves, and inflorescences of plants. Silica that forms phytoliths is carried up from ground water as monosilicic acid (Jones and Handreck 1967) and is deposited in epidermal and other cells of the growing plant, eventually forming bodies composed of opaline silica ($SiO_2 \cdot nH_2O$) (Jones et al. 1963). In some plant taxa, distinctive forms that retain cell shape after organic tissue has decayed or burned are formed (Blackman 1971) (Fig. 5.1). In other taxa, silica deposition produces less distinctive masses, ones difficult to characterize by shape or size (Fig. 5.2). Research into distribution in the plant kingdom of diagnostic silica bodies is ongoing; in general, as more taxa are studied, more distinctive phytoliths are found; it is clear that such phytoliths are much more widely produced than previously thought.

Many monocotyledons are known to produce abundant, distinctive phytoliths. Volumes in the series *Anatomy of the Monocotyledons* (Metcalfe 1960, 1971; Tomlinson 1961, 1969) yield frequent descriptions of silica inclusions in epidermal and subepidermal cells.

One abundant phytolith producer among monocots is the grass family. Phytoliths produced in grass epidermis can be divided into two broad morphological classes, bodies in long cells and bodies in short cells (Metcalfe 1960) (Fig. 5.3). Long cell silica bodies are variable in shape but tend to be rectangular with sinuous, often interlocking borders (Fig. 5.4). Cells of the epidermis are also silicified in other

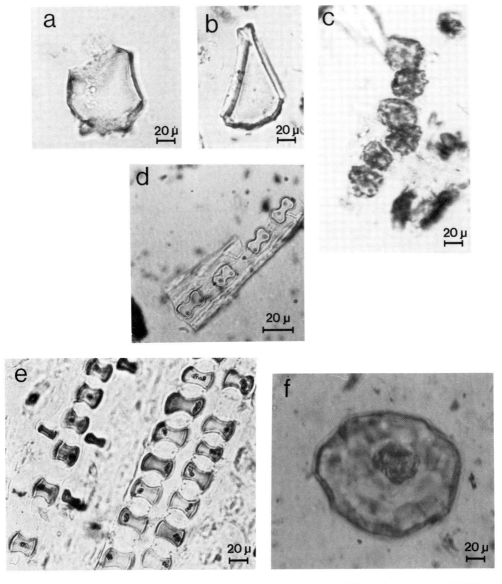

Figure 5.1 Examples of distinctive phytoliths: (a) keystone bulliform from a raised field soil sample; (b) keystone bulliform from *Oryza sativa*; bulliforms are diagnostic of grasses; (c) angled and folded spheres from *Canna edulis*; (d) dumbbell-shaped short cells from *Phragmites australis*; dumbbells are diagnostic of panicoid grasses; (e) saddle-shaped short cells from *Bambusa australe*; saddles are diagnostic of bamboos and chloridoid grasses; (f) scalloped sphere from *Lagenaria siceraria*. Photographed at 400×

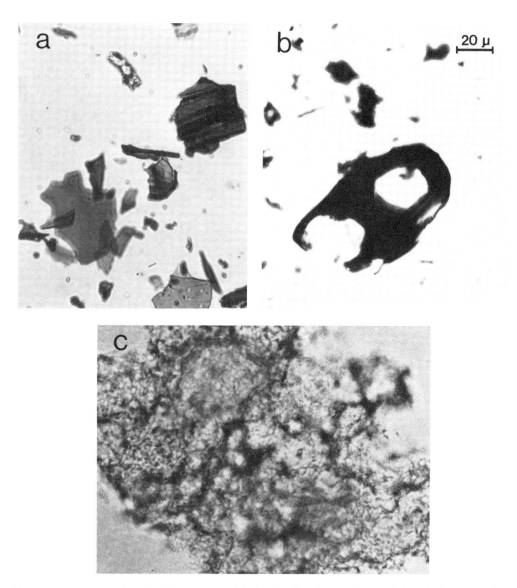

Figure 5.2 Less distinctive silica bodies: (a) classic platelike epidermal phytolith from *Echinochloa crus-galli*; (b) carbon-occluded epidermal material from *Typha domingensis*; (c) classic honeycomb material from the Asan site, Guam. Photographed at 400×.

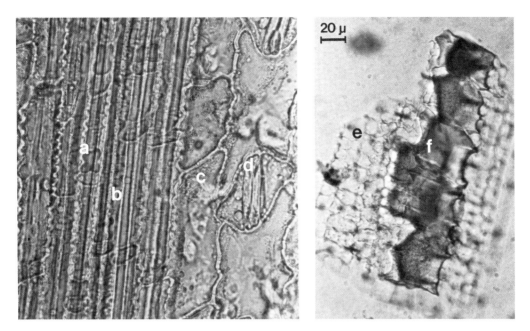

Figure 5.3 Typical cells of grass epidermis: maize epidermal peel (a) short cells, (b) vascular tissue, (c) long cells, and (d) stomata; Yap surface sample (e) long cells and (f) bulliform cells. Photographed at 400×.

groups, for example, in a variety of woody and herbaceous dicotyledons. Epidermal appendages (edge spines, prickles, and hairs) are also often silicified. Long cells vary in their usefulness in distinguishing among plant taxa. In grasses, some forms are produced widely in the family. Other long cells are more limited in occurrence and therefore more useful in identification. In some taxa, presence of spines, hairs, or other epidermal projections on long cells gives them a very distinctive appearance. For example, rice glumes have very characteristic long cells with unicellular hairs, which allowed Watanabe (1968) to identify spodograms (ashed "skeletons" of the epidermis) of this cultivar archaeologically. Distinctive long cells of another form also occur in rice leaf epidermis (Pearsall 1986a) (Fig. 5.5). Not all such distinctive silica bodies survive in soil, however. Analysis of modern surface soil samples from a rice-processing area, granary, and rice field yielded none of these distinctive silica bodies (Pearsall 1986a). I return to differential survival of phytoliths and other sources of bias in phytolith assemblages below. Silica may also be deposited in stomata, bulliform cells, and vascular tissue of grass epidermis, as well as in subepidermal cells, forming easily recognizable bodies (Fig. 5.3). Some of these may be used in combination with short cells to indicate presence of grasses in a deposit.

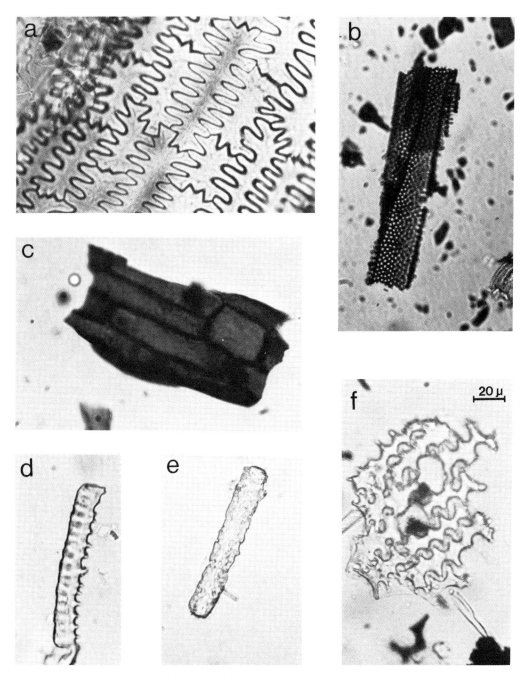

Figure 5.4 Various types of epidermal long cell: (a) highly serrated-edged cell from *Oplismenus hirtellus*; (b) serrated-edged cell with perforations from *Saccharum spontaneum*; (c) smooth-surface, smooth-edged cells from *Rhynchelytrum repens*; (d) very long cell with one sinuous edge, one serrated edge, and a striated surface from a Yumes raised field soil sample; (e) cell with sinuous edge, and uneven surface from a Yumes soil sample; (f) very highly serrated cells from *Sorghum bicolor*. Photographed at 400×.

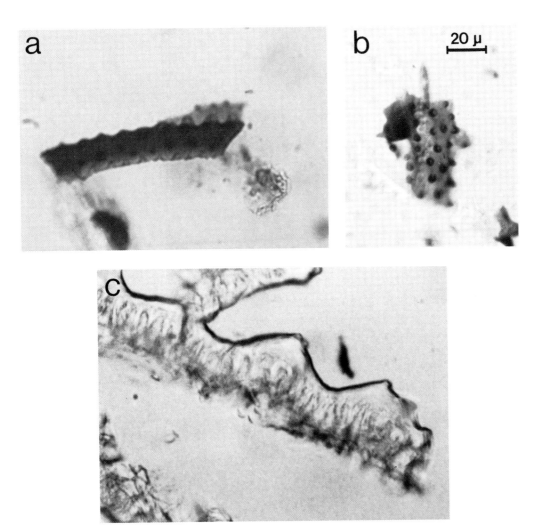

Figure 5.5 Distinctive cells in rice, *Oryza sativa*: (a, b) long cells with conical projections, from the leaf epidermis; (c) distinctive glume epidermal cells. Photographed at 400×.

The second major class of grass silica body is the short cell (Metcalfe 1960). Short cell phytoliths lie across the veins of the leaf and leaf-derived tissues (e.g., glumes, husks) and may also occur in cells lying between the veins (Fig. 5.3). In general, short cells can be classified into three distinctive classes: the festucoid class (occurring in Festuceae, Hordeae, Aveneae, and Agrostideae tribes), characterized by circular, rectangular, elliptical, acicular, crescent, crenate, and oblong shapes; the chloridoid class (occurring in Chlorideae, Eragrosteae, and Sporoboleae tribes), characterized by

318 • Chapter 5 Phytolith Analysis

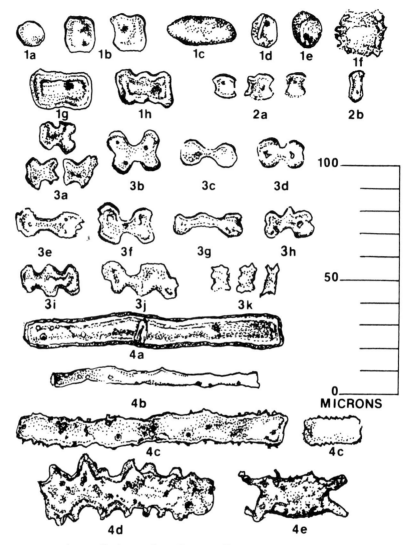

Figure 5.6 Grass silica cell-shape classification (from Twiss *et al.* 1969:111).

I Festucoid Class
    1a. Circular
    1b. Rectangular
    1c. Elliptical
    1d. Acicular, variable focus
    1e. Crescent, variable focus
    1f. Circular crenate
    1g. Oblong

    1h. Oblong, sinuous

II Chloridoid Class
    2a. Chloridoid
    2b. Thin Chloridoid

III Panicoid Class
    3a. Cross, thick shank
    3b. Cross, thin shank
    3c. Dumbbell, long shank

saddle-shaped bodies; and the panicoid class (occurring in Andropogoneae, Paniceae, Maydeae, Isachneae, and Oryzeae), characterized by cross-, dumbbell-, and crenate-shaped phytoliths (Twiss et al. 1969). The Twiss et al. (1969) short cell classification (Fig. 5.6), modified in various ways, has been widely used by phytolith analysts working in archaeological sediments (e.g., Lewis, 1978, 1979, 1981; Mulholland 1986a; Pearsall 1979, 1981, 1982b, 1984a, 1985a, 1986a, 1986b, 1987a, 1987b; Pearsall and Trimble 1983, 1984; Piperno 1979, 1983, 1985a, 1985b, 1985c; Rovner 1983b, 1983c). A short-cell soil assemblage, the total range of types occurring in a sample expressed quantitatively (Pearsall 1982a), can be characterized as predominantly festucoid, panicoid, or chloridoid based on percentage occurrences of short cell forms. Figure 5.7, for example, illustrates the relative abundance of festucoid, panicoid, and chloridoid phytoliths at the Hudsen-Meng site (Lewis 1981). How such an assemblage is interpreted depends on the floral setting, as we shall see.

The Twiss et al. short cell classification system was developed using grass species dominant in prairie formations of the Great Plains of the United States. Grasses occurring in short grass prairie were found to produce predominantly chloridoid phytoliths, species in tall grass prairie predominantly panicoid phytoliths, and grasses preferring locally moist habitats festucoid short cell phytoliths. Seventeen species were analyzed in this pioneering study. Twiss et al. demonstrated that phytoliths recovered from soil could be used to characterize the grass component of vegetation. In areas such as the Great Plains, where grass-dominated vegetation formations occur, characterizing grass vegetation gives considerable insight into basic parameters of climate; abundance of chloridoid phytoliths signifies warm, dry conditions, panicoid warm, moist conditions, and festucoid cool, moist habitats (Lewis 1978, 1979, 1981; Pearsall 1981; Robinson 1980).

As phytolith analysis is applied to a wider range of settings, the limitations of the original Twiss et al. system are becoming apparent. Problems arise when species and genera of grasses not included in the original study are present in the flora, or when grasses do not dominate vegetation and phytoliths not described in the Twiss

---

3d. Dumbbell, short shank
3e. Dumbbell, long shank, straight or concave ends
3f. Dumbbell, short shank, straight or concave ends
3g. Dumbbell, nodular shank
3h. Dumbbell, spiny shank
3i. Regular, complex dumbbell

3j. Irregular, complex dumbbell
3k. Crenate

IV Elongate Class (no subfamily characteristics)
4a. Elongate, smooth
4b. Elongate, sinuous
4c. Elongate, spiny
4d. Elongate, spiny with pavement
4e. Elongate, concave ends

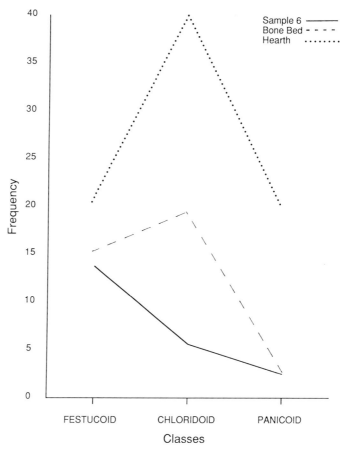

Figure 5.7 Application of the Twiss *et al.* (1969) classification of short-cell assemblages to material from the Hudsen-Meng site (from Lewis 1981:179).

et al. system predominate. An example from my own experience serves to illustrate both points.

During an analysis of soil samples from Hawaii Island, studied as part of the Waimea-Kawaihae project (Clark and Kirch 1983; Pearsall and Trimble 1983, 1984), a variety of phytoliths not previously seen at the Missouri lab were found to occur commonly in archaeological samples (Fig. 5.8). These phytoliths were of short cell size but were not included in the Twiss *et al.* classification. Study of comparative grasses provided by the project, and reference to phytolith literature, suggested that a number of these were from grasses. We added six commonly occurring types, then referred to as horned towers, flat towers, regular spools, irregular spools, angles, and half-rotated, to the sum used for phytolith quantification, but we could not inte-

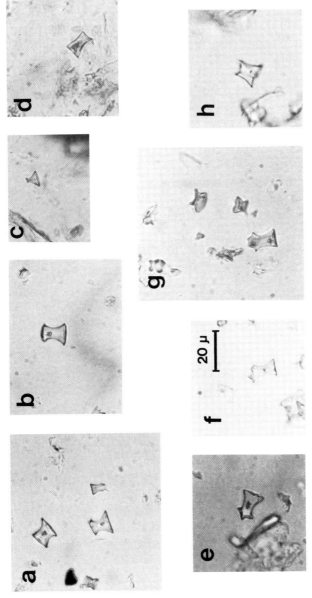

Figure 5.8 Some short cells encountered for the first time at the Missouri lab in materials from the Waimea-Kawaihae project, Hawaii: (a, b) spools from *Schizostachyum glaucifolium*; (c, d) flat towers from *Pennisetum clandestinum*, showing the range of variation; (e, f) irregular spools from *Schizostachyum glaucifolium*; (g) two horned towers and an irregular spool from *Schizostachyum glaucifolium*; (h) horned towers. Photographed at 400×.

grate these into the Twiss *et al.* panicoid–festucoid–chloridoid system. In samples where these "other" grass types were common, it was difficult to characterize grass vegetation and to draw inferences about moisture and temperature. A number of samples in this same study were found to contain very few short cells (i.e., a 200-count could not be obtained using one standard slide mount). When we analyzed comparative soil samples collected from modern vegetation formations, a similar situation was observed in samples collected from tree and shrub-dominated formations. If we scanned additional slides of such samples, the grass short cell class that "dominated" could be characterized following Twiss *et al.*, but how were we to characterize relative abundance of grasses, trees, shrubs, and forbs? In this case, grasses were not the most important component of vegetation, and a classification system incorporating other phytoliths was clearly needed.

These limitations of the Twiss *et al.* grass classification system should be taken as a caution that regional classification systems cannot always be used for interpreting archaeological phytolith assemblages from other areas. In some cases, such as the example discussed above, new classification systems must be developed by analyzing phytolith assemblages produced by known natural and human-altered vegetation formations. Such activity forms an integral part of archaeological phytolith analysis.

Brown (1984) has developed a broad-based key for identification of grass phytoliths. His key is based on 112 grass species common to central North America and is designed for use in the midcontinental grasslands of North America. Eight major classes of phytolith shapes are defined: plates, trichomes, double outline, saddles, trapezoids, bilobates, polylobates, and crosses (Fig. 5.9). This classification retains some of the basic shapes of the Twiss *et al.* system while restructuring other shape categories. Each major category contains a variety of types based on minor shape characteristics. Figure 5.10, for example, shows subclassification of the saddle category. A major advantage to this approach is that the key is easily expanded. If new comparative grass species are found to produce new variations of a shape, these can simply be added to the key. Another key is being developed by Mulholland and Rapp (1988).

My focus to this point has been on grasses, a group of early interest in phytolith research. Grasses are not, however, the only group of monocotyledons forming distinctive phytoliths. Early work on other monocot groups includes Mehra and Sharma's (1965) study of dome-shaped or hat-shaped silica bodies which have systematic value for the Cyperaceae. This silica type occurred in 60 species studied by these authors. I studied *Canna edulis*, the South American cultivar *achira*, and described phytoliths that appeared to be diagnostic to at least the family level (Cannaceae) (Pearsall 1979). Wilson (1985) reports presence of diagnostic phytoliths in the genus *Musa*, the banana group. Although it may prove impossible to separate

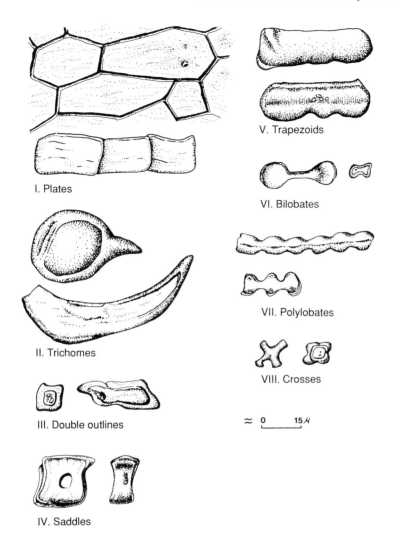

Figure 5.9 The eight basic divisions of grass phytolith types in Brown's classification system (from Brown 1984:362–364).

the three sections of the genus (*Eumusa*, *Australimusa*, and *Ingentimusa*), making it difficult to distinguish cultivated species from disturbance species, Wilson's work has clear application for identifying agricultural activity and for documenting spread of bananas outside their native range. Distinctive phytoliths have also been described for gymnosperms such as Pinaceae (Klein and Geis 1978; Rovner 1971) and for pterdophytes such as *Equisetum* (Kaufman *et al.* 1971).

The most comprehensive work to date on the distribution and nature of phy-

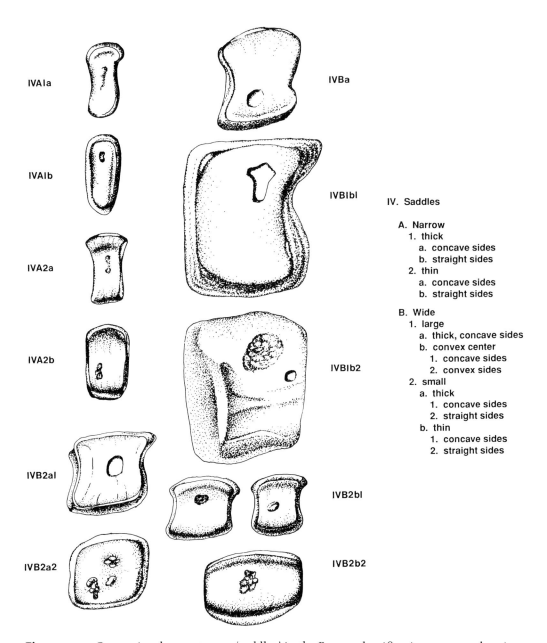

Figure 5.10 One major shape category (saddles) in the Brown classification system, showing distinctions based on minor shape characteristics (from Brown 1984:363).

tolith production in monocots other than Gramineae has been carried out by Piperno (1983, 1985d, 1988a). Piperno's work, which is focused on tropical taxa, includes detailed study of 105 species of wild monocots in twenty families. Piperno used plant anatomy tests (Metcalfe 1971; Tomlinson 1961, 1969) to choose silica-accumulating families for study. She then selected species occurring in her study area (Panama) for initial analysis. Piperno's research revealed that, while a number of monocot families did not produce silica bodies, others produced abundant and taxonomically significant phytoliths. In the latter group are the Cyperaceae, Commelinaceae, Marantaceae, Orchidaceae, Palmae, and Strelitziaceae (Piperno 1985d). Among the significant findings of this research was the discovery that *Heliconia* (Strelitziaceae) produces genus-specific phytoliths, and that at least family-specific

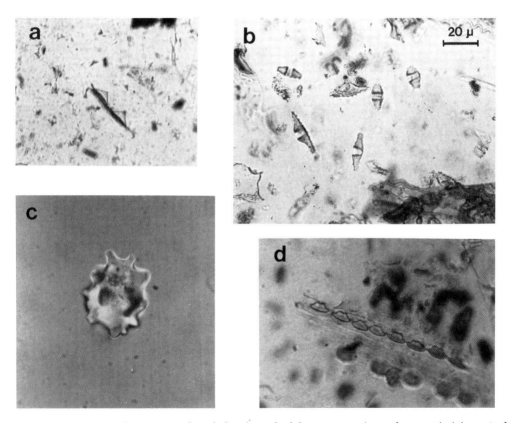

Figure 5.11 Some diagnostic phytoliths described by Piperno (1985d, 1988a): (a) conical phytoliths produced by Cyperaceae; (b) genus-specific bodies with troughs from *Heliconia*; (c) family-specific seed phytoliths from Marantaceae; (d) hat-shaped phytoliths from Palmae. Photographed at 400×.

forms are present in Palmae, Marantaceae, Cyperaceae, and Comelinaceae (Fig. 5.11). Among these taxa are important ecological indicators, which, when used in conjunction with grass and distinctive dicotyledon phytoliths, allow considerable information to be gained on seed-cropping systems, wild plant usage, landforms, and vegetation reconstruction (Piperno 1985a, 1985b, 1985c).

Piperno's research is also significant in demonstrating that diagnostic phytoliths are produced in a variety of dicotyledon families. Prior to her research, little systematic work had been done on characterizing phytolith production in dicots. Amos (1952), Ter Welle (1976), and Scurfield et al. (1974) studied phytoliths in various woody taxa; Geis (1973) investigated a number of deciduous angiosperms; and Wilding and Drees (1968, 1971), Karstrom (1978, 1980), Postek (1981), and Rovner (1971) studied limited numbers of taxa. Bozarth (1986, 1987) has identified diagnostic phytoliths in bottle gourd (*Lagenaria siceraria*) and the New World squashes (*Cucurbita* spp.) (Fig. 5.1). Piperno's research on dicotyledons included 260 species in thirty families (Piperno 1985d). She found that abundant, taxonomically significant phytoliths were produced by members of Acanthaceae, Boraginaceae, Compositae,

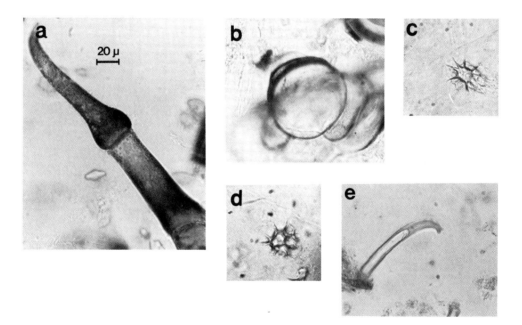

Figure 5.12 Silicified hairs and hair bases—non-grass: (a, b) jointed hairs and hair cell bases from *Cucurbita*; (c, d) hair cell bases from manioc, *Manihot esculenta*; these also occur in *Acalypha diversifolia*; (e) hair cell from lima bean, *Phaseolus lunatus*. Photographed at 400×.

Cucurbitaceae, Dilleniaceae, Loranthaceae, Moraceae, Piperaceae, Podostemaceae, Urticaceae, and Rosaceae. A number of other families were found to produce silica bodies that were not useful in distinguishing among taxa. Piperno (1985d) found that hair cell phytoliths are the most abundantly produced type of silica body in dicot leaf tissue, and that these, with hair bases, can be useful diagnostics (Figs. 5.12 and 5.13). Her most recent work has demonstrated that diagnostic phytoliths are also produced

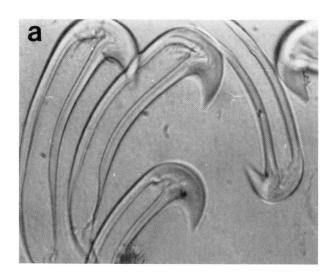

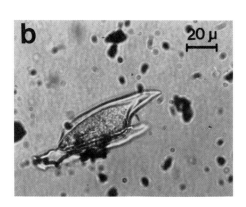

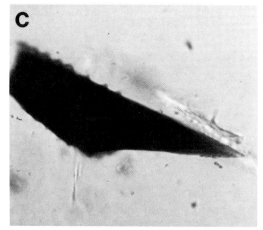

Figure 5.13 Silicified hairs and hair bases—grass: (a) hair cells from the grass *Pharus glaber*; (b) edge spine from *Saccharum spontaneum*; (c) edge spine from rice, *Oryza sativa*. Photographed at 400×.

in seed epidermis (Piperno 1988a); this opens up a new avenue of inquiry into phytolith production in plants, since most previous work has focused on vegetative tissue.

This brief overview of the nature and occurrence of phytoliths is not exhaustive of what is known about phytolith production in pteridophytes, gymnosperms, monocotyledons, and dicotyledons. As I discuss in the next section, much early work in phytolith analysis centered on grasses, where silica accumulation is high. Only now is a more balanced picture of phytolith production in the plant kingdom emerging. One strength of the phytolith technique is its application for distinguishing grasses below the family level, but this is far from being its only strength or application. It is now possible to identify a number of cultivated or useful plants, such as maize, squash, gourd, banana, arrowroot, achira, palms, and sedges, and research continues on identification methods for wheat, rice, and other cultivars.

Phytoliths and Archaeology: A Brief History

Study of phytoliths from soil of archaeological sites is a recent addition to paleoethnobotanical research in the New World. Use of this technique in the Old World has considerable antiquity, however. Among early uses of phytoliths to identify cultivated grasses is the research of Netolitzky (1900, 1914), Schellenberg (1908), and Edman and Soderberg (1929). Later, Watanabe (1968, 1970) used spodograms to identify the presence of rice and millet archaeologically.

Application of phytolith analysis to identification of New World crops dates to the 1960s with investigations at the Kotosh site, Peru (Matsutani 1972). With publication of Rovner's (1971) seminal article, interest in the technique began to grow. The late 1970s and 1980s have seen a dramatic increase in soil phytolith studies, among them Carbone (1977), Dunn (1983), Lewis (1978, 1979, 1981), Miller (1980; Miller-Rosen 1987), Mulholland (1986b), Mulholland *et al.* 1982, Pearsall (1978, 1979, 1980, 1981, 1982b, 1984a, 1984b, 1984c, 1985a, 1985b, 1986a, 1986b, 1987a, 1987b), Pearsall and Trimble (1983, 1984), Piperno (1979, 1980, 1981, 1983, 1985a, 1985b, 1985c, 1988), Piperno and Husum-Clary (1984), Piperno *et al.* (1985), Robinson (1980), Rovner (1975, 1980, 1983a, 1983b, 1983c), Starna and Kane (1983), and Veintimilla *et al.* (1985).

The history of development of soil phytolith analysis in archaeology is tied closely to two important research foci of the 1970s and 1980s: origins and intensification of agriculture, and reconstruction of past environments.

Investigation of the potential of phytolith analysis for identifying New World crops began with a search for a method of identifying maize. (*Zea mays* L.). Two techniques to identify maize were developed: the cross-shape size technique (Pearsall 1978, 1979, 1982a) and the three-dimensional morphology technique (Piperno

1983, 1984). I describe application of these techniques later in this chapter; here I review the history of their development.

The technique of identifying maize by cross-shaped phytolith size was developed during analysis of plant remains from the Real Alto site in Ecuador (Pearsall 1976, 1978, 1979, 1982a). The idea of analyzing ash and soil from Real Alto for phytolith content came from Donald Lathrap, who drew Matsutani's (1972) analysis of ash samples from Kotosh to my attention. Although results of that study were inconclusive, further checking on the technique showed me that phytoliths had been used successfully to identify cultivated grasses archaeologically in Europe, the Middle East, and Japan (Helbaek 1959, 1961; Netolitsky 1900, 1914, Schellenberg 1908; Watanabe 1968, 1970). Since one hypothesis we hoped to test at Real Alto was whether maize had been used at the site, a portion of every soil sample taken for flotation was reserved for phytolith analysis.

I began working on the problem of distinguishing maize by analyzing leaves of seven wild grasses and two varieties of Ecuadorian maize collected during fieldwork. This initial work revealed that, in comparison to the wild grasses, maize was characterized by an abundance of cross-shaped bodies (Pearsall 1976). I defined the cross-shape as a panicoid short cell with indentations on at least three of four sides which was no more than two eyepiece micrometer units longer than wide (Pearsall 1976, 1982c). Converted to absolute size, crosses are square to rectangular in outline and no more than 9.16 microns longer than wide (Fig. 5.14). This artificial classification, formally creating the cross category, allowed me to separate crosses and dumbbells (bilobates) and to create two clearly defined types. This was the key to the success of the identification technique. The cross-shape type occurred in several of the wild grasses I studied, but at much lower frequencies than in maize. I found that, while some maize crosses overlapped in size with those of the other grasses, the largest crosses occurred in maize. I grouped crosses of like size by width—that is, by the shorter side (Fig. 5.14). Crosses larger than 16 microns on the shortest side were observed only in maize. This initial phase of development of the size technique

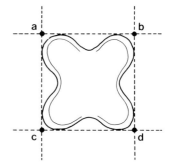

Figure 5.14 Definition and measurement of cross-shaped phytoliths. Length and width are measured by visualizing a rectangle around the cross. The longer sides are length, the shorter are width. In this example, all four sides (ab, cd, ac, bd) are of equal length. Length and width are measured by aligning the cross with a microscope eyepiece micrometer. A panicoid phytolith with three or four lobes can by definition be no more than 9.16 microns longer than wide to be a cross. Crosses are classified into size groups by width (i.e., shorter side). The largest outline (heavy line) is measured.

eventually included tests of ten genera of wild grasses and nine races of Ecuadorian maize, grown in Illinois from seed collected in Ecuador (Pearsall 1978).

I selected seventeen soil samples from Real Alto for a preliminary test of the identification procedure. I processed 5 g of soil from each sample using the Rovner (1971, 1972) procedure with some modifications. All samples contained phytoliths of various types, and twelve contained cross-shaped phytoliths. Crosses falling in the large category (longer than 16 microns on the shortest side) occurred in ten samples (Table 5.1). I concluded that "although cross-shaped archeological phytoliths of the small and medium size range cannot be identified definitely on the basis of size as maize, the archaeological phytoliths in the large category can be distinguished as maize rather than wild grass" (1978:178).

Following this preliminary stage of development and testing of the cross-shape size maize identification technique, I carried out rigorous tests (Pearsall 1979). I was concerned about the possibility that wild grasses on the Ecuadorian coast produced crosses overlapping in size with maize. I received permission from the herbarium of the Field Museum of Natural History in Chicago to collect leaves and floral parts from grasses in their collection. Using Metcalfe's (1960:601–602) list of grasses that tend to form cross-shaped phytoliths, I collected broken leaves and floral parts of all

Table 5.1 • Percentage of Cross-Shaped Phytoliths by Size Category

Material	Size category[a]			
	Small	Medium	Large	Extra-large
Wild grasses				
Tripsacum	25	75	0	0
Cenchrus	20	80	0	0
Eleucine	100	0	0	0
Races of maize				
Mischa	12	63	25	0
Mischa-Huandango	0	46	52	2
Sabanero	7	81	12	0
Morochon	9	50	41	0
Purple flint	77	23	0	0
(C) Patillo	6	79	15	0
Canguil	12	74	14	0
(O) Patillo	0	50	48	2
Cuzco	0	25	62	13
Maize, average	13	55	30	2
Archeological	12	52	36	0

[a]Size categories: small, 6.87–11.40 microns; medium, 11.45–15.98 microns; large, 16.03–20.56 microns; extra-large, 20.61–25.19 microns.

Source: Pearsall (1978;178) © American Association for the Advancement of Science; reproduced with permission.

available species in a genus listed by Metcalfe and known from Acosta-Solis (1969) to occur in the coastal zone of Ecuador (Esmeraldas, Manabi, and Guayas provinces). I also sampled all grasses considered dominants of the coastal flora. In all, 62 wild grass species, chosen to represent grasses most likely to overlap with maize, were included in the final study. After studying these specimens, I found only five taxa that produced cross-shaped phytoliths larger than 16 microns (Table 5.2). These produced only a low frequency of larger crosses, yielding an overall phytolith assemblage distinct from most maize races studied. Additionally, none of these species were ecological dominants on the coast (Pearsall 1979, 1982a).

I subsequently analyzed 95 archaeological soil samples from Real Alto and found that the percentage distribution of crosses in the four defined size categories (small, medium, large, and extra-large) corresponded closely in many samples to typical size distribution in maize rather than to that of wild panicoid grass (Table 5.2). By contrast, panicoid phytolith assemblages in five modern surface samples strongly resembled those of wild grasses studied. These data, and the presence of cross-shaped phytoliths larger than 16 microns on the shortest side in many archaeological soil samples, led me to conclude that there was a grass with abundant large and some extra-large cross-shaped phytoliths present at Real Alto that was not present in the modern grass population. I suggested that this grass was *Zea mays* L. (Pearsall 1979).

The technique of identifying maize by three-dimensional morphology of crosses (variant technique) grew out of Piperno's research in Panama and Ecuador. In her initial Panamanian work, Piperno (1979) analyzed soil samples from the Monagrillo and Aguadulce sites. A major aim of this research, directed by Anthony Ranere, was to collect data relevant to investigations of food production in the late preceramic and early ceramic periods (5000–1000 B.C.) in Panama. When analysis of pollen and charred botanical samples proved unproductive, phytolith analysis was undertaken in an attempt to augment data on subsistence for these early periods. Using the Pearsall (1978) criteria for classifying cross-shaped phytoliths, Piperno demonstrated that large crosses were present in ceramic levels at Aguadulce (2500–1000 B.C.). Few phytoliths were recovered from Monagrillo samples. Piperno concluded that presence of maize was the best interpretation of the Aguadulce phytolith data, but that positive identification of these phytoliths as maize required further study of the factors that influence their size and study of phytoliths of other wild grasses not yet tested.

In her subsequent (1980–1981) analysis of soil samples from the preceramic Las Vegas site, Ecuador (Piperno 1986; Stothert 1985, 1986), Piperno again used cross-shape size to test for presence of maize. My final study of Real Alto samples (Pearsall 1979) had recently been completed, and data on phytolith distribution in grasses native to coastal Ecuador were available. Piperno's analysis revealed that the high-

Table 5.2 • Size Distribution of Cross-Shaped Phytoliths from Ecuadorian samples

Number	Name	No. measured[a]	% Small[b]	% Medium	% Large	% Extra-large
In maize and wild grasses						
563	Oplismenus rariflorus Presl.	33	33.3	60.6	6.1	0
564	Olyra latifolia L.	63	19.0	76.2	4.8	0
578	Tripsacum dactyloides L.	1	0.	0	p[c]	0
591	T. laxum Nash.	54	22.2	74.1	3.7	0
594	Pennisetum nervosum Trin.	15	13.3	80.0	6.7	0
643	Zea mays L., sabanero	97	7.2	53.6	38.1	1.0
645	Z. mays L., canquil	97	7.2	86.6	6.2	0
662	Z. mays L., patillo	101	3.0	29.7	61.4	5.9
663	Z. mays L., mischa-chillo	100	0	26.0	66.0	8.0
664	Z. mays L., blanco blandito	106	1.9	38.7	57.5	1.9
In soil samples						
Archaeological						
96	Feature 274, Valdivia I	17	0	23.5	58.8	17.6
122	FS 1291, vessel, Valdivia II	17	23.5	35.3	35.3	5.9
117	Structure 20, Valdivia III	20	10.0	25.0	50.0	15.0
70	Feature 214, Valdivia III	26	0	30.8	61.5	3.8
64	Feature 198, Valdivia VI	28	28.6	53.8	21.4	0
Modern						
129	Xerophytic forest/open grassland	40	35.0	57.5	7.5	0
130	Xerophytic forest/open grassland	4	p	p	0	0
131	Open grassland	2	0	p	p	0
132	Scattered shrub	8	p	p	0	0
133	Deciduous trees	17	76.5	23.5	5.9	0

[a] All crosses on slides, up to ca. 100, were measured.
[b] For size categories, see Table 5.1.
[c] Percentage not calculated for occurrence < 15; p = presence.
Source: adapted from Pearsall (1982a:865).

est numbers of crosses, and those of largest size, were recovered from samples dating to the Late Las Vegas phase (ca. 8000–6600 B.P.) (Piperno 1986). Frequencies of large crosses recovered from earlier levels (Early Las Vegas, 9800–8000 B.P.) were lower, more similar to levels observed in the few wild Ecuadorian grasses known to produce large crosses rather than to maize. Number of crosses recovered was also lower in earlier levels. These results led Piperno to conclude that maize was present only in Late Las Vegas, and that lower percentages of large cross-shaped phytoliths in Late Las Vegas samples, as compared to samples from the Formative period Real Alto Site, could be interpreted as weaker evidence for maize. An alternative explanation, that more primitive varieties of maize might produce an assemblage of overall smaller cross-shaped phytoliths, was not investigated at that time. A subsequent reanalysis of Real Alto and Las Vegas samples using the variant identification method supports presence of maize at these sites (Pearsall and Piperno 1986). I discuss this study below.

Piperno's applications of the cross-shape size method of identifying maize led her to investigate ways of solving the problem of overlap between maize and wild grasses that produce large crosses. Such a technique would be especially important for applying the method in the hearth area of maize domestication, where maize would overlap geographically with closely related grasses. Realizing that three-dimensional grain characteristics were useful in distinguishing pollen, Piperno decided to investigate three-dimensional phytolith structure (Piperno 1983, 1984).

Piperno's study of three-dimensional characteristics of cross-shaped phytoliths revealed that crosses differ in three-dimensional structure, and that crosses of different types were sometimes found in the same species (Piperno 1983). Cross-shaped phytoliths can be visualized as having two broad sides, or tiers, in planar view. One, the "type" tier, is by definition cross-shaped; the other, the "non-type" tier, may or may not be cross-shaped (Fig. 5.15). In a thin slide mount, crosses usually lie with the type or non-type tier upward; views from the narrower "sides" are rarer. Piperno established eight variants of crosses based on morphology of the non-type tier (Piperno 1984) (Fig. 5.16):

Variant 1 plain; the non-type tier is also cross-shaped. It looks like a smaller cross within the cross.

Variant 2 tentlike; the non-type tier is arched into a tentlike shape aligned down the long axis of the phytolith. It appears as a line down the long axis.

Variant 3 large nodules on each corner; the non-type tier looks like four "legs" on the bottom of the type tier. Nodules usually appear as lines in the lobes of the cross.

Variant 4 thin, elongated plate, which appears as a narrow rectangle lying along the long axis.

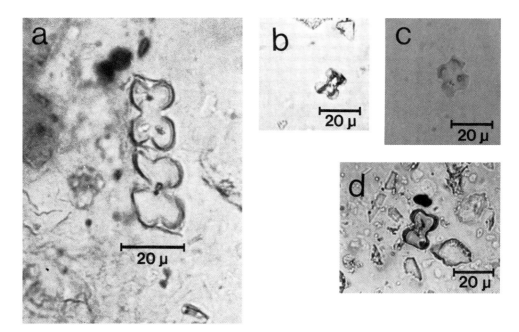

Figure 5.15 Type and non-type tiers of cross-shaped phytoliths. Type tiers on all examples shown are cross-shaped. (a) Variant 1 cross from *Zea mays*; non-type tier is also cross-shaped; (b) Variant 5 cross from a soil sample from the San Isidro site, Ecuador; non-type tier appears as lines paralleling the long axis; (c) Variant 3 cross from a soil sample from the Peñón del Rio site, Ecuador; non-type tier is composed of four nodules; (d) Variant 7 cross from a soil sample from the Peñón del Rio site; non-type tier is dumbbell-shaped. Photographed at 400×.

Variant 5 two elevated pieces of silica along the long axis which form the outline of a near dumbbell. These appear as lines paralleling the long axis.

Variant 6 irregularly trapezoidal to rectangular; the non-type tier appears as a squarish or irregularly angled body enclosed within the cross. Note that Variants 5 and 6 are now combined, as in Figure 5.16.

Variant 7 dumbbell shape; the non-type tier appears as a squat to slightly elongated dumbbell within the cross.

Variant 8 conical projection at each corner; the non-type tier has cone-shaped projections on each lobe of the cross. These do not lie flat but are angled to one side, and they appear as curved lines in the lobes.

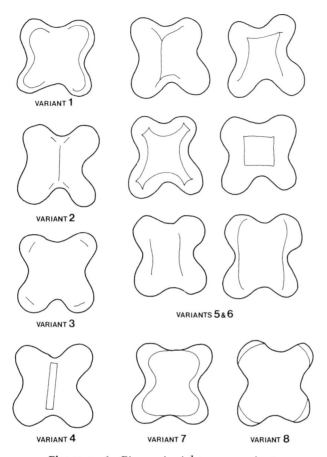

Figure 5.16 Piperno's eight cross variants.

From her study of three-dimensional cross structure in maize and wild grasses (21 races of maize, 6 races of teosinte, 40 species of other wild grasses; see Piperno 1984: Table 1), Piperno was able to demonstrate that by combining size and morphological criteria maize could be securely identified. Almost every race of maize studied was characterized by high percentages of Variant 1 crosses and low quantities of Variants 2 and 6. By contrast, many teosinte races were dominated by Variant 2 and Variant 6 crosses. Most other wild grasses tested also had three-dimensional morphologies quite distinctive from maize, with a marked tendency for high frequencies of Variant 6 crosses and low occurrence of large crosses. Piperno concluded that none of the wild grasses combined (1) high percentages of large, cross-shaped phytoliths, (2) three-dimensional configurations like maize, and (3) low ratios of dumbbell- to cross-shaped phytoliths, and that phytolith assemblages that satisfied (1) and (2) alone could be confidently identified as maize.

Piperno subsequently applied the variant identification method to analysis of archaeological soils from four sites in the Pacific watershed of central Panama (Piperno 1983, 1984, 1985a). Sites spanned the period from 6500 B.C. to A.D. 500, encompassing preceramic, early ceramic, and later ceramic phases. Phytolith analysis of the earliest deposit studied (6610 B.C., Cueva de los Vampiros) revealed a cross-shaped phytolith assemblage characteristic of wild grasses (small crosses, high dumbbell–cross ratio). Analysis of samples from deposits at Cueva de los Ladrones, dated from 4910 to 1820 B.C., and Aguadulce, dated from 2100 to 1000 B.C., gave very different results, however. High percentages of large crosses were recovered, most samples were dominated by Variant 1 crosses, dumbbell–cross ratios were low, and some extra-large Variant 1 crosses, observed only in maize, were recovered. A clear reading of maize was indicated.

Piperno's study of three-dimensional cross morphology has ramifications beyond refining the technique for identifying maize. Her work illustrates the importance of viewing phytoliths as three-dimensional rather than as planar bodies. Distinctive three-dimensional structures also exist in dumbbells, for example, and three-dimensional attributes of other classes of short cells, such as elongates, may prove important for distinguishing taxa (Mulholland 1986b). The successful development of maize identification techniques also illustrates the importance of utilizing multiple approaches in phytolith analysis. For maize, quantitative characterization of complete short cell assemblages (to illustrate percentage of crosses, ratio of dumbbells to crosses) and focus on size and structure of a selected diagnostic (cross) were important for separating the cultivar from wild grasses.

While research on identifying maize was in progress, other analysts were investigating the utility of phytolith analysis for identifying other crops and cropping systems. For Gramineae, for example, Miller-Rosen and Kaplan and colleagues investigated phytolith production in cultivated wheat species. Because wheat species differ from one another in chromosome number, it is likely that the size of certain epidermal cells, and phytoliths associated with those cells, also differ. Preliminary results of this research indicate that size differences are present and could form part of the basis for distinguishing cereal grain phytoliths of different species in archaeological soils (L. Kaplan, personal communication, 1982; Miller-Rosen 1987).

Another promising application of phytolith analysis to investigation of plant domestication is the discovery by Miller (1980; Miller-Rosen 1987) that silica content of grasses can be used to distinguish remains of irrigated grasses from those that were grown using only rainfall. Miller observed multiple-cell phytoliths in contexts where crops were grown in warm, moist conditions (floodwater fields in Egypt; high rainfall areas in Belize). In regions where rainfall agriculture was the rule, grasses absorbed less silica, forming single-cell phytoliths. Application of this technique allowed Miller to evaluate the agricultural techniques of a second-millennium site in northern Syria.

In addition to analyses of soil samples taken from site occupation areas (e.g., house floors, storage pits, crop-processing areas), recent phytolith analyses focused on agriculture have also applied the technique in nonhabitation contexts, specifically former cropping surfaces (Pearsall 1987a, 1987b; Pearsall and Trimble 1983, 1984; Piperno 1985b) and geological cores taken for regional vegetation analysis (Piperno 1985c; 1988b).

Recovery of phytoliths and other ethnobiological data through archaeological study of field or garden areas is a means of gaining direct information on subsistence, cropping techniques, and other aspects of human interaction with the environment. If an agricultural field or garden plot can be identified by means of surface features, then subsurface soil layers can be tested to identify buried cultivation surfaces to confirm or reject a hypothesis of agricultural use. Analysis of samples from old field areas may yield more direct data on cropping systems than samples from habitation contexts, since decay or burning of crop residues, which may include silica-rich vegetative portions of cultivars as well as field weeds, often occurs in fields rather than at habitation sites.

In a study of phytolith samples from archaeological sites and adjacent agricultural fields on the island of Hawaii, Pearsall and Trimble (1983, 1984) were able to identify in many cases when native vegetation was altered and cropping, fallow, and other agricultural activities began. Post-agricultural interpretations proved more difficult, since episodes of erosion and loss of soil, in part revealed by the phytolith analysis, followed field abandonment. One of the foci of the Peñón del Rio Archaeological-Agricultural project, Guayas, Ecuador, was study of a system of raised fields, or *camellones*, associated with the Peñón del Rio site (Martinez 1987; Muse and Quintero 1987; Pearsall 1987b). The Peñón raised field complex is part of more than 50,000 ha of raised fields identified in the lower Guayas basin (Denevan and Mathewson 1983; Parsons 1969). Analysis of phytolith samples from the Peñón *camellon* complex and habitation site is still under way, but testing for presence of maize has been carried out. Samples taken from early, buried fields and later, extant fields contained maize phytoliths, indicating that maize was part of the crop assemblage (Pearsall 1987b; and see pp. 418–424, below). Phytolith research is also under way on prehistoric agricultural fields in the Valley of El Dorado, Calima region, Colombia (Bray *et al.* 1985; Piperno 1985b, 1988a). Two types of fields occur in this zone: ditched fields and raised fields. Soils analyzed for phytoliths came from cultivation zones in both types of fields. Piperno reports a rich and diverse assemblage of silica remains from field contexts. It appears that maize was grown throughout the complex, and that shifts in land usage and environments can be documented, in part, by phytolith analysis.

Analysis of phytoliths from noncultural, nonagricultural contexts can also play a role in documenting evolution and spread of agricultural systems. As I discussed in Chapter 4, pollen analysis has traditionally been a major tool in paleoenvironmental reconstruction. It is often possible to identify episodes of prehistoric land clearing from the pollen record of lakes or bogs. To test the utility of phytolith analysis for vegetation reconstruction, Piperno (1985c) analyzed sediments from a series of cores taken in Gatun Lake, Panama. These had previously been analyzed for pollen by Bartlett and Barghoorn (1973). Comparison of pollen and phytolith data revealed that recognizable assemblages representing mature tropical forest, marine swamp, freshwater swamp, and disturbed forest could be documented. Evidence for agricultural and forest clearance activities appeared in the phytolith record approximately 1000 years earlier than documented in the pollen sequence (see pp. 424–428, below).

The complementarity of pollen and phytolith analyses has been explored by several other researchers. Kurman (1981), for example, conducted a study of pollen and phytolith occurrence in modern and paleosoils in Kansas. Schreve-Brinkman (1978) recovered phytoliths during a study of pollen from the El Abra rockshelter in Colombia and investigated the relationship between pollen and phytolith occurrence. Both studies concluded that phytolith data aided paleoenvironmental reconstruction and complemented data available from pollen analysis.

The second major focus of soil phytolith research, reconstructing past environments, was an important focus early in the development of phytolith analysis. In the Americas, much of this work has been centered in North America. Much research was done in the 1960s and 1970s by soil scientists interested in documenting changes in prairie–forest boundaries. Such studies used phytoliths to distinguish grassland from forested environments by using bulk density of silica in soil to indicate the presence or absence of grass cover (e.g., Bonnett 1972; Verma and Rust 1969; Wilding and Drees 1968, 1971; Witty and Knox 1964). Phytolith density has also been used to document direction and intensity of trade winds and to indicate extent of continental aridity (Folger *et al.* 1967; Maynard 1976; Parmenter and Folger 1974). Environmental reconstructions done in conjunction with archaeological projects, and using phytolith assemblages rather than bulk density to characterize vegetation, are more recent.

In an early study from Wyoming, MacDonald (1974) analyzed phytoliths from two archaeological sites, the Worland Mammoth site (ca. 9000 B.C.) and the Medicine Lodge Creek site (ca. 5000 B.C. to present), and interpreted relative abundances vegetation zones. Recent work has shown that phytoliths are produced much more widely in the plant kingdom than previously thought, making them useful in a wide variety of environmental settings.

3. Many plants of cultural significance can be identified through phytolith analysis; among these are cultivated or utilized plants such as maize, squash, gourd, arrowroot, *achira*, banana, sedges, and palms. Manioc produced distinctive hair cell bases, which were observed in only one wild species of 28 Euphorbiaceae tested, but because these are rarely silicified it is unlikely that they can be recovered from archaeological soils (Piperno 1985d). Arrowroot (*Maranta arundinacea*) and *achira* (*Canna edulis*) show more promise (Pearsall 1979; Piperno 1985d). Continued work with grain crops such as wheat, barley, rice, and millet should increase the roster of culturally significant plants securely identifiable through soil phytolith analysis.

4. Taxa important as ecological indicators can be identified through phytolith analysis; among these are subfamily groupings of the Gramineae, families such as Cyperaceae, Palmae, Marantaceae, and Compositae, and genus-specific forms such as *Heliconia* (Strelitzeaceae) and a number of Compositae genera (Piperno 1985d). Dicotyledon leaf hair cells and epidermal cells of fruits and seeds are sources of diagnostic phytoliths in many groups.

5. Soil phytolith assemblages usually represent localized deposition of silica. Although wind-borne dust can carry phytoliths from open topsoil considerable distances (Folger *et al.* 1967; Parmenter and Folger 1974; Twiss 1983; Twiss *et al.* 1969), the quantity of silica released through insitu decay of vegetation makes wind-borne silica of negligible importance in most contexts. Since no specialized conditions (cf. burning activity for macroremain preservation) are required for phytolith preservation, considerable potential exists for discerning discrete activity areas. Most research to date suggests that illuviation, a downward movement of phytoliths in soil, is not a major problem in most soils (see Rovner 1986 for a review of soil science literature on movement of silica in stable soils). There are many studies which show that phytoliths stay in the A horizon, or at most in the top of the B horizon, with few found lower in soils. No matter what the absolute quantity of phytoliths in soil, there is a clear boundary between the A and lower horizons, which argues against movement of phytoliths during normal water percolation in soils. Phytoliths stay in the soil horizon where decay occurred. Vertisols may be an exception. Rovner's (1986) main point, that phytoliths behave like other soil constituents, argues against any exceptional mobility on the part of silica

bodies. One should, of course, always evaluate the stability of deposits and avoid sampling disturbed, redeposited, or eroded locations. Comparison of phytolith assemblages of surface soil with known plant cover to that of subsurface archaeological assemblages is one way to control for disturbance (Pearsall 1979, 1982a). Comparing phytolith distribution to the distribution of other materials can also serve as a check on movement of particles (Piperno 1985c).

In summary, phytolith analysis has been shown to provide independent evidence in paleoethnobotany and paleoecology as well as to complement analysis of macroremains and pollen. Many paleoenthnobotanists and archaeologists would agree with Piperno that "it is clear that phytoliths should no longer be considered the data base of last resort. . . . Phytolith analysis can be entered in its own right as a reliable and important contributor of paleoecological data" (1985c:265).

In spite of the optimistic view taken by experienced phytolith researchers, the technique has not met with universal acceptance by archaeologists. Dunn (1983), for example, criticized Pearsall's (1978, 1979, 1982a) maize identification technique because of potential overlap between maize and some grasses that form large cross-shaped phytoliths. However, as discussed above, Piperno (1984) subsequently demonstrated that the two species mentioned, *Tripsacum dactyloides* and Johnson grass, neither of which are native to Ecuador, could be easily separated from maize by three-dimensional morphology of crosses, as can other wild grasses.

Studies have also been critical of "missing" phytoliths, soil samples lacking phytoliths altogether or lacking the expected phytolith assemblage (Dunn 1983; Starna and Kane 1983). I have not encountered problems recovering phytoliths from floodplain soils, such as reported by Starna and Kane (1983) for the Street site in upper New York state, in my research in the Mississippi (Pearsall 1982b), Guayas (Pearsall 1987b), or Daule (Pearsall 1987a) floodplains. Starna and Kane acknowledge that they experienced some difficulties during the soil extraction process. Given success in recovering phytoliths from sites in diverse floodplain settings elsewhere in North and South America, it would be premature to generalize from the New York case.

Dunn (1983) has been cited in support of the claim that phytoliths move readily in the environment and are therefore very difficult to tie securely to given archaeological contexts (Roosevelt 1984:10). In her discussion of phytolith analysis in archaeology, Dunn reports finding no phytoliths in a test column in a garden plot where maize was grown and allowed to decay over a winter, and no phytoliths in soil samples from prehistoric fields and irrigation canals in the Moche Valley, Peru. Pollen and diatoms were present in the latter samples.

There are two problems in using Dunn's study to generalize that phytoliths cannot be tied to specific archaeological contexts. First, there are alternative explanations for the "missing" phytoliths. For example, in the gardening test column, maize residue was only allowed to "decay" over one winter. Very little silica would be released to the soil in that short time unless the field were burned. Phytoliths are bound in organic material not only while it is still recognizable as crop residue but also for a considerable period in the soil humus layer. In the Moche Valley study, presence of pollen and diatoms (but no phytoliths) indicates that downward movement or horizontal removal could not be the explanation for "missing" phytoliths, since all these microfossils are similar in size and density and would be expected to move together. Absent phytoliths could be due to lack of deflocculation or incomplete removal of organic material during soil processing. Piperno (1988a), for example, reports recovering no phytoliths from samples processed without deflocculation; full processing subsequently yielded abundant silica bodies. Similarly, soil organic material can impede phytolith flotation. Inexperience of the analyst could easily be the explanation for Dunn's results.

Second, other analysts working in garden and agricultural field deposits have recovered abundant phytoliths from such settings. (Pearsall 1987a, 1987b; Pearsall and Trimble 1983; 1984; Piperno 1985b, 1988a). Cases of "missing" phytoliths should be considered in the light of the success of such studies. Piperno (1985c) and Rovner (1986), in reviewing evidence for phytolith illuviation, both conclude that it is of negligible effect in stable soils.

Although final judgment on the potential of phytolith analysis in paleoethnobotany cannot yet be made, results of much research carried out by experienced analysts is encouraging. It is no longer possible to ignore the findings of phytolith analysis, nor to dismiss the techniques as of no more strength than indirect data on subsistence such as maize kernel impressions, depictions, or presence of metates (Lippi et al. 1984).

Weaknesses

1. Many plants do not deposit silica in vegetative tissue or inflorescences, or they deposit very little; such taxa must be considered "invisible" in the phytolith record. Although a number of important food plants, including root crops such as manioc, sweet potato, and yam, fall into this group, many important taxa remain to be tested and should not be assumed to be nonaccumulators until material is examined.

2. Redundancy of phytolith production exists for a number of forms. Similar silica bodies may be produced in widely divergent plant groups. For example, silica deposited in epidermal and mesophyll leaf tissues looks similar in many taxa. At times only very broad characterization (e.g., dicotyledon vs. grass) can be based on phytolith assemblages.

3. Not all phytoliths are preserved equally well in soil. Although differential preservation is not a serious problem, as it is with organic remains, several factors can affect phytolith preservation. Low silica content in soil can result in "lightly" silicified bodies that do not hold up well under soil pressure. Phytoliths may lose delicate hairs; surface features may erode with time. Larger silicified cells may break, making identification more difficult. High pH may lead to dissolution. These factors do not, however, affect the majority of phytoliths or depositional contexts.

4. Differential silica production among plant taxa leads to problems of bias in phytolith assemblages. Like pollen data, some taxa will always be under- or overrepresented in a phytolith assemblage. Quantitative comparisons among diverse taxonomic groups must take this factor into account.

Problems for Research

1. *Phytolith taxonomy.* Many advances have been made in this area, both within the Gramineae and in other taxonomic groups, especially in the tropics, but vast numbers of plants await systematic study. We lack a standardized vocabulary for describing siliceous forms; methods of characterizing absolute or relative abundance of forms in plants differ greatly; measurements are not always included in descriptions; diverse plant extraction procedures are utilized. Published studies of comparative materials and development of keys are high priorities in the field.

2. *Processing.* A number of different processing procedures are currently in use for recovering phytoliths from soil matrix (see "Laboratory Analysis," below). No systematic testing has been carried out to indicate whether one procedure extracts more silica than another or if soil type may affect processing and silica recovery. All chemicals in use are toxic to one degree or another; tests for less toxic alternatives would be useful.

3. *Quantification.* I discuss quantitative approaches used in phytolith analysis under laboratory procedures. Efforts to quantify phytolith assemblages are a recent development in the field. A number of problem areas exist, including how to deal with under- and overrepresentation of taxa that produce different quantities of silica and how to characterize abundance of silica bodies that cannot be meaningfully counted. Standardization of mounting, scanning, and counting procedures are needed before quantitative results of research can be compared and replicated. Absolute counting procedures, modeled after methods used in pollen analysis, are just beginning to be explored.

4. *Impact of environment on phytolith production.* How is silica deposition affected by abundance of silica in groundwater and by quantity of

rainfall? Do these and other environmental factors affect phytolith quantity, size, and type? Research to date on these issues is sparse. Some data exist linking phytolith size to quantity of rainfall (Yeck and Gray 1972), but studies of maize and its relatives suggest that phytolith size and morphology are under considerable genetic control (Pearsall 1979; Pearsall and Piperno 1986; Piperno 1983, 1984, 1988a). Depending on their function and location in tissue, all classes of phytoliths may not be equally affected by environmental factors. Replicate studies of phytoliths from the same species grown under different conditions should clarify what kinds of phytoliths are most affected by environmental conditions. The work of Pearsall (1979) and Piperno (1983) included many of the same species of grasses, most of which were found to have very similar short cell assemblages. Piperno (1988a) has found strong evidence for stable patterns of silification in many taxa in diverse ecological settings.

My goal in this section has been to summarize the status of phytolith analysis as a paleoethnobotanical technique. It is premature to attempt a final assessment of the role of phytolith analysis in archaeology; the technique is still in its developmental stage in many respects. Like pollen analysis in the 1960s, phytolith analysis must prove itself both by application in real archaeological settings and through advances in basic research in the laboratory. I would argue that neither focus is sufficient without the other. Development of identification techniques or keys and studies of production and morphology should be focused on ultimate application in soil analysis. By the same token, archaeologists must be willing to support the basic research necessary for development of techniques. Unrealistic claims about the potential of the technique are damaging, but even more damaging is unwarranted pessimism and an unwillingness to let a new approach find its role in archaeological research.

Field Sampling

In this section I discuss procedures for collecting phytolith samples in the field. There are many similarities between phytolith and pollen sampling (see Chapter 4); phytolith sampling techniques are, to a considerable extent, derived from techniques developed for pollen research. I discuss two aspects of field sampling: sampling soil and sampling modern vegetation.

Sampling Soil

A question I am frequently asked by archaeologists headed for the field is how to take phytolith samples during excavation. "How" in this case usually means both how and where: what are procedures for actually taking samples, and what areas should be sampled? To address these related issues, I divide soil sampling into three types: (1) sampling in cultural contexts (e.g., habitation sites, agricultural features), (2) sampling natural strata (e.g., soil horizons, lake deposits), and (3) sampling modern surfaces.

Cultural Contexts

The key to successful phytolith sampling during excavation of a habitation site, former agricultural field, or other locus of human activity is to fit the sampling design to the questions to be answered. Sampling to identify activity areas within structures is different from sampling to identify crops under cultivation, for example. Although it is always possible to select samples for actual processing from a larger body of samples taken in the field, inappropriate sampling cannot be remedied after the fact. Following a brief discussion of the mechanics of taking samples, I give a number of examples of sampling to address specific research goals.

Supplies and equipment needed to take soil samples for phytolith analysis include the following:

1. clean medium- to heavy-weight small plastic bags
2. trowel or other hand tool for scrapping or cutting soil
3. wash bottle or cleaning cloth
4. labels, masking tape
5. field notebook, indelible pen
6. nails, tape measure, string

As mentioned above, sampling for phytolith analysis is modeled after archaeological pollen sampling. Let us imagine two common situations: taking samples from a profile in a garbage midden, and sampling a horizontal surface such as a house floor.

I discussed collecting soil samples from vertical profiles in some detail in Chapter 4, since this is a common situation for pollen sampling. After a fresh profile surface has been cut with a trowel, upper and lower limits of each stratum to be sampled are identified. Limits can be marked with nails, or by outlining with a sharp trowel or knife. If a thick stratum is being subdivided, clearly mark the limits of each arbitrary division as well. Position a tape measure adjacent to the sampling area, so that depth of each sample from the surface can be read accurately. Sample from the bottom of the profile toward the top, so that soil dislodged during sampling falls only on previously sampled areas. Using a sharp trowel, knife, or sampling scoop, remove a portion of soil from the stratum by cutting straight back into the profile. Thin strata must be carefully sampled so that soil is not incorporated from adjacent levels. Take samples from each side of a stratum boundary rather than one sample incorporating soil on both sides. As was discussed for pollen sampling, there are advantages to sampling continuously, rather than at separated intervals, within strata. Changes may occur in soil assemblages which are not associated with a color or texture change. I generally recommend taking samples of approximately 50 g; this gives enough soil for a number of replicate processing runs, should these be necessary.

Place soil in a plastic bag, which should be clearly labeled with all appropriate provenience information, including depth of sample. Clean your trowel or other sampling tool between samples, either by washing or thorough wiping. Close and seal bags immediately. An easy way to label and seal plastic bags is to write provenience information directly on the bag with an indelible pen, shake the sample to the bottom of the bag, roll the bag up from the bottom, and tape up the resulting cylinder with masking tape. To facilitate reading provenience information without unrolling the bag, label the sample again using a tie-on label or by writing on the masking tape. This also serves to double tag the sample.

It is useful to list samples taken each day in a field notebook, noting at the same time any problems encountered and information on soil color, texture, and other features. A quick sketch of the profile, with sampling loci marked, is also helpful. Such a sketch is not a substitute, of course, for a detailed profile drawing of the sampled area.

It is not necessary to freeze, dry, or add fungicide to phytolith samples, since silica is not destroyed by microbial growth. However, if soil samples are to be divided later for parallel phytolith and pollen processing, it is important to follow procedures outlined in Chapter 4 for preserving pollen and to take larger samples.

Sampling a horizontal surface, such as a house floor, requires a slightly different procedure. Make sure that the surface to be sampled is free of overlying soil layers or fill. If there is a clear boundary between layers, remove overlying ones and scrape the surface of the desired layer clean just before sampling. If floor deposits are not readily divided into discrete layers or are indistinguishable from fill without artifactual analysis, take phytolith samples from each arbitrary level used to excavate the floor. Data from phytolith analysis may help determine depositional boundaries.

Once the layer to be sampled is exposed, areas to be covered by each sample should be delineated. For a large floor, this will probably involve griding off the surface into sampling units, each of which is given a number or letter designation (as in Fig. 2.34). The smaller the grid used, the finer the distinctions that can be drawn among areas. Features (hearths, pits) in the floor should be numbered for separate sampling.

Sample unit by unit across the floor. Scrape each unit just before sampling to guard against transfer of soil between units or contamination with soil on shoes or clothing. Within each unit, take soil from the entire area of the unit, to give a composite sample, rather than from only one or two points. Care should be taken to cut into the floor to a uniform depth in each unit. When sampling the floor toward the bottom of arbitrary levels, it is important not to cut into the next level. Measure the depth of the sampling surface below the site surface. Bag and label soil samples as described above. A sketch map of the floor, with sampling units clearly labeled, is useful and an easy way to keep records of depths and sample numbers.

As mentioned above, the real key to successful phytolith sampling, as in sampling for macroremains or pollen, is fitting sampling design to research questions. As an example, consider the difference between sampling for identifying foods and sampling for identifying household activity areas. For identifying foods, sampling should be focused on areas of food preparation, cooking, garbage or crop residue disposal, and storage—in other words, areas where concentrations of foods were present, or repeated activities involving foods were carried out—maximizing recovery of phytoliths deposited through decay or burning of subsistence items. Grid sampling a house floor to identify activity areas might lead to recovery of phytoliths from food plants (especially in food-preparation areas), but remains may be absent or masked in samples from other activity areas by quantities of silica from nonfood plants such as those released from bedding, floor mats, fallen roofing thatch, basketry, or weaving materials. In this case, sampling a number of food-preparation or storage contexts would probably yield better data on subsistence.

Similarly, although sampling pit fill may give data on garbage thrown into the pit after it is no longer used for storage, only samples scraped carefully from the bottom and sides of the pit may reflect its original function. Even these samples may be contaminated by phytoliths from later fill, unless a clear boundary is visible between the fill and residue of the original contents. Excavating large pit features in sections, so that profiles can be studied to guide sampling, improves chances of obtaining botanical samples representing original function.

Because phytoliths are released into soil by either decay or burning of plant tissue, sampling need not be restricted to ashy deposits. Much ash recovery at any archaeological site is the product of cooking fires, especially ash recovered in situ in a hearth or in small localized deposits representing hearth cleaning. In many cases such deposits contain concentrations of phytoliths from woody plants used as fuel, which may mask occurrence of a few phytoliths from subsistence plants. More extensive ash lenses, such as might result from burning garbage or crop residue, may be good contexts for sampling for food plants. Sampling can also be guided by the presence of certain types of artifacts. Sampling the floor around an in situ milling stone may be quite productive for recovering phytoliths from foods, for example.

I have so far dealt with sampling in a large excavation, where many more potential sampling contexts exist than funds or time to analyze samples. In such a situation, a balance must be achieved between sampling a wide diversity of contexts and replicative sampling of a narrower group of contexts chosen according to research questions. But what about the small-scale excavation? Should phytolith samples be taken from test pits where cultural context may be unknown?

If testing is being done in advance of more extensive excavations, for example to determine natural stratigraphy or to choose areas to excavate, then soil sampling in test pits is useful to assess types of phytoliths likely to be encountered later or to obtain a few samples from an area that is not going to be investigated further.

Collecting soil samples for phytolith analysis, as well as for flotation and pollen analysis, from a small-scale excavation can yield valuable paleoethnobotanical data. It is often possible to obtain data on the presence or absence of plants, as well as a general idea about nature of vegetation surrounding a site. It is important that the area sampled be free of post-depositional disturbances, however. If animal burrows, pot-hunter holes, or other disturbances are evident in the unit, sampling should be restricted to clearly undisturbed areas. If the site is very shallow, with extensive root penetration, or if most of it has been reworked by plowing, phytolith analysis is usually unproductive, since ancient and modern materials may be mixed substantially. One way to determine whether extensive mixing has occurred is to sample the surface of the site prior to excavation. Depth to which subsurface samples are identical to surface control samples is an indication of the extent of mixing.

Sampling residues in or on artifacts for phytoliths may yield valuable data on artifact function or lead to identification of utilized plants. I describe how to do pollen washes of ground stone artifacts in Chapter 4. This technique can also be used to recover phytoliths from grinding tools. Scrapings from the interior of cooking vessels can also be an excellent source of phytoliths of food plants. At the Real Alto site, for example, analysis of scrapings from an inverted Valdivia II vessel yielded one of the clearest indications of maize from that site (Pearsall and Piperno 1986). Phytoliths have also been successfully extracted and identified from scrapings of ungulate teeth (Armitage 1975).

Before considering sampling in natural strata, I should say a few words about phytolith sampling in ancient agricultural fields. Phytolith analysis is a paleoethnobotanical technique well suited to investigating agricultural field features. Analysis of samples from old field areas can yield direct data on cropping systems, since decay or burning of crop residues often occurs in fields rather than at habitation sites. An ideal situation for applying phytolith analysis to ancient field soils occurs when former cultivation surfaces are buried and not mixed with modern surfaces by deep plowing or erosion. A profile through such an idealized field would show the modern surface, old buried surfaces representing cultivation surfaces, episodes of fallow, flooding, mucking, and the like, and lower soil horizons. Samples taken from each discrete layer should allow description of phytolith assemblages and identification of specific plants or vegetation formations present.

Sampling from an exposed vertical profile of a field gives "point" data—phytolith assemblages of soil at each location. In many traditional agricultural systems, it is not uncommon for a variety of crops to be grown in one plot. In this situation, sampling a series of closely spaced locations on an exposed horizontal surface (as described above), or creating a composite sample of such a surface by collecting pinches of soil from all over it, increases chances of recovering phytoliths from the full range of crops present. Pinch sampling is described in more detail under "Modern Surfaces," below.

Natural Contexts

Analysis of phytolith samples from cultural contexts, whether house floors, pits, vessel scrapings, or agricultural fields, yields information on human interaction with the plant world. Such sampling contexts do not yield an unbiased picture of the natural environment, although plants chosen for use or accidentally included in the archaeological record may reflect resources available near a site. Phytolith analysis for the purpose of reconstructing past environments requires sampling in contexts free of direct disturbance. Human activities that alter the makeup of local or regional vegetation may be reflected in samples taken from natural soil depositions, but sampled areas should be away from direct human disturbance (occupation sites, old field areas).

Strategies for sampling such deposits, which I refer to as natural contexts to distinguish them from cultural contexts, are similar in some ways to selection of loci for pollen sampling. As I discussed in Chapter 4, pollen analysis of sediments from lake or ocean cores allows reconstruction of regional vegetation patterns through time. Wind-borne pollen falls on the water's surface, sinks, and becomes part of bottom deposits. Pollen extracted from such deposits reflects broad patterns of vegetation, since pollen is carried in from considerable distances as well as from local vegetation. Disturbance of the environment, such as deforestation due to clearing for agriculture, is often reflected in the depositional record.

Lake or ocean bottom deposits are also excellent contexts for phytolith sampling. Like pollen, phytoliths can be carried considerable distances as a component of wind-borne dust from open topsoil. As discussed earlier, the dynamics of phytolith rain are not well understood. Unlike pollen grains, which are most readily preserved under anaerobic conditions, phytoliths are preserved under a broader range of conditions, including dry soil deposits. Thus it is possible to core or profile dry, as well as waterlogged, natural contexts for phytolith sampling. This can be especially useful for paleoenvironmental study of dry regions, where the only contexts appropriate for palynological study may be very distant.

Procedures described in Chapter 4 for taking pollen cores and sampling sediments work equally well for phytolith sampling of waterlogged deposits. Phytolith samples can be taken from cores intended for pollen analysis. Contamination can occur during coring, so soil samples should be taken from the interior of the core. Small hand corers, such as those used for probing archaeological sites, are inappropriate for phytolith or pollen sampling, since sampled soil is easily contaminated when cores are removed from the ground and the small core diameter makes taking a sample from the interior difficult. Such cores should be used only to test for presence of phytoliths in deposits. Before extensive phytolith sampling was done on raised fields associated with the Peñón del Rio site, for example, we extracted two cores from a field using a hand corer and analyzed several samples to determine whether phytoliths were present and what types should be expected.

Phytolith samples useful in paleoenvironmental reconstruction can be taken from a variety of natural contexts, depending on what undisturbed contexts are available and what aspects of paleoenvironment are of interest. As mentioned above, analysis of phytoliths from wind-blown soil deposited in still water may yield a composite view of vegetation on a regional scale as well as document local plant communities. Soil deposited on land surfaces, by the action of either water or wind, can be of similar value. For example, phytolith analysis of naturally deposited soils in Natural Trap Cave, Wyoming, yielded information on changes in local grassland vegetation over a 100,000-year period (Pearsall 1981). Silt carried into the cave covered remains of animals that had fallen into the natural trap created by the narrow (12' × 15') opening of the 85' deep cave (Martin and Gilbert 1978).

Analysis of phytoliths from buried paleosols, former land surfaces, can also give valuable insight into past vegetation patterns. A buried A horizon in a soil profile should be sampled as described above for profile sampling in archaeological contexts. Considerable work has been done in North America on distinguishing former grassland soils from those formed under forest cover. Because grasses are abundant phytolith accumulators, grassland soils are composed, by weight, of a much higher quantity of silica than forest soils. The assemblage of phytolith types present also differs considerably. Shifts in boundaries between grassland and forest can be investigated through phytolith analysis of paleosols.

A phytolith sequence from noncultural stratigraphic contexts contemporary with an archaeological site occupation can serve as control for archaeological phytolith analyses. Comparison of the "natural" phytolith assemblage of an undisturbed buried surface to assemblages recovered from archaeological strata can help one distinguish between "background noise" and phytolith patterns that reflect human selection of plants.

Modern Surfaces

Phytolith analysis of modern surface soils carried out in conjunction with archaeological or paleoenvironmental phytolith investigations serves two basic purposes. First, as mentioned above, sampling the modern surface of a site can serve as a means of detecting downward movement of phytoliths into archaeological strata. Unless human occupation of an area was so ephemeral that little disturbance of natural vegetation or introduction of foreign elements occurred, marked differences should exist between regrown surface vegetation and the record of plants burned or decayed on the site during and immediately after its occupation. Homogeneity in surface and subsurface phytolith assemblages may be a signal of soil mixing. Second, analyzing soil phytolith assemblages produced by known vegetation serves as a means of establishing correspondences between vegetation and phytolith assemblage, a valuable interpretative tool for archaeological or paleoenvironmental phytolith analysis. I focus on the latter aspect of surface sampling for the remainder of this section.

Modern surface soil can be sampled in several ways. When taking samples from a profile, always take a sample from the modern soil surface. This type of sample is a point sample, which should reflect the character of vegetation immediately above the profile. Take the sample right at the surface to ensure this correspondence. During analysis of profile samples from the Waimea-Kawaihae project (Clark and Kirch 1983), lack of correspondence between surface vegetation and both phytolith and pollen assemblages of soil taken in the first 5 cm below the current surface alerted us that recent top soil had been removed by erosion (Pearsall and Trimble 1983, 1984).

Another approach to phytolith sampling of modern surfaces is pinch sampling, a technique borrowed from pollen analysis. As I discussed in Chapter 4, interpreting pollen data depends on establishing correspondences between pollen assemblages and vegetation. This is how over- and underrepresentation of pollen in soil is documented. Pinch samples are taken by collecting several small samples from surface soil in a bounded area, such as a 10- × 10-m square, and combining them. This creates a composite sample, which characterizes the mix of vegetation in the square rather than plants at any one point. Pinch sampling for phytolith analysis is carried out in the same way.

To get the most benefit from such samples, make a detailed description of vegetation in the sampling area. I usually estimate percentage cover of plants by canopy to describe vegetation growing on a surface control square. Table 5.3 is an example of such a description, made during research at Peñón del Rio (Pearsall 1984b). Correct identification of plants is essential. If determinations cannot be made in the field, collect an example of each species for later identification (see Chapter 3 for collecting techniques).

Once the vegetation cover is described and quantified and the soil analyzed to determine corresponding soil phytolith assemblages, the extent of over- and underrepresentation of plants in the phytolith record can be gauged and the phytolith

Table 5.3 • Vegetation Description Accompanying Surface Soil Sample Peñón del Rio, Ecuador

A. Trees   none
B. Shrubs   none
C. Dicotyledonous Herbs

Lab #	Identification	Cover
980	*Vigna luteola* (Leguminosae)	20%
—	Acanthaceae	20%
—	Malvaceae and Leguminosae	20%
1219	*Momordica charanta* (Cucurbitaceae)	very scattered cover, small
1279	*Vigna cf. sinensis* (Leguminosae)	20%

D. Monocotyledonous Herbs

1278	*Chloris cf. radiata* (Gramineae)	very scattered cover, small
1193	*Thalia geniculata* (Marantaceae)	isolated cover, small
979	Cyperaceae	very scattered cover, small

Sample 1983-1: 7- × 10-m area, the top of a cleared camellon; herbaceous cover, 100%.

"signature" of the vegetation formation characterized. In our Waimea-Kawaihae work, soil phytolith assemblages of surface pinch samples from each vegetation zone (Fig. 5.17) were compared to actual percentage cover of categories of plants (Table 5.4). This helped us interpret soil assemblages from ancient fields.

Looking at the large-cell data in Table 5.4, going from the dry coastal zone (I) to the most mesic zone (VII), changing vegetation is reflected in corresponding changes in percentages of grass and non-grass phytoliths in surface soils. Zone I has the lowest grass soil phytolith occurrence (3.1%). Relative abundance of grass large cells in soil increases steadily from Zone I through Zone VI. Grass large-cell levels of Zones V, VI, and VII are all between 44 and 53%. These three zones have no canopy tree cover and little bare ground. Zones I, II, and III show a correlation between canopy cover, barer ground, and lower grass large-cell occurrence. These data suggest that in the coastal zones production of platelike and honeycomb-like phytoliths by canopy trees and large shrubs contributes more silica to soil than that produced by scattered grasses. Silica production by smaller herbs does not appear to alter ratios significantly. In inland zones, lack of large producers of platelike and honeycomb-like phytoliths, and denser grass stands give high grass large-cell occurrence.

We used these data as a guide for interpreting archaeological phytolith as-

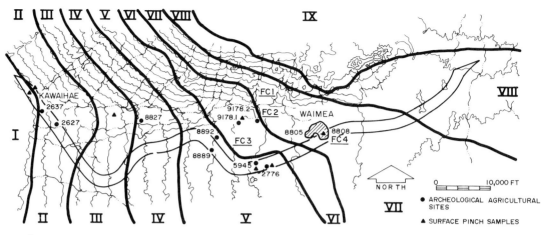

Figure 5.17 Tested archaeological sites and vegetation zones from the Waimea-Kawaihae study area (from Pearsall and Trimble 1984:123).

semblages. Large-cell grass phytolith percentages of 43% and above indicated grass-dominated vegetation, occurrence of very low large-cell grass counts (<5%) indicated forest or large shrub cover, and occurrence in between (5–40%) signaled transitional formations. The value of conducting comparative analyses of this type as a routine part of any archaeological phytolith project should be evident.

Table 5.4 • Soil Phytolith Assemblages (surface pinch samples) and Actual Percentage Vegetation Cover, Waimea-Kawaihae project

	Percentage by zone					
	I	II	III	V	VI	VII
Actual vegetation cover						
canopy	60–80	20–40[a]	5–25	—	—	—
native shrub	5–25	5–15	5–15	5–25	25–50	—
exotic shrub	5–25	—	5–15	5	5	—
native herb	—	—	5–15	—	—	5–25
exotic herb	—	—	—	25–50	50–75	—
exotic grass	20–50	50–75	50–75	25–50	50–75	90–100
bare ground	50–75	25–50	5–25	5–25	5–10	—
Soil phytolith assemblage						
large cells, grass	3.1	17.0	25.2	44.0	52.9	43.6
large cells, non-grass	94.9	81.9	74.8	56.0	47.1	56.4
panicoid short cells	I[b]	63.7	32.7	17.2	16.0	30.9
festucoid short cells	I	33.5	58.6	69.9	74.9	49.5
chloridoid short cells	I	0	1.9	4.3	3.3	8.6

[a]From surrounding slope.
[b]I = insufficient count to calculate percentage.
Source: adapted from Pearsall and Trimble (1984:125).

Sampling Vegetation

Describing plants growing on plots sampled for modern surface phytolith analysis is clearly one way study of present-day vegetation aids archaeological phytolith analysis. Another way is investigating the nature of phytolith production in plant species. Earlier I described recent advances in phytolith taxonomy which are outgrowths of applied phytolith research. This work has not only advanced our knowledge of the nature and distribution of phytoliths in the plant kingdom but has also aided interpretation of archaeological phytolith assemblages and identification of utilized plants. Collecting comparative plant material and describing phytoliths produced by plants are important components of archaeological or paleoenvironmental phytolith research.

Ideally, an archaeological project incorporating analyses of botanical remains, whether macroremains, pollen, or phytoliths, should include field collection of wild and cultivated plants for comparative purposes. In reality, the paleoethnobotanist may not be able to spend time in the field, or there may be insufficient time or funds to do a full collection. In the latter case, comparative work should focus on obtaining materials not readily available from herbaria or other existing collections and on establishing the nature of the flora in the study area. This might include field identification or collection of ecological dominants, collection or purchase of crop plants, and collection of wild plants known to have been utilized.

As I discussed in Chapter 3, it is important to collect specimens that are identifiable, representative, and usable. When collecting for phytolith analysis, this means that specimens must be in good flower, so that they can be identified, and should include portions of the plant that would have been used and portions that could have been discarded as part of on-site processing. This is especially important for phytolith comparative specimens, since phytolith production differs among different portions of the plant. I try to collect whole specimens whenever possible—or, in the case of trees, good leaf and branch material and an example of wood, bark, and fruit.

One can often augment field collections by obtaining permission to remove broken pieces from herbarium voucher specimens. In some cases, this may be preferable to collecting specimens. For example, for my research during the Real Alto project it was necessary to test whether phytoliths thought to be diagnostic for maize occurred in any of the numerous wild grasses native to coastal Ecuador. My own collections of grasses were incomplete and not all specimens were identifiable. Using material from the Field Museum of Natural History herbarium allowed me to test grasses from a much broader range of ecological zones and to use only professionally identified materials.

I describe procedures for extracting phytoliths from comparative plant material in the next section. Several specimens of the same species should be collected or sampled if variation in phytolith production among individuals is to be investigated.

As is necessary in all comparative plant-collecting work, notes should be made on plant habit, flower color, plant size, range of habitats, associated plants, and common names and uses (refer to Chapter 3 for details on how to collect, press, and dry plant specimens).

Laboratory Analysis

In this section I present laboratory procedures for recovering phytoliths from soil and comparative plant materials. None of these procedures is overly difficult, but prior training in introductory chemistry laboratory techniques is recommended. Noxious chemicals are used in processing; exercise caution at all times, work in a fume hood where indicated, and wear proper protective clothing, gloves, and goggles. There are no standard processing procedures in phytolith analysis; procedures described below are those currently in use by experienced analysts. This discussion is organized into four sections: (1) phytolith laboratory, (2) recovering phytoliths from soil samples, (3) extracting phytoliths from comparative samples, and (4) scanning and counting procedures.

Phytolith Laboratory

Equipment and glassware needed to process and analyze soil or comparative phytolith samples are similar to those used in pollen analysis (refer to Chapter 4). Other components have been borrowed from sedimentary soil analysis. Essential equipment, although more extensive than that needed for macroremain analysis, is quite minimal by standards of modern analytical laboratories. There are, however, several sophisticated extras that can speed data analysis.

Phytolith analysis is a two-stage process: recovery of phytoliths from soil or comparative material, and examination of recovered remains. The first stage of analysis requires a processing laboratory, the second a microscopy laboratory. As in pollen analysis, these two aspects of research are best housed separately, since chemical fumes can damage microscopes, as well as bother the researcher, and the processing area must be closed off from other activities to minimize dust movement and possible contamination.

Processing Laboratory

A phytolith processing laboratory requires the following items of permanent equipment:

Fume hood with acid cup drain and acid-resistant working surface. I cannot understate the importance of carrying out all procedures involving noxious or corrosive chemicals in a properly installed and maintained fume hood. Soil phytolith analysis should not be attempted without access to a fume hood facility.

A fume hood certified for standard laboratory situations (presence of toxic, offensive, corrosive, and flammable fumes produced by routine laboratory operations such as evaporation, distillations, digestions) is sufficient for phytolith work. Although Rovner's (1971) original procedure called for use of perchloric acid in comparative processing, he (Rovner 1972) as well as other researchers now strongly advise against using this acid, which produces explosive residues, requiring use of a fume hood specifically certified as safe for perchloric acid.

The inside work surface of the hood should be made of material resistant to acid spills. A small cup drain with glass-lined pipes inside the hood facilitates decanting of acids. The hood should also have running water, several electrical outlets, and good internal light. Gas outlets for a bunsen burner are useful but not essential. The hood should be large enough that a tabletop centrifuge can be placed inside, with sufficient space left for working with samples.

Tabletop centrifuge. A sturdy, multipurpose tabletop centrifuge is essential for soil phytolith analysis. Since centrifuging includes spinning of heavy liquids (specific gravity 2.3) and 50-ml volumes, the centrifuge must be capable of handling heavy loads and be large enough to spin 50-ml test tubes. It is useful to have interchangable rotors: a 6- or 8-place rotor for 50-ml test tubes and a 12-place or larger rotor for 15-ml test tubes. Adapters that fit inside 50-ml rotor slots, allowing centrifuging of 15-ml tubes, can also be purchased. Since capacity of the centrifuge limits the number of samples processed per processing run, it is cost efficient to purchase a large-capacity centrifuge.

Hot water bath. A hot water bath is needed to hold samples during HCl and organic removal steps. General-purpose water baths can be purchased, or water can be heated on a hotplate or over a bunsen burner. I use a bath that consists of a hotplate, a metal basin to hold water, and a metal rack to hold tubes. (Fig. 5.18). The bath is used inside the fume hood and should be large enough to hold all samples being processed at the same time.

Sieving apparatus. In some procedures, samples are washed through a sieve to remove large soil particles and other material. The sieving apparatus used in my lab consists of a large funnel into which is fitted a 250-micron screen. Samples are washed through the sieve into a series of 50-ml test tubes held upright in a test tube rack (Fig. 5.19).

Low-temperature oven. A tabletop oven has several uses in phytolith processing. Some soil extraction procedures call for drying phytolith residue; an oven can also be used for drying mounted slides, moist soil samples, or leaf material that has been washed prior to processing.

Muffle furnace. Phytoliths can be extracted from comparative plant specimens either chemically or by ashing. If the ashing method is chosen, a muffle furnace or kiln is needed to heat specimens. I have processed much of my comparative collec-

358 • Chapter 5 Phytolith Analysis

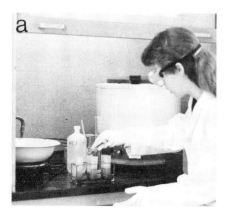
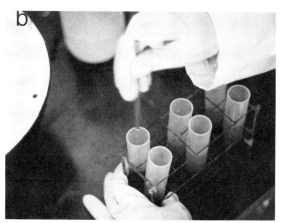

Figure 5.18 Missouri lab hot water bath setup; this photograph sequence shows treatment to remove carbonates: (a) about 30 ml of dilute HCl are added to each sample; (b) samples are mixed thoroughly; (c) samples are placed in hot water bath for 15–20 minutes, or until reaction ceases; (d) cooled samples are centifuged.

tion with a muffle furnace, which we also use for charring comparative specimens for macroremain analysis. Since only 500°C is needed for either purpose, a relatively inexpensive furnace can be used.

Balance. A balance precise to 0.01 g is needed for mixing heavy liquid used in soil processing. The lab should also have a triple beam or similar balance precise to 0.1 g for weighing out soil for processing and leaf material for ashing or chemical oxidation.

In addition to the permanent equipment listed above, the phytolith processing lab should have a sink for washing glassware, generous counter space with a rack for

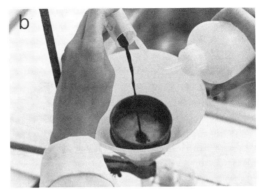
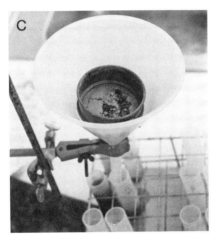

Figure 5.19 Missouri lab sieving setup; samples are passed through a 250-micron sieve to remove larger soil particles: (a) sample is poured into sieve; (b) all material is washed from the test tube with a distilled water jet; (c) after soil is thoroughly washed, large residue remains on the sieve and is discarded.

drying glassware and room to carry out procedures not requiring use of the fume hood, storage cabinets for chemicals, glassware, and samples, a rolling cart for moving the ashing oven to and from the fume hood, and access to a source of distilled water. The lab should also have a fire extinguisher, a first aid kit, and an emergency eyewash station. Inexpensive wall-mounted eyewash units that require no plumbing connections are available from scientific supply houses.

I use the following glassware and other expendable supplies during comparative and soil phytolith processing:

1000-ml glass beakers
10-ml graduated cylinder
1000-ml graduated cylinder
glass stirring rods
acid-resistant lab apron
acid-resistant gloves
heat-resistant gloves
safety goggles
polypropylene storage bottles
distilled water carboy
polyethylene washing bottles
beaker brush
test tube brush
round-bottom 50-ml polypropylene centrifuge tubes
conical 15-ml polypropylene centrifuge tubes
crucibles
wax pencils
pipettes
general-purpose tongs
clear flexible tubing
pycnometer
1-dram storage vials

Before leaving equipment, I should add a word about centrifuge tubes. Which are preferable, glass or polypropylene tubes? Glass centrifuge tubes are easily cleaned and do not become scratched or pitted, which happens eventually with polypropylene tubes. After prolonged submergence in boiling water baths, polypropylene tubes can become brittle and crack in the centrifuge. This is more of a problem in some wet oxidation procedures, rather than in routine soil processing, where polypropylene tubes hold up well. The major problem with glass tubes is breakage, especially if an angled, rather than swing-out, centrifuge head is used. In general, I recommend using high-speed polypropylene test tubes for all processing procedures, because breakage and loss of samples are more serious problems than contamination from pits or scratches in tubes. Breakage during centifuging contaminates all samples in a set and can cause injury from chemical spattering. The latter is an excellent reason to be sure the centrifuge lid is always closed when the instrument is running. Contamination from pitted tubes can be avoided simply by replacing all tubes periodically. If wet oxidation is carried out by prolonged boiling, tubes used in that procedure should be replaced frequently.

Microscopy Laboratory

After processing soil or comparative leaf material to extract phytoliths, work moves to the microscopy laboratory. This laboratory can be as simple as a desk with an optical microscope and supplies for mounting and storing specimens or as elaborate as a facility with scanning electron microscope and digital image analysis system. We start here with the basics; a good optical research microscope with camera.

Routine scanning and photographing of phytoliths requires magnifications of 200–400×. Higher magnifications are useful for detailed study of phytolith structure but are impractical for scanning and counting. A standard research grade microscope equipped with 10× to 15× eyepieces and a range of objectives (such as 10×, 25×, 40×, and 100× oil immersion) offers good versatility. My lab microscope has a planachromatic lens on the "scanning" objective (25×). This type of lens flattens the field of view, eliminating curvature at the edges of the field and minimizing eye strain. Polarized light and dark field capability are useful features for a phytolith microscope, as is a movable mechanical stage equipped with index and vernier scales (for determining the x and y coordinates of any object on a slide). An eyepiece micrometer is essential for measuring phytolith size.

A camera mounted on the microscope by means of a trinocular head is the surest means of documenting nature and variety of phytolith occurrence in samples. Camera systems vary considerably in complexity and price, but good results can be obtained by using a 35-mm camera with built-in automatic light meter. The camera setup illustrated in Figure 5.20 has a 10× lens in the photo tube. We photograph using the 40× objective, giving photographs at 400×. By having a camera exclusively for phytolith work, photographs can be taken during routine scanning work, providing documentation of new types, photographs for publication, and so on. At the Missouri lab we have developed a desktop key using black-and-white photographs of phytoliths (Dinan and Pearsall 1988). This is invaluable in maintaining consistency in classification.

As more distinctive phytoliths are recognized and described, and as work becomes more quantitative, storage and manipulation of raw data becomes increasingly time consuming and difficult. My first phytolith scans were recorded in handwritten form in a few lines in a notebook; now the lab uses three different

Figure 5.20 Missouri microscopy lab: (a) material is scanned under a research microscope with camera mount; (b) to mount specimens, Canada Balsam is placed on a clean slide, sediment is added and stirred in, and a cover slip is added and pressed down.

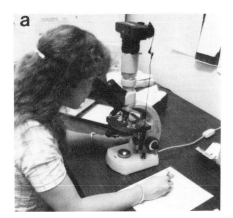

printed forms. Compiling and analyzing data recorded in this way is time consuming. Computer image analysis may offer an alternative to this type of record keeping (Russ and Rovner 1987).

An image analysis system allows visual data to be quantified, stored, manipulated statistically, and evaluated. By mounting a video camera on the microscope, phytoliths or other microscopic bodies in a microscope field can be projected onto a computer monitor screen. Using a video overlay module and a digitizer tablet, direct on-screen measurements of projected microscopic images can be made and stored in the computer. Phytolith counts or area estimations could be similarly stored. Application of image analysis in phytolith research is only beginning, but it holds much potential.

Another tool already proven useful in phytolith research is the scanning electron microscope. Scanning electron microscopy allows detailed study of phytolith morphology; SEM photographs of phytoliths give detailed, three-dimensional images not possible using light microscopy (Fig. 5.21). Piperno (1983) used SEM to study the minute structure of cross-shaped phytoliths. This research resulted in definition of variants within the cross-shape type. Although it is impractical to use SEM for routine scanning and counting, it is a valuable tool for basic research in phytolith morphology.

Figure 5.21 SEM photograph of *Zea mays* cross-shaped phytolith.

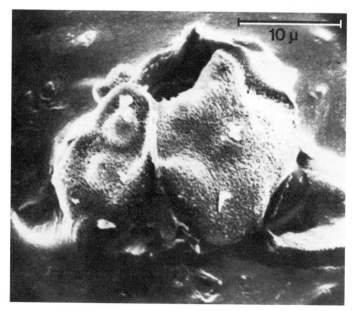

For the basic phytolith microscopy laboratory, the following items of small equipment and expendable supplies are necessary for routine work:

Laboratory counter. Standard counts are part of most phytolith scanning procedures. A simple hand-held counter is useful for keeping a running total of short-cell counts. A manual blood cell counter, with multiple recording keys and a totaling unit, can be used to tally common phytolith types or subgroups within a standard count.

Microscope slides, cover slips, mounting media, and related supplies. Phytoliths are mounted for viewing under light microscopy on glass microscope slides. Mounting procedures are described below; all involve use of slides, glass cover slips, and a liquid mounting medium such as Canada Balsam or Permount. Each slide should be labeled with laboratory number and provenience information. Disposable pipettes (wet mounting) or probe and spatula (dry mounting) are also necessary. Alcohol is used to clean hand tools. A slide warmer or low-temperature oven is useful for drying slides after mounting, although some mounting media, such as Canada Balsam, air dry overnight without heating.

Slide storage boxes. Mounted slides should be stored horizontally to prevent the mounting medium from "running" over time.

Soil-Processing Procedures

As mentioned earlier, there is no standard procedure for extracting phytoliths from soil; that is, no systematic tests have been conducted to determine whether different flotation media are better for some soils than others, whether different deflocculation schedules or procedures should be used on soils of different clay content, whether organic material should be removed, whether results differ if samples are fractionated or processed without fractionation, and so on. However, only a limited number of procedures are currently in use by experienced analysts. These have been found to work for a variety of soils and seem to give comparable results. Until systematic work on refining processing procedures is done, the beginner can choose any of the procedures described below and have reasonable expectation of success.

Most procedures currently in use are variations of three procedures published in the late 1960s and 1970s—those of Twiss et al. (1969), Rovner (1971), and Carbone (1977). I begin by presenting these procedures and then discuss more recent procedures developed from these, namely those used by myself (Pearsall 1979, 1984a), Piperno (1983), and Mulholland and Rau (1985). All of these have in common the basic technique for separating phytoliths from soil matrix: flotation using a heavy liquid medium. Differences include whether or not soil is fractionated at the beginning of the procedure, order in which carbonates, organic material, and clays are removed, and whether organic material is destroyed as part of processing. Removing carbonates and organic material is included in processing to remove all non-mineral elements from soil prior to heavy liquid flotation. Carrying out these removals prior to deflocculation (dispersing clays) may give better clay dispersal and cleaner removal, since clay can bind with organic material. This in turn may give cleaner silica flotation, since clays interfere with phytolith removal. Piperno (1988a) advises removing organic material, noting that it interferes with flotation. See Powers and Gilbertson (1987) for a sandy sediment procedure which does not use flotation.

Twiss et al. (1969) Procedure

Included in this important article on morphological classification of grass phytoliths is an abbreviated procedure for extracting phytoliths from soils (Twiss et al. 1969:11). This procedure, which is not detailed enough to follow without considerable background knowledge, is as follows:

1. Samples are fractionated by gravity sedimentation according to Jackson (1956); the fraction ranging from 10 to 25 microns is retained.

2. Clay aggregates are dispersed by adding 50 ml of a solution of Calgon (102 g/l) per 1000 ml soil suspension. The processing of settling is repeated 10 to 19 times, depending on concentrations of organic matter and clay in the samples. Sample weights are between 150 and 200 g.

3. After samples are dry, phytoliths are separated from silt by centrifuging in a suspension of bromoform and tetrachloromethane with a specific gravity of between 2.232 and 2.319. Specific gravity is checked using specific gravity glass cubes. To increase dispersion, about 2 ml of acetone per 1.5 g of silt is added to 6 ml of heavy liquid solution. The acetone is allowed to evaporate for 24 hours; the suspension is then centrifuged for 5 minutes at about 3000 RPM.

4. Floating material is decanted, filtered, and washed several times in acetone. After the phytolith residue has dried, it is mounted in Cadex on glass microscope slides.

This procedure begins with soil fractionation. Fractionation is the process of separating soil into particles of like size through gravity sedimentation. If a soil sample is suspended in water, particles of like size settle at the same rate; larger particles settle before smaller ones. Sedimentation rates for given size grades of minerals have been established by laboratory test, and tables of settling times published (Tables 5.5 and 5.6). Settling velocities of mineral material are governed by Stokes' law:

$$V = KD^2$$

where V is velocity of settling in cm/minute, K is a constant based on particle density and water temperature, and D is particle diameter in millimeters. As Table 5.6 illustrates, clay (0.002 mm, or 2 microns) is the largest grade particle in suspen-

Table 5.5 • Mineral Size Grades for Determining Settling Velocities

Fraction	International (mm)	USDA (mm)
Coarse sand (CS)	2.0–0.2	2.0–0.2
Fine sand (FS)	0.2–0.02	0.2–0.05
Silt (Z)	0.02–0.002	0.05–0.002
Clay (C)	< 0.002	< 0.002

Source: adapted from Smith and Atkinson (1975: 116).

Table 5.6 • Settling Velocities in Water, from Stokes' Law

Settling time[a]	Settling velocity (cm/sec)	Max. particle diameter[b]
4 min 48 sec	0.0347	0.0200*
20 min	0.00833	0.0098
1 hr 15 min	0.00222	0.00506
8 hr	0.000347	0.00200†
25 hr	0.000111	0.00113

[a]Time required to settle through 10 cm water at 20°C.
[b]Maximum diameter in suspension at 10 cm depth (mm c.s.d.). *international silt plus clay; †international clay.
Source: adapted from Smith and Atkinson (1975: 116).

sion after 8 hours of settling in a 10-cm water column; silts and clays are both still in suspension after 4 minutes 48 seconds. Depending on when liquid is poured off, different size particles are present in supernatant and residue. Pouring off water after 8 hours removes clay from soil. Note that only the 10–25-micron fraction was analyzed by Twiss et al. (1969). Although this is sufficient to recover many grass phytoliths, a broader range of sizes should be analyzed if the full range of silica present in the soil is to be documented.

Rovner (1971) Procedure

Rovner's (1971) procedure is designed to avoid the problem of excluding some phytoliths from consideration on the basis of size, so no fractionation is carried out. Rovner also eliminates the use of filters to avoid possible contamination from this source. Distilled water is used throughout the procedure. Rovner's procedure is as follows:

1. Place 2 to 5 g of soil in 50 ml of distilled water and 1–2 g of Calgon and leave to disaggregate overnight. After decanting the liquid, refill samples with distilled water, allow to settle, and decant. Rinse the samples a second time.

2. Place 1–2 g of soil in a 15-ml test tube and add about 10 ml dilute HCl. Test tubes may be placed in a hot water bath to accelerate the reaction, which removes organic material and carbonates that can bind phy-

toliths with other material in the soil. Excessively high temperatures should be avoided in the water bath.

3. Centrifuge the samples at moderate speed for 20 to 30 minutes and decant and discard the liquid. Refill the samples with distilled water, agitate, centrifuge, and decant. Repeat this procedure until the supernatant liquid is clear.

The original procedure called here for repeating this step using a nitric-perchloric acid oxidation mixture to remove organic material from soils with high organic content. Rovner (1972) later advised against using this potentially explosive mixture. Treatment with Schulze solution (3 parts concentrated nitric acid, 1 part saturated potassium chlorate) or concentrated nitric acid may be substituted.

4. Remove water from samples by drying in a low-temperature oven at 50° to 60°C or by adding absolute ethyl alcohol to samples, centrifuging as in Step 3, and decanting the supernatant liquid.

5. Prepare a heavy liquid mixture by adding tetrabromoethane to absolute ethyl alcohol until a specific gravity of 2.3 is reached. Specific gravity glass cubes can be used to determine when correct specific gravity has been reached.

6. Add approximately 10 ml of heavy liquid to each dried sample. Agitate samples to float phytoliths from heavier sediments. Centrifuge the samples for 30 minutes and decant the liquid carefully into a 40- or 50-ml centrifuge tube. Repeat this step until phytolith yield is negligible.

7. The supernatant liquid containing suspended phytoliths can also be removed by freezing the sediment in an ethanol and dry ice bath before decanting, or by using a pipette to transfer the liquid. These methods may produce a cleaner separation. If a pipette is used, it should be flushed with absolute ethyl alcohol into the large centrifuge tube to ensure that the entire phytolith sample is removed.

8. Add absolute ethyl alcohol to each sample to reduce the specific gravity below 1.5. Specific gravity can be checked with a heavy liquid hydrometer. Phytoliths will collect at the bottom of the tube. Centifuge and decant the liquid. The tetrabromoethane–alcohol mixture can be reclaimed in a separatory funnel. This is done by adding distilled water to the used mixture. Water combines with alcohol to form a mixture immiscible with the tetrabomoethane, which can then be drawn off for reuse.

9. Wash the phytolith sediment with absolute ethyl alcohol to remove any remaining heavy liquid. Agitate the sample, centrifuge, and decant. Repeat to ensure a complete rinse. Dry samples in a low-temperature oven.

10. Mount the sediment in Canada Fir Balsam.

I used a modified version of this procedure to process more than one hundred samples from the Real Alto site, Ecuador, with excellent recovery of phytoliths. The modifications I incorporated are described below.

My procedure also includes substitution of cadmium iodide and potassium iodide for tetrabromoethane. After prolonged exposure to tetrabromoethane, both myself and my lab assistants experienced discomfort from fumes produced by this chemical, in spite of using a fume hood for all steps involving the heavy liquid. I switched to the cadmium iodide–potassium iodide mixture after reading of Piperno's (1983) excellent results with that solution. I have found it to give very clean separations and have not experienced any difficulties using it. Mulholland and Rau (1985), as well as Rovner himself, have adopted the use of zinc bromide rather than tetrabromoethane. At the present time, tetrabromoethane is not the chemical of choice of most analysts.

Carbone (1977) Procedure

Carbone's procedure was developed as part of research into the usefulness of phytoliths as paleoecological indicators in Paleoindian and early Archaic sites in the middle Shenandoah Valley. A considerable part of the research effort was aimed at finding an alternative to the commonly used mixture of tetrabromoethane and nitrobenzene, which the author considered too costly and hazardous, and at devising an economical procedure. The cadmium iodide–potassium iodide solution mentioned above was the result. This procedure includes soil fractionation; only the silt fraction (5–50 microns) is processed. The decision to process only the silt fraction came as the result of a series of whole soil mounts of different soil fractions (size fractions), which showed most phytoliths to be concentrated in the silt fraction. The Carbone procedure is as follows:

1. Boil samples in water for several hours with approximately 30 ml of 30% hydrogen peroxide; this destroys organic matter in soil. After boiling, remove the hydrogen peroxide solution by means of ceramic filters. Wash and filter the samples three times and place them in an oven to dry overnight.

2. Weigh the dried samples and transfer them to shaking bottles. Add Calgon to aid in soil dispersion; place the bottles in a mechanical agitator overnight.

3. Following dispersion, fractionate the samples. Remove the sand fraction by wet sieving through a 53-micron (270 mesh) sieve. Separate the clay fraction less than 5 microns in size by repeated dispersions and decantations at specified time periods for particles of 5 micron size (i.e., Table 5.6).

Further subdivide the silt fraction into fine silt (5–20 micron) and coarse silt (20–50 micron) fractions. Dry the silt fractions.

4. Suspend 1 g of dried sample in a saturated solution of potassium iodide and cadmium iodide which has been boiled down to 2.3 specific gravity. In order to minimize loss of the heavy liquid, the following special procedure was devised. Pour the liquid into a piece of flexible plastic tubing, which is folded into a U shape. Introduce the sample into the liquid and agitate. Insert the plastic tube, still folded, into a test tube for centrifuging. After centrifuging, clamp off the tube just below the layer of floating phytoliths, and decant the phytoliths onto a filter; this economizes loss of liquid and does not disturb the settled soil. Repeat this step five times, until all phytoliths are removed.

5. Rinse the phytolith residue on the filter with acetone, dry it and weigh it.

6. Mount the dried samples in Cadex on microscope slides.

Pearsall (1979, 1986a) Procedure

I began phytolith processing with a modified version of the Rovner (1971) procedure. This initial procedure (Pearsall 1979) was replaced by a procedure incorporating use of cadmium iodide and potassium iodide as the flotation medium (Pearsall 1986a). The current Missouri laboratory soil extraction procedure, run on six samples at a time, is as follows:

0. (optional step prior to Step 1) To remove organic material from highly organic soil, place 5 g of soil in a beaker and add 100 ml of 30% hydrogen peroxide. Reaction can be vigorous; use eye protection. Leave for 24 hours, or until the reaction ceases. Decant the liquid and proceed with normal deflocculation (Step 1). Concentrated nitric acid, with the addition of potassium chlorate crystals, is also reported to give good organic removal (Piperno 1988a).

1. Place 5 g of dry soil in a beaker, and add distilled water to a height of 10 cm. Add approximately 1 tsp. dry automatic dishwasher detergent (active ingredient, sodium hexametaphosphate), stir to mix, and allow to settle for 8 hours. Decant the liquid with a siphon hose (hose should be thoroughly rinsed between samples). Repeat until the decanted liquid is clear, usually 4–5 times in all. This step, deflocculation, removes the clay fraction (particles <2 microns) from the soil, leaving the silt (2–50 microns) and sand (>50 microns) fractions. Use more soil if silica content is low.

2. Begin heating water for heated bath (see Steps 4, 6).

3. Wash the sediments that remain after Step 1 from beakers with

distilled water into 50-ml centrifuge tubes. Centrifuge for 10 minutes at 3000 RPM; decant.

4. Add 30 ml of dilute (10–15%) HCl to each sample, stir to mix, and place in a heated bath for 15 to 20 minutes or until the reaction ceases (see Fig. 5.18). Stir the samples periodically to stimulate reaction. This step should be carried out under the fume hood. Carbonates are eliminated.

5. After the reaction ceases and the tubes cool slightly, centrifuge for 10 minutes and decant. Refill with distilled water, centrifuge, and decant.

(optional step after Step 5) To remove humic colloids from highly organic soils, add 30 ml of 10% potassium hydroxide to each sample. Stir to mix, and place in a heated bath for 5 minutes (a longer heating period can etch silica bodies). This step is carried out under the fume hood. Stir periodically. Cool tubes slightly, centrifuge for 10 minutes, and decant.

6. Set up screening apparatus (large funnel with 250-micron sieve [60 mesh] in mouth, twelve test tubes, test tube rack) (see Fig. 5.19). Wash each sample through the 250-micron sieve with a distilled water jet. Wash the sediment until two centifuge tubes are filled for each sample. Discard the residue on the screen. Wash the screen and funnel and rinse with distilled water between samples. Centrifuge and decant. Combine each pair of tubes into one. Centrifuge and decant.

7. Add 10 ml of heavy liquid (cadmium iodide and potassium iodide solution, specific gravity 2.3) to each sample; stir to float phytoliths. This step is carried out under the fume hood. Centifuge for 20 minutes and decant the supernatant liquid into a clean centrifuge tube. Repeat once for each sample. (Repeat several more times if silica content is low.)

8. Discard the residue and save the supernatant liquid. Add 25 ml of distilled water to each of the tubes with supernatant liquid. Stir vigorously to precipitate phytoliths (drops specific gravity below 1.5). Centrifuge for 20 minutes. Pour the liquid into a chemical storage container; save the solid phytolith precipitate.

9. Add distilled water to wash the phytolith precipitate. Centrifuge for 10 minutes and decant carefully. Repeat once.

10. Add acetone to the samples, centrifuge for 10 minutes, and decant carefully.

11. Transfer the phytolith sediment to storage vials and allow to dry. Mount the residue in Canada Balsam.

I added a multiple-stage deflocculation step (Step 1), also used by Carbone, Piperno, and Mulholland and Rau, to the beginning of this modified Rovner procedure to remove most of the clay fraction (particles <2 microns) from samples.

This gives much cleaner final samples, since if clay is left in sediment it can float in the heavy liquid solution along with phytoliths. Samples are settled long enough for the silt fraction (particles 2–50 microns) to settle; clays still in suspension are siphoned off. Since some clay particles also settle from lower in the water column, this step is repeated until the supernatant liquid is clear, indicating removal of most clay. Care must be taken not to let samples settle too long, since this allows more clays to settle out and lengthens the time needed for deflocculation. For water at 20°C, all silt settles in a 10-cm column of water in 8 hours (Smith and Adkinson 1975).

Removing organic matter from samples can also produce better flotation and cleaner residue. We found that in highly organic phytolith soil samples, such as those processed during the Waimea-Kaiwaihae project (Pearsall and Trimble 1983, 1984), clays were not removed during routine deflocculation but remained in samples. Organic material acts as a cementing agent, causing clay particles to stick together and settle faster (Kilmer and Alexander 1949). Treating samples with a strong oxidizer (hydrogen peroxide) before deflocculation not only removes organic material but also releases more clays. Potassium hydroxide, found effective at removing humic colloids from peaty soils (Piperno 1983), was inadequate for the Hawaiian sediments. Hydrogen peroxide can also be used to aid dispersion in highly clayey soils where sodium hexametaphosphate may be ineffective. Removal of organics is done prior to deflocculation. Removing carbonates (HCl step) prior to deflocculation might also enhance clay removal. The order given by Kilmer and Alexander (1949) for processing steps before mechanical analysis is (1) treatment with HCl, (2) treatment to remove organics (hydrogen peroxide), and (3) dispersion, using an agent such as sodium hexametaphosphate. Samples are then separated into size fractions using gravity sedimentation.

Although I agree with Rovner (1971) that processing unfractionated samples allows easy recovery of phytoliths of all sizes, I have added one sieving step to the procedure (Step 6). Passing samples through a 250-micron sieve (60 mesh) removes several sand fractions (very coarse, coarse, and medium; 2.0–0.25 mm diameter particles) (Kilmer and Alexander 1949) but leaves the fine sand and very fine sand fractions (0.25–0.05 mm particles), which may contain larger phytoliths such as those found outside the Gramineae. We have encountered no difficulty in flotation using samples of mixed particle sizes, provided that the clay fraction is removed before flotation.

The heavy liquid mixture of cadmium iodide and potassium iodide is prepared at the Missouri lab following the procedure in Piperno (1983, 1988a) rather than the boiling procedure described in Carbone (1977). Piperno gives the following formula for a 2.3 specific gravity solution of cadmium iodide–potassium iodide: 470 g cadmium iodide and 500 g potassium iodide in 400 ml distilled water. Gently warming the solution helps dissolve the dry chemicals. Quantities in this formula are only approximate; specific gravity of each batch of solution should be checked with a pycnometer and analytical balance. Weigh the pycnometer, add the solution, weigh both, and then determine the net weight of the solution. Since the specific gravity of a fluid is the weight, in grams, of 1 ml of the fluid, 1 ml of heavy liquid solution should weigh 2.300 g, 2 ml should weigh 4.600 g, and so on. If the specific gravity is less than 2.3, add more potassium iodide and cadmium iodide in small increments in the ratio given in the formula.

Piperno (1983) Procedure

The procedure used by Piperno (1983) for her study of archaeological and geological sediments from Panama was adapted from the procedures outlined by Twiss et al. (1969), Rovner (1971), and Carbone (1977). Piperno's procedure is as follows:

1. Deflocculate soil samples by repeated stirring in a 5% solution of either Calgon or sodium bicarbonate.

2. Pass wet soil samples through a 53-micron sieve (270 mesh) to separate the sand fraction from the silt and clay fractions.

3. Remove clay by gravity sedimentation. Place the soil samples in large beakers and add water to a height of 10 cm. Stir the samples vigorously, and after 1 hour carefully pour off the supernatant liquid. This process leaves behind particles from 5 to 50 microns. Repeat this step seven or eight times to remove most of the clay from samples.

4. To fractionate the silt fraction into two sizes, fine (5–20 microns) and coarse (20–50 microns), place samples in 100-ml tall beakers and add water to a height of 5 cm. Stir, allow to settle for 3 minutes, and pour off the supernatant liquid into a 1000-ml beaker. Repeat, allowing the sediment to settle for 2 minutes 20 seconds, and pour off the supernatant liquid, adding it to the beaker. Repeat this step seven or eight times to complete fractionation.

5. Place 1–1.5 g of each silt fraction and the sand fraction into 16 × 100-mm test tubes and wash with distilled water.

6. Add a 10% solution of hydrochloric acid to each sample, centrifuge at 500 RPM for 3 minutes, and decant the liquid. Repeat until no reaction is observed on addition of HCl. Wash the samples with distilled water. This step removes carbonates.

7. Add either concentrated nitric acid or a 3% solution of hydrogen peroxide to the samples, place centrifuge tubes in a boiling water bath, and

heat until reaction ceases. Repeat once. This step removes organic material. A mixture of potassium chlorate and nitric acid has also been used to remove organic matter (Piperno 1988a).

(optional step for removing humic colloids from peaty samples) Before heavy liquid flotation, add a 10% solution of potassium hydroxide to the samples and place them in a boiling water bath for 5 minutes. This step removes humic colloids, which can bind to phytoliths, preventing flotation.

8. Add 10 ml of a heavy liquid solution of cadmium iodide and potassium iodide, specific gravity 2.3, to each sample (this solution is mixed as detailed above). Mix thoroughly and centrifuge for 5 minutes at 1000 RPM. Remove the floating phytolith fraction using a Pasteur pipette and transfer it to another test tube. Remix the sample, centrifuge, and remove additional floating phytoliths. Remix and recentrifuge samples until most phytolith material is removed.

9. Add distilled water to the liquid containing the phytolith fraction in a ratio of 2.5 to 1. Phytoliths will settle to the bottom of the centrifuge tube. Centrifuge at 2500 RPM for 10 minutes and decant. Repeat twice to remove all heavy liquid from phytoliths.

10. Wash the phytolith fraction in acetone. Mount on slides in Permount.

Notice that the sedimentation time for separating clays from silts given in Step 3 (1 hour) is considerably shorter than the 8 hours given in the Pearsall procedure. This is due to a difference in boundary between clay and silt fraction. If this boundary is placed at 5 microns, silt sediments settle in 1 hour from a 10-cm height; if placed at 2 microns, 8 hours are required for complete settling of silts. The next procedure discussed also uses a boundary of 5 microns.

Mulholland and Rau (1985) Procedure

Mulholland and Rau, working in collaboration with Rapp at the Archaeometry Laboratory of the University of Minnesota–Duluth, have developed detailed procedures for extracting phytoliths from soil samples and plant tissue. Their soil extraction procedure uses zinc bromide as the heavy liquid flotation medium. Samples are fractionated in this procedure. A condensed version of the Mulholland and Rau (1985) procedure is presented below:

1. Split each sample into two subsamples weighing between 2 and 3 g. Save one subsample for duplication; weigh the other and place it in a plastic bottle. Add 30 ml of 0.005N sodium hexametaphosphate and leave overnight. This step disaggregates the sample.

2. Sieve the sample through an 88- or 90-micron sieve using an aspirator. Larger mesh sieves can be used for recovery of larger phytoliths. Transfer the sand fraction to a beaker and save. Pour the sieved sample into 600-ml beakers.

3. Check the pH of the silt–clay fraction. If carbonates are present (pH <7) add 10% HCl to dissolve carbonates and disaggregate any clay lumps. Add HCl 1 ml at a time, stirring thoroughly, until pH equals 7. Disaggregation of carbonates will be complete. If necessary, bring the pH back to 7 with drops of $3N$ NaOH.

4. Settle the silt fraction from the clay fraction (fractionation). Bring the volume of each beaker up to about 10 cm depth with distilled water. Add 5 ml of $0.005N$ sodium hexametaphosphate to each beaker, stir thoroughly, and let settle for 1 hour. Carefully pour off the supernatant clay suspension and save. For each sample, combine the remaining sediment into one beaker, refill to 10 cm with distilled water, add 5 ml of dispersant, and repeat the settling process. Repeat this step until the supernatant liquid is essentially clear after the 1-hour settling period. At that point, silt-size particles 5 microns and larger (including phytoliths) are the only sediment size left in the beaker.

5. Transfer each silt fraction to a 15-ml centrifuge tube, centrifuge for 15 minutes at a moderate setting, and decant. Repeat as often as necessary to transfer the entire sample.

6. Wash the sample with 10 ml of 10% HCl. Mix thoroughly with glass stirring rods, centrifuge, and decant. Repeat once.

Steps 7–13 must be done under the fume hood. Zinc bromide ($ZnBr_2$) is poisonous as well as very corrosive when mixed with HCl and water. Wear goggles, gloves, and a lab apron or coat when handling this chemical.

7. Prepare the $ZnBr_2$ heavy liquid to a specific gravity of 2.3 as follows. Stir 20 ml of concentrated HCl into 215 g of $ZnBr_2$. Heat will be generated. Slowly add 40 ml of distilled water in small amounts and stir thoroughly. Use extreme caution. Allow to cool to room temperature. Test specific gravity using a 2.3 sink/float standard. To adjust the specific gravity, add small amounts of distilled water or $ZnBr_2$ (makes approximately 125 ml of solution).

8. Add 10 ml of heavy liquid to each sample. Mix thoroughly with glass rods. Centrifuge for 15 minutes at moderate speed. All mineral grains with a specific gravity greater than 2.3 spin to the bottom (heavy fraction); phytoliths and other light particles remain suspended in the heavy liquid mixture. Decant the supernatant heavy liquid into a clean, dry 50-ml test tube, leaving the heavy fraction behind.

9. Add 19 ml of 10% HCl to the supernatant liquid and mix thoroughly with glass rods. This lowers the specific gravity to 1.5, causing phytoliths to sink to the bottom. Centrifuge and decant the supernatant (light fraction) into a beaker or 250-ml flask, retaining the phytolith fraction in the test tube. To ensure complete phytolith recovery, repeat separation of the heavy fraction twice (Step 8); each time, decant the resulting supernatant heavy liquid into the previously obtained phytolith sample, add 19 ml 10% HCl, mix, centrifuge, and decant the light fraction (combine with the previous light fraction). Cover and store the final heavy (mineral residue) and light (1.5 specific gravity liquid) fractions as a check on procedure.

10. Rinse the recovered phytolith fraction with distilled water. Centrifuge and decant. Repeat rinse to ensure removal of all heavy liquid. Rinse with 10 ml of 95% ethanol. Centrifuge and decant.

Concentrate and store the various fractions as follows.

11. (*Phytolith fraction*) Rinse 1-dram glass vials with 95% ethanol. Using disposable pipettes, transfer phytolith samples to vials. Cover each with a square of parafilm and cap. Label tops and sides with provenience or other pertinent data.

12. (*Heavy fraction*) Rinse twice with distilled water and once with 95% ethanol. Transfer from test tube to cleaned 1-dram vial. Cap with parafilm and label.

13. (*Light fraction*) Dilute by adding 20 ml of distilled water. Transfer to 50-ml test tube and centrifuge. Decant supernatant liquid into a dark bottle. Repeat until the entire sample has been centrifuged and decanted. Label the dark bottle "used, unfiltered $ZnBr_2$" and save for cleaning. Rinse the residue in the bottom of the centrifuge tube with 95% ethanol. Transfer to cleaned 1-dram vials; cap with parafilm and label.

14. (*Sand fraction*) centrifuge supernatant in 50-ml test tubes, decanting and repeating until all excess water has been processed. Using fine sprays of 95% ethanol, wash the residue into cleaned 1-dram vials. Add sediment from test tube. Cap with parafilm and label.

15. (*Clay fraction*) Let each container of clay suspension settle for at least 56 hours. Decant excess water and consolidate the sediment into one bottle. Let this settle for 56 hours and decant. Centrifuge the remaining water and sediment in 50-ml test tubes, decanting and repeating until the entire sample is processed. Rinse with 95% ethanol. Transfer the sediment to a clean 1-dram vial. Cap with parafilm and label.

Rovner and Noguera (1986) have suggested changing the procedure for preparing zinc bromide heavy liquid described in the Mulholland and Rau (1985) processing procedure. Contrary to normal laboratory procedure, in the Mulholland and Rau (1985) procedure water is added to acid (full strength HCl is first added to zinc bromide granules, then distilled water is added to the mixture). These authors explain that this is done to preclude formation of zinc precipitates and caution that, because water floats on acid and heat is generated at the interface of acid and water, water should be added slowly, in small amounts, and stirred in thoroughly (Mulholland and Rau 1985). Rovner and Noguera suggest the following modification. Add measured amounts of HCl to 40 ml of distilled water until all 20 ml of acid has been stirred in. Slowly add 215 g of zinc bromide powder; stir with a magnetic stirring device.

As I pointed out in the beginning of this section, no one procedure has emerged as standard in phytolith processing. Whether samples should be fractionated is one area of difference, as is sequence of removal of clays, carbonates, and organic material. Zinc bromide and a mixture of potassium iodide and cadmium iodide are currently in widest use as flotation media. The field would benefit from side-by-side tests of different procedures using a variety of soil types.

Processing Comparative Plant Material

Another laboratory procedure essential to phytolith research is the extraction of phytoliths from comparative plant specimens. Although a number of aids for identifying phytoliths have been published, such as Twiss *et al.* (1969), Brown (1984), and Mulholland (1986a) for Gramineae and Rovner (1971), Pearsall (1979), Piperno (1985d, 1988a), and Bozarth (1986) for other taxa, there is as yet no comprehensive reference text for phytolith identification. The beginner must not only familiarize himself or herself with the available literature on phytolith classification but must also prepare a reference collection containing ecological dominants, utilized plants, and crops from the study area.

There are two widely used techniques for extracting phytoliths from plant tissue: chemical oxidation and dry ashing. A third technique, preparation of epidermal peels, has limited uses. Many analysts use wet, or chemical, oxidation procedures for phytolith extraction. In these procedures, a strong oxidizing agent is used to digest the organic component of a sample, leaving silica and other inorganic materials. Rovner (1971, 1972) advocated wet oxidation over the more traditional technique of incineration for extracting phytoliths, citing studies such as Jones and Milne (1963), which reported distortion (warping, twisting), blurring of detail, or carbon occlusion in phytoliths extracted by dry ashing. The Jones and Milne (1963) study also reported that dry ashing led to incorporation of accessory elements such as sodium, potassium, calcium, and magnesium into silica bodies. Water is released from heated plant tissue at 60–150°C and again at 570–670°C. At 1150°C, opal silica crystallizes into critobalite and tridymite (Jones and Milne 1963).

Other authors have reported comparable results using wet oxidation and dry ashing. Labouriau (1983), for example, reports no differences between silica bodies extracted from *Casearia grandiflora* using the combined dry ashing and wet oxidation procedure described below and the wet oxidation procedure recommended by Jones and Milne. I observed no differences in quantitative assemblages of short cell phytoliths in maize leaves processed by dry ashing and by epidermal peel (Pearsall 1979). However, comparison of cross-shaped phytolith measurements revealed that some shrinkage of phytoliths did occur in dry ashed samples, presumably due to loss of water during heating (Table 5.7). Samples presented in Table 5.7 were ashed at

Table 5.7 • Size Distributions of Cross-Shaped Phytoliths among Ashed and Peeled Maize Leaf Samples

			No. crosses	Percentage by size			
				Small	Medium	Large	Extra-large
Modern grasses (ashed)							
563	*Oplismenus rariflorus* Presl.		33	33.3	60.6	6.1	0
564	*Olyra latifolia* L.		63	19.0	76.2	4.8	0
578	*Tripsacum dactyloides* L.		1	0	0	p	0
591	*T. laxum* Nash.		54	22.2	74.1	3.7	0
594	*Pennisetum nervosum* Trin.		15	13.3	80.0	6.7	0
Modern maize (ashed)							
643	*Sabanero*		97	7.2	53.6	38.1	1.0
645	*Canguil*		97	7.2	86.6	6.2	0
646	*Canguil*		101	1.0	76.2	22.8	0
662	*Patillo*		101	3.0	29.7	61.4	5.9
663	*Mishca-Chillo*		100	0	26.0	66.0	8.0
664	*Blanco Blandito*		106	1.9	38.7	57.5	1.9
Modern maize (peeled)							
E1	*Mishca*		56	12.5	62.5	25.0	0
E2	*Mishca-Huandango*		50	0	46.0	52.0	2.0
E3	*Sabanero*		58	6.9	81.0	12.1	0
E4	*Morochon*		58	8.6	50.0	41.4	0
E5	Purple Flint		44	77.3	22.7	0	0
E6	*Patillo*		52	5.8	78.8	15.4	0
E7	*Canguil*		43	11.6	74.4	14.0	0
E8	*Patillo*		48	0	50.0	47.9	2.1
E9	*Cuzco*		53	0	24.5	62.3	13.2
Ashed vs. peeled maize							
160-5	*Cuzco*	ashed	49	4.1	42.9	53.1	0
	upper leaf	peeled	47	2.1	38.3	57.4	2.1
101-1	*Morochon*	ashed	65	1.5	47.7	50.8	0
	lower leaf	peeled	66	1.5	43.9	54.5	0
190-5	*Chillo*	ashed	64	3.1	54.7	40.6	1.6
	upper leaf	peeled	67	1.5	44.8	29.9	23.9

Source: adapted from Pearsall (1979).

500–600°C, following the procedure described below. Differences in size distribution of cross-shape types between ashed samples of wild grass and maize were clear, even with some shrinkage. This observation, and the low cost, safety, and speed of dry ashing, led me to continue using the technique for the Real Alto research (Pearsall 1979). It continues to be my procedure of preference, but I now use 400–500°C for ashing, which keeps samples below the temperature at which Jones and Milne (1963) report a second water release.

As is the case for soil-processing procedures, procedures for extracting phytoliths from leaf material need systematic testing and evaluation. All procedures described below have been used successfully by experienced analysts, although, as Rovner and Noguera (1986) report, wet oxidation procedures using Schulze solution do not work equally well on all plants. After a brief description of procedures for dry ashing and making epidermal peels and spodograms, I review several bulk wet oxidation procedures currently in use.

Dry Ashing (Incineration) Techniques

Twiss *et al.* (1969: 111) utilized the following procedure for extracting phytoliths from leaf material:

1. Wash two or three leaves from mature plants several times in distilled water and dilute HCl to remove any mineral particles on the surface and to soften mineralized tissue.
2. Place leaves in a ceramic crucible and ash for at least 6 hours at 500°C.
3. Wash ashed residue and mount in Cadex or Canada Balsam.

This procedure produced samples that contained spodograms (silica "skeletons" of leaf tissue, showing all silicified cells in growth position) as well as disarticulated silica bodies. Mounts of each type of silica material could be made.

The procedure we use at the Missouri lab (Fig. 5.22) is very similar to that published by Twiss *et al.* We cut material to be processed into small pieces, wash it in distilled water to remove surface dust or other deposits, and dry it in a low-temperature oven to remove excess water. We then place dry material in aluminum ashing boats (formed from aluminum foil) and ash at 400–500°C. Most samples are completely ashed in 2 hours; some require longer. Samples are finished when the residue is whitish-gray in color. Tissue first blackens. If samples are not heated long enough, these carbon deposits are not burned away, obscuring silica.

Labouriau (1983:6–7) describes a comparative-material processing technique combining dry ashing and chemical oxidation which has been used successfully for several years at the Instituto Venezolano de Investigaciones Científicas. This combined technique was developed specifically for removing alkaline and alkaline-earth metals, in the form of their soluble chlorides, from samples in order to avoid fusing

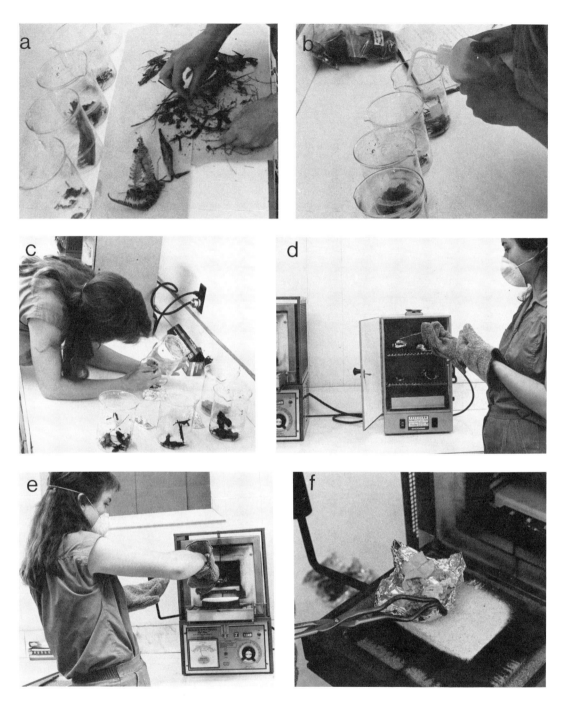

Figure 5.21 Dry ashing comparative material: (a) material for each sample is selected and placed in a beaker; (b) material is washed with distilled water to remove surface dust and other debris; (c) washed material is placed in aluminum ashing boats; (d) material is dried in a low-temperature oven; (e) dried material is transferred to a muffle furnace and heated at 400–500°C for several hours; (f) samples are finished when residue is a whitish-gray color.

silicates during ashing. Fusing distorts the shape of silica bodies or cements them together. The procedure is as follows:

1. Wash material with distilled water and dry at 100°C in a labeled porous clay crucible (disposable crucible).
2. Close the crucible with another crucible and char contents at 200°C for 2 hours.
3. Transfer the sample residue to a 250-ml beaker and boil for 10–30 minutes in 100 ml of $5N$ HCL.
4. Filter the HCl solution into a Buchner funnel over ashless filter paper and wash with water until no more chloride ions are detected in the washings. Presence of chlorides is tested with 1% $AgNO_3$.
5. Wrap the washed material in the filter paper disk and return to the crucible. Dry and char it in a closed crucible for 20 minutes.
6. Open the crucibles and heat for 2 hours at 800°C.

Epidermal Peels and Spodograms

Preparing epidermal peels is a traditional botanical technique for exposing the epidermal cells of leaf tissue in situ. In this technique, after several days of soaking in glycerol, dried leaves are sufficiently softened so that the upper epidermis and mesophyll tissue of the leaf can be peeled away with a razor blade, leaving the lower epidermis and vascular tissue intact. Phytolith short cells, which form across the veins, are visible for measurement and counting after peels are mounted in Canada Balsam. To use this technique, it is necessary to hold the specimen down firmly while cutting the epidermis away. Small leaf fragments, husk fragments, glumes, stems, and irregularly shaped material cannot be processed easily this way (Pearsall 1979).

Kaplan and Smith (1980) have developed several procedures for processing phytolith reference materials. I describe their procedure for bulk extraction of cell-free phytoliths below; here I present their procedure for producing spodograms from plant tissue.

Kaplan and Smith describe spodograms as "the cleared cellulose framework of the epidermis with the phytoliths in situ" (1980:1). Unlike the Twiss et al. (1969) ashing procedure presented above, which can result in ash spodograms, the Kaplan and Smith (1980) procedure is designed to clear plant parts of cytoplasmic contents without heating and to detach the epidermal layer with its associated phytoliths for study. The procedure is as follows:

1. Trim plant parts with a razor blade into 0.5-cm square pieces. Cut leaf squares so that the leaf margin is included on one edge; for larger stems

cut the epidermis away from the inner tissue; use the whole stem for smaller plants. Detach flowers from the inflorescence; remove fruits.

2. Place tissue squares in petri dishes containing 50% commercial Clorox (5.35% sodium hypochlorite) and allow to stand overnight or until tissues appear clear or milky. Clorox, an oxidizing agent, clears the tissues of their cytoplasmic contents.

3. Remove cleared tissues from the Clorox solution and place in dishes of distilled water to rinse off the oxidizing agent. These may be stored under refrigeration in water for several weeks.

4. Remove upper and lower epidermal layers for examination by agitating the solution containing the tissue squares, tapping the dish, and gently prodding with a glass rod or forceps until epidermal tissues separate from other tissues. Very thin leaves do not require separation.

5. Dehydrate and infiltrate specimens by processing through an ethanol dehydration series. Float specimens onto slides and inundate (using a pipette) with the following series of solutions: distilled water; 10% ethanol; 50% ethanol; 75% ethanol (2 changes); 95% ethanol (3 changes); 100% ethanol (3 or more changes); xylene (2 changes); dilute Canada Balsam in xylene.

If samples cloud upon addition of xylene, repeat the 100% ethanol washes. After adding mounting medium, place slides on a slide-warming table until the medium hardens.

Wet Oxidation Procedures

The wet oxidation processing procedure suggested by Rovner (1971, 1972) in his seminal article on phytolith analysis is used in various forms today. I first present the original Rovner procedure, then modifications of it by Rovner and Noguera (1986), Piperno (1983), and Rattel (1983). All use concentrated Schulze solution for oxidation. I then describe another procedure currently in wide use, that of Kaplan and Smith (1980). The Rovner (1971, 1972) Schulze procedure is as follows:

1. Soak a 1-g plant sample overnight in 50 ml of water with detergent. Flush with distilled water and decant. Add dilute HCl and allow to soak overnight. These procedures remove material adhering to the plant surface. Transfer material to a small metal screen and flush with distilled water to remove acid. Transfer to a centrifuge tube.

2. Prepare a Schulze solution by combining three parts concentrated nitric acid (HNO_3) and one part saturated potassium or sodium chlorate ($KClO_3$ or $NaClO_3$). These are strong reagents and should be handled very

carefully, under a fume hood. *Do not use concentrated nitric–perchloric acid mixture as published in Rovner (1971).*

3. Add 10 ml of the acid mixture to each sample and place in a heated water bath under a fume hood until digestion of organic material is complete; this takes from 2 to 4 hours. Place a glass stirring rod in each test tube to keep samples from overflowing while digestion takes place.

4. Centrifuge samples at moderate speed for 30 to 60 minutes. If considerable undigested material continues to float, remove about 5–7 ml of acid by suction pipetting from the area between the sediment and the suspendant. *Never pipette acids by mouth.* Add fresh Schulze solution and return to the water bath. A second digestion is usually necessary to ensure maximum disaggregation of phytoliths.

5. Centrifuge and decant. Remove acid from the sample by washing with distilled water. Agitate the sample, centrifuge, and decant. Repeat distilled water wash twice to remove all acids.

6. Dry sample in a low-temperature oven. Mount.

A number of analysts have encountered difficulties using this procedure. The main problem seems to be a resistance of some plants to digestion by the Schulze solution. In addressing this issue Rovner and Noguera (1986) note that a frequent problem is aging of the mixed Schulze solution; only small quantities of solution should be mixed at one time, so that the acid mixture is always fresh when used. They also report that pretreatment of plant tissue improves digestion. Among pretreatment steps they have found effective are the following (Rovner and Noguera 1986:6): (1) Cut or tear the plant tissue into small pieces; this exposes more surface area for digestion. (2) Boil the plant material in water to remove waxes, resins, and similar compounds. (3) Soak boiled material in an acetone or ether bath to reinforce the results obtained by boiling and to further soften tissues. (4) Soak material in chloryl hydrate solution to disaggregate fibers.

Rovner and Noguera (1986) report significant improvement in digestion of plant tissue found resistant to processing using the original Schulze procedure, but they emphasize that these pretreatment steps need further testing. Which treatments are most effective, or what combination is optimal for certain plant groups, is not known.

Piperno (1983:46) presents a very abbreviated procedure for wet oxidation which utilizes a mixture of nitric acid and potassium chlorate. This is one combination of acids that gives a Schulze solution. This procedure is very similar to Rovner's, except that plant material is placed in the acid solution before treatment with HCl. After organic material is dissolved, the phytolith residue is washed in distilled water, then washed in a 1N solution of hydrochloric acid. Samples are then washed again in water, and finally in acetone. The phytolith extract is allowed to dry and is mounted in Permount. Piperno notes that some plants were very difficult to digest (e.g., Euphorbiaceae). Tissue resisted digestion even after repeated additions of new solution. Piperno found that solid potassium chlorate granules added to the samples quickly digested the remaining organic matter.

Cheryl Rattel, of the University of Minnesota Archaeometry Laboratory, has developed a wet oxidation procedure that also uses concentrated Schulze solution (Rattel 1983, revised by Mulholland and Rau 1985):

1. Weigh out 0.25 g of each plant part to be processed and place in a labeled 50-ml beaker. Record plant and part numbers, part description, and any deviation from standard weight in a logbook.

2. Add a sufficient quantity of concentrated Alconox solution to immerse each sample (about 25 ml per sample). Concentrated Alconox solution is mixed by adding 1 g of Alconox to 50 ml of distilled water. Allow to soak at least 10 hours, but no more than 24 hours or mold may develop. Remove the plant parts with forceps and place them in a 500-micrometer Coors filtering crucible positioned on top of a widemouth Ehrlenmeyer flask. Flush the material with distilled water and return to a rinsed beaker.

3. Add sufficient dilute (10%) HCl to cover each sample. Soak overnight. If necessary, plants can remain in HCl up to a week. Again using the Coors crucible, flush the plant material thoroughly.

4. Allow samples to dry; this improves digestion. Samples may be dried in a low-temperature (100°C) oven or hot water bath, or by air evaporation. Place dried material in labeled 15-ml centrifuge tubes.

The remaining steps must be carried out under a fume hood with use of protective garments.

5. Prepare a Schulze solution by adding three parts HNO_3 to one part saturated $KClO_3$ or $NaClO_3$. Allow 15 to 20 ml per sample for the first day of digestion, 10–12 ml for the second day. The solution should be made fresh for each day of plant digestion.

6. Add 10–12 ml of Schulze solution to each tube using a manostat. Place a glass stirring rod in each tube and stir to remove air pockets. Place tubes in a hot water bath, leaving room for additional water. The bath should be at approximately 95°C, but not boiling, when samples are placed in it. Distilled water should be added to maintain water levels as high as possible throughout digestion, since low water levels can cause bumping of test tubes and possible sample contamination. Monitor the samples carefully for the first hour, since convection currents often cause undigested

material to rise to the top of the tube or to overflow. After the first hour, maintain acid level in the tubes at the 15-ml level. To accelerate digestion, break up parts with the stirring rod. Stir frequently. If plant parts are not completely digested after 10 hours, remove samples from the bath, rinse stirring rods with excess acid and remove, and centrifuge for 15 minutes at moderate speed. If material still floats, return the tube to the water bath without decanting. If all material sinks, decant acid, add fresh solution, and return to the water bath. Continue digestion until it is complete. Most samples require at least one centifuging cycle to facilitate complete digestion. Remove the completed samples from the bath.

7. *Wear gloves for the first wash in this step. Do not mix ethanol waste with Schulze solution, or a violent reaction will occur.* Add enough distilled water to the tubes to equalize the liquid level. Stir well. Rinse the stirring rods with distilled water while removing them from the tubes. Centrifuge for 15 minutes. Decant acid. Refill with distilled water, stir, centrifuge, and decant. Repeat, using wooden stirring rods to mix the samples. Use wooden rods for the remaining washes. Decant, add 8 drops of 1M KOH, and fill with distilled water. Centrifuge and decant. Rinse with distilled water, centrifuge, and decant. Repeat. Rinse 1-dram storage vials with distilled water and label. Using repeated fine sprays of 95% ethanol, transfer material from tube into vial. If any solid material remains in the tubes, use a wooden rod to assist transfer. Rinse tube and rod twice, or until the vial is full. If the vial becomes full before rinsing is complete, insert an empty vial below the storage vial in a centrifuge sleeve, centrifuge, and remove the liquid with a disposable pipette. Complete the transfer. Label vial, cover with parafilm to prevent evaporation, and cap.

Another wet oxidation procedure currently in use is that developed by Kaplan and Smith (1980) of the Department of Biology of the University of Massachusetts–Boston. This procedure is a bulk extraction technique like the Schulze procedures described above, designed to obtain free phytoliths by removal of all organic material from samples. The procedure is as follows:

1. Chop up plant material. Place each sample in a beaker with distilled water and sonicate to remove most of the surface dust and pollen. Decant water. Dry the samples overnight at approximately 57°C to dehydrate tissues.

2. Add 0.1 g of dehydrated plant tissue to 40 ml of commercial Fisher brand chromic–sulfuric acid solution; this solution is a strong oxidizing agent used routinely to remove organic material from laboratory glassware. Allow samples to stand overnight. Oxidation should be complete.

3. Swirl the resulting solution and divide among four 15-ml tapered glass centrifuge tubes. Centrifuge for 5 minutes at a low to moderate setting (clinical centrifuge with swing-out head). Decant the supernatant liquid.

4. Wash the precipitate, which consists of phytoliths and a small amount of organic debris, by resuspending it twice in distilled water with the aid of a vortex agitator. Follow the distilled water washes with a wash of 95% ethanol. Decant the alcohol.

5. Resuspend the alcohol-washed precipitate in 95% ethanol and store in glass vials.

I have used the Kaplan and Smith procedure with good results. The steps involving the chromic–sulfuric acid solution should be carried out under a fume hood. A protective lab apron and gloves must be worn. Samples may overflow early in the digestion process, so it is advisable to place a stirring rod in each sample (I use 50-ml centrifuge tubes for the digestion) and to monitor the samples for at least the first hour.

Scanning and Counting Procedures

In this final section on laboratory procedures I discuss approaches for scanning and counting phytoliths extracted from plant material or soil samples. This is the heart of phytolith analysis: observing types of silica bodies which occur in samples, classifying or describing them, and quantifying abundance of forms. From these raw data come archaeological and paleoenvironmental interpretations.

I begin by discussing how this process is carried out at the University of Missouri laboratory. This includes slide mounting, scanning and classifying phytoliths, quantifying data, and dealing with new types. I also describe procedures used by other analysts.

Missouri Laboratory Procedures

Every laboratory procedure I use currently has evolved during the years I have conducted archaeological phytolith analyses. Scanning and counting procedures have undergone marked changes. These have come about mostly in response to problems encountered during applied research rather than as directed inquiry into the best way to do things. This is not the ideal way to develop a corpus of procedures, but it reflects the state of the art of phytolith analysis.

Once phytolith material has been extracted from plant or soil samples, work moves from processing laboratory to microscopy laboratory. Extract from soil samples is allowed to air dry completely in covered test tubes, then is transferred to 1-dram (4-ml) glass storage bottles. Ashed samples are similarly stored. All samples are labeled with provenience information or specimen identification and lab number. I keep one master inventory of all soil samples; most comparative materials are numbered by specimen voucher number and stored alphabetically by family and genus.

When I first began phytolith analysis, I mounted enough dry extract on a microscope slide so that there were a variety of bodies to examine, but not so many that fields were cluttered and difficult to examine. "Enough" and "not too many" are not very exact, but most slides were fairly consistent. Later, however, it became clear that a standard slide-mounting procedure was needed. This would not only make it easier to teach new lab assistants how to mount slides but would also permit more precise comparison of results among projects. Creating standard mounts became especially critical when we began to quantify phytolith occurrence using methods other than counting (see below). I recommend that slides be prepared using a known quantity of phytolith material, whether mounted as dry extract or liquid suspension.

We developed a standard dry mounting procedure by crafting a small scoop which when filled contains the amount of extract "normally" used in slide mounting (i.e., the amount of extract that creates easily read slides). By repeatedly filling the scoop and weighing extract, we determined that on average we mount 0.1 g extract per slide. The scoop is used to mount samples without weighing out each one.

The slide-mounting procedure now in use at the Missouri lab for both soil and comparative phytolith material is as follows:

1. Label a clean, glass microscope slide (3" × 1") with project name and sample lab number (soil samples) or plant name and voucher number (comparative samples). White adhesive paper labels are used.
2. Using a spatula cleaned with alcohol, place a small quantity of Canada Balsam on the slide. Canada Balsam tends to thicken over time; it can be thinned with xylene. Xylene should be handled carefully and used in a well-ventilated room.
3. Using a small scoop, cleaned with alcohol and wiped dry, remove a standard amount of phytolith extract from the storage bottle and shake onto Canada Balsam. Using a probe, cleaned with alcohol and wiped dry, stir extract completely into balsam. Scoop and probe should never enter the Canada Balsam bottle.
4. Place a clean glass cover slip (22 mm × 22 mm) over the mount and press down with a probe handle until the mounted material spreads out evenly, extending no more than 1/8" beyond the edges of the cover slip. Let slides air dry flat. Most air bubbles escape as the mounting medium

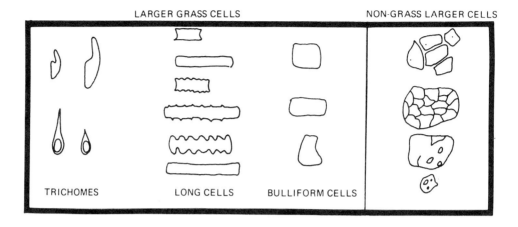
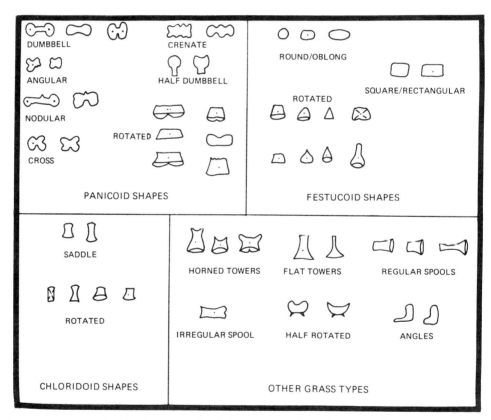

Figure 5.23 Modified Twiss et al. (1969) short-cell phytolith classification system used during the Waimea-Kawaihae project: top, long-cell silica bodies; bottom, short-cell silica bodies (from Pearsall and Trimble 1983:476).

dries. Store slides flat so that the balsam does not run. Clean all hand tools with alcohol between specimens. Prepare empty Canada Balsam mounts periodically to check for contamination of the mounting medium.

A slide drier is handy for speeding the drying process. Air-dried samples are most easily scanned if left to dry several days; otherwise, air bubbles interfere with scanning and counting. Mounting dry phytolith extract in Canada Balsam is the procedure recommended by Rovner (1971); it produces excellent mounts with good contrast. Slides that I prepared in 1976 are still stable and easy to read.

Once slides are mounted, scanning begins. We currently use three different scanning procedures for identifying and quantifying phytolith assemblages: short-cell scanning, quick-scanning, and 60-field scanning. I describe each of these procedures, and the circumstances for their use, below.

Every time a slide is scanned, phytoliths must be tallied by type. This process of classification is critical; each phytolith must be classified in a way that is consistent and replicable. Because my first work in phytolith analysis dealt with maize, I initially focused on short-cell counting. We still often tally short cells by named type using a modified Twiss *et al.* (1969) classification (Fig. 5.23) devised during the Waimea-Kawaihae project (Pearsall and Trimble 1983, 1984):

Panicoid. dumbbell, nodular, crenate, angular, cross, rotated, half dumbbell
Festucoid. round/oblong, rectangular/square, rotated
Chloridoid. saddle, rotated
Other Grass. horned towers, flat towers, regular spools, irregular spools, angles, half-rotated

This classification lumps a number of distinguishable shapes into larger form categories—dumbbell, nodular, and so on. This is adequate precision for distinguishing among different grassland formations, or, with cross measurements, for identifying maize.

When we need more detailed data on variety of phytoliths present in samples, we use a hierarchical classification in which each phytolith shape has a discrete number. There are currently ten major categories in our system (Table 5.8). Within each category, distinctive phytoliths are assigned unique number codes, which indicate how similar each is to other phytoliths in the category. Figure 5.24 illustrates a section of category 10, epidermal quadrilaterals. Long cell type 10ICa100, for example, is a two-dimensional rectilinear long cell (10I) with two serrated edges (10IC), no projections or perforations (10ICa), highly serrated (10ICa1), with rounded serrations (10ICa100). This type is most similar to 10ICa101, which is identical except for having pointed serrations, and is quite dissimilar to 10ICb2, a long cell with conical surface projections and highly serrated edges.

388 • Chapter 5 Phytolith Analysis

Table 5.8 • Major Categories of Phytolith Types in the Missouri Classification System

10	epidermal quadrilaterals (large cells of the epidermis, quadrilateral, square to elongated, often very two-dimensional in appearance, variable surface and edge characteristics)
20	epidermal, non-quadrilateral (large cells of the epidermis, irregular in shape, usually two-dimensional in appearance)
30	short cells (Gramineae, produced in the leaf vein area)
40	dermal appendages (prickles, hairs, edge spines, including base cells)
50	bulliform cells (located in the mesophyll or epidermis, usually three-dimensional in appearance, highly silicified, shapes variable: rectilinear, rounded, keystone)
60	anatomical origin unknown (a category for plant silica of unknown origin in the plant)
70	honeycombed material (amorphous silicified material, usually of irregular shape, lightly silicified, breaking easily)
80	spheres (mesophyll cells of spherical-spheroidal shape)
90	other biogenic silica (diatoms, sponge spicules, and the like)
100	cystoliths (silicified outgrowths of cell walls, occurring in the epidermis and subepidermis, variable shape, usually nodular surfaces)

Although this system may seem cumbersome, it is easy to use, since each type is documented by photograph and detailed description in a desktop key. All terms used in describing phytoliths are defined in a lab dictionary. Scanning with this classification system generates data such as the quick-scan results illustrated in Table 5.9, which shows some of the epidermal quadrilaterals encountered in samples from raised fields in the Yumes area of the Daule River, Ecuador (Pearsall 1987a). A list describing types accompanies tables.

The Missouri phytolith classification system is still in its developmental stage. We currently have over 300 numbered types in ten categories. My point in describing our system here is to illustrate how a high level of consistency in classification can be maintained. However phytoliths are named or numbered, the same criteria for classifying silica bodies must be used from one sample to another, one project to another.

Unless we are interested only in documenting grass occurrence or testing for maize, the process of examining soil phytolith slides at the Missouri lab begins with a "quick-scan." Otherwise, only a 200-count of short cells is done (see below). Quick-scanning, a term coined by Irwin Rovner, is the process of rapidly scanning through an entire slide, noting what phytoliths types are present and roughly in what quantities. We developed a quick-scan form (Fig. 5.25) on which is tallied each type of phytolith observed and its abundance using the following scale:

    VR   very rare; one or two specimens per slide
     R   rare; one specimen every several fields, 3–15 per slide
     M   moderate; one specimen every 2–5 fields, 15–40 per slide

A   abundant; one specimen in almost every field
VA very abundant; several specimens in every field

Numbers on the far left of the quick-scan form (10, 20, ..., 100) refer to major categories of phytoliths in our classification system. All 10s are listed together, for example; additional pages are used if any categories are represented by numerous types. "Number" refers to individual phytolith number (e.g., 10ICa100) and "Description" is added for easier interpretation of results. The "Tally" column is used to keep track of observations; after scanning is completed the tallies are translated

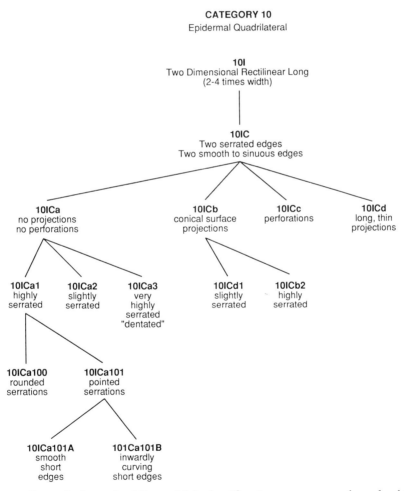

Figure 5.24 Example from the Missouri lab classification system; one branch of Category 10, epidermal quadrilateral, is illustrated.

Table 5.9 • Example Quick-Scan Report: Archaeological Raised Field Samples, Yumes Area, Ecuador (from Pearsall 1987a)

			IA100A	IA100B	IA101	IA2	IAb	IAc	IBa1	IBa200	IBb	IBc	IBd	ICa100	ICa101	ICa2	IIAa2	IIAb100	IIAb101	IICa	IID	IIEa	IIIAa2	IIIBa100	IIIBa101	IIIBa102	IIIBa300	IIIBa301	IIIBb100	IIIBb101	IIICa	IV	V	VIAa2	VIAb1
		Category 10																																	
	Raised Field 4																																		
582	#41 West Profile, Dep. 3		R			R						R	R																						
583	#44 West Profile, Dep. 10							R				M																				VR			
584	#47 West Profile, Dep. 11		VA								M																								
585	#50 West Profile, Dep. 14						R																												
586	#53 West Profile, Dep. 16		M																																
587	#56 West Profile, Dep. 17		R									R				R							VR	R				M	M						
555	Feature 1		VR VA	VR		VR R	R			M VR	A	R		M VR		A	M	VR M	R VR	VR	VR	VR		VR VR		VR VR	M	M		R VR				
556	Feature 5		VA	M R		VR VR								VR																					
	Raised Field 9																																		
557	#11 Feature 1							M										VR			VR	M	R						R VR						
558	#1 Feature 3																														VR				
	Raised Field 12																																		
588	#5 Feature 1		A			M		M	M	VR VR															VR								R		

QUICK SCAN FORM

Project______________ Scanner______________

taxon______________ Date______________

slide #______________ Rows scanned______________

Number	Description	Tally	Abundance Code
10			
20			
30			
40			
50			

Figure 5.25 Missouri lab quick-scan form.

into abundance codes listed above. Scanning is carried out at 250×. In general, a good estimate of abundance of types is gained without scanning the entire slide (22 non-overlapping rows). At some point, forms begin to repeat; redundancy is reached. The rest of the slide is then spot-checked to catch any rare forms.

We began quick-scanning only recently because the lab comparative collection is only now large enough that soil slides can be scanned with few new phytoliths (i.e., phytoliths not in the photograph collection) observed per slide. Phytoliths are thus quickly tallied by lab number, with only a few having to be described, drawn, photographed, and assigned a number. While we were compiling our photograph collection and desktop key, quick-scanning would not have been any quicker than examining a predetermined number of fields or phytolith count. Quick-scanning is a useful way to see the variety of phytoliths present in a soil or comparative sample and to determine the presence or absence of diagnostic phytoliths. It is not, however, an adequate basis for quantitative analysis.

Another scanning procedure routinely carried out on soil samples is a 200-count short-cell scan. Short cells, phytoliths produced in the vascular region of grasses, are among the easiest phytoliths to recognize. They are found in the silt size fraction (2–50 microns). Because relative abundance of short cells from different groups of grasses can give data on environmental conditions, most analysts quantify short-cell occurrence in some fashion. Using a standard count is the most common method, and the one we use.

I originally chose 200 as a standard short-cell count because this was the number often used in pollen analysis to give a representative sample of pollen on a slide. That seemed like a good place to start. Later I ran a simple test using soil slides from the Waimea-Kawaihae project, calculating percentage occurrences of short cells based on the first 100, 150, 200, 250, and so on, up to 400 short cells. Patterns of occurrence calculated using a 200-count remained fairly consistent even when more phytoliths were counted. Although some analysts recommend higher counts, I am satisfied with the reliability of 200-counts, which also allow rapid determination of short-cell assemblages.

We record short-cell occurrences using the modified Twiss *et al.* (1969) system illustrated in Figure 5.23 and a standard lab form. Figure 5.26 shows the latest version of this form, which is designed to facilitate both raw data recording and percentage occurrence calculations.

Note that the top sections of our lab forms do not have space for detailed provenience information. Only the project name (site or archaeologist's name) and slide number are recorded. This ensures that the scanner has no information about the sample, such as stratigraphic position or assumed feature function, which might influence how phytoliths are classified. Working blind is an easy way to guard against bias. "Test" refers to number of rows scanned. Starting and stopping points

```
                    SHORT CELL COUNTS
    Project_____      Scanner_____
    Slide #_____              Date_____
    Taxon_____      Hours_____
    Test_____ Starting Point_____ Objective_____
                    Stopping Point_____
    Panicoid                                                Count | Percent
        Dumbbell_____
        Nodular_____
        Crenate_____
        Angular_____
        Cross_____
        Half-Dumbbell_____
        Rotated_____
                                        All Panicoid
    Festucoid
        Round/Oblong_____
        Rect/Square_____
        Rotated_____
                                        All Festucoid
    Chloridoid
        Saddle_____
        Rotated_____
                                        All Chloridoid
    Other Gramineae
        Horned Tower_____
        Flat Tower_____
        Regular Spool_____
        Irregular Spool_____
        Angles_____
        Half-Rotated_____
        _____
        _____
        _____
                                        All Other
                                    Total Short Cells
```

Figure 5.26 Missouri lab short-cell counting form. On side two of the form (not shown here), measurements of all crosses are recorded, rotated forms are sketched and listed, and phytoliths not categorized on side one are recorded.

are recorded in x and y coordinates. The scanner's initials, date of the scan, and total hours spent on the scan help record the progress of the analysis. Scanning is carried out at 250×.

Scanning proceeds by rows, beginning either at the top or the bottom of the slide. Every short cell encountered is examined and either classified as a type included in the 200-count (those types printed on the front of the lab form) or drawn on side 2 of our form as "Other," a phytolith of short-cell size which does not fit established categories. "Other" phytoliths are not included in the 200-count. The category "Other Gramineae" are phytoliths that have been observed in grasses but do not fit into the Twiss et al. classification. These are included in the 200-count. Also included in the count are "rotated" phytoliths that can be classified as panicoid, festucoid, or chloridoid. Rotated phytoliths are short cell phytoliths that are not oriented on their planar surface in the mount. Although it is often impossible to classify rotated short cells precisely, major groups can often be determined. Because distinguishing among rotated forms can be difficult, the scanner draws all rotated short cells on side 2 of the form.

Each short cell is recorded by placing a tally mark next to the appropriate type name. Total count is tallied with a hand counter. When 200 short cells have been observed, or the entire slide has been scanned, the scanner counts up tally marks for each type and enters totals in the count column. Subtotals are calculated for panicoid, festucoid, chloridoid, and other Gramineae, and the final sample total is determined (this should be 200, unless few short cells occur in the sample). I calculate percentage figures after checking classification of rotated forms and phytoliths not included in the 200-count.

Because occurrence of cross-shaped short cell phytoliths is used to test for maize, detailed information is recorded for all crosses encountered during short-cell scanning (on side 2 of the form). In addition, if the entire slide has not been scanned during the 200-count, the remaining portion is quickly checked for additional crosses. These are not added to the 200-count, but they are included in any discussion of cross occurrence.

As I discussed earlier, panicoid phytoliths must meet two criteria to be classified as crosses, proportion of sides and number of indentations (see Fig. 5.14). Once a phytolith is determined to be a cross, the scanner records its size in micrometer units and variant type, following Piperno (1984). Location in x and y coordinates is recorded and a photograph is often taken to document the occurrence, especially if there is any question about variant type. I convert micrometer measurements to microns and place crosses in size categories based on length of the shortest side.

The third scanning procedure used at the Missouri lab is the 60-field scan. This recently developed scanning procedure serves primarily to quantify occurrence of "large cells," a category that excludes short cells and short cell–like bodies and includes silicified epidermal cells, bulliforms, epidermal appendages, honeycombs, platelike and similar larger bodies and fragments of these. As the name implies, in the 60-field scan sixty microscope fields are examined at 250×.

My first efforts to quantify these non–short cell phytoliths utilized a standard count, much in the same way I quantified short cells. We counted large cell phytoliths in a separate 400-count tally (see Fig. 5.23). I found, however, that counts varied considerably among lab personnel. This was not a problem in short-cell counting. Establishing a minimum size for counting silica bodies that occurred in fragmented form (honeycomb and platelike, especially) helped standardize counts, but it was still necessary to have one assistant count all slides in a study for comparable results.

The Waimea-Kawaihae project crystallized a number of difficulties with large-cell counting procedures and, in general, classification of these phytoliths. When we compared soil phytolith assemblages of surface soil samples to percentage vegetation cover (see Table 5.4), phytolith assemblages correlated fairly well with plant cover, but a number of discrepancies were observed. The most obvious discrepancy was apparent overrepresentation of non-grass phytoliths as measured by relative abundance of non-grass large cells (honeycomb and platelike). Honeycomb and platelike phytoliths were described by Rovner (1971) as characterizing many forbs and trees. We found that even in grass-dominated formations up to 56% of large cell silica was "non-grass" (Pearsall and Trimble 1984).

We hypothesized that there were two probable sources of error for these results: our method of quantifying large-cell occurrence, and the presence of "non-grass" silica bodies in grasses. Because honeycomb, platelike, and other large epidermal silica bodies break up into masses of varying size, counts are not an entirely appropriate method of determining abundance of these classes of phytoliths. A new method of quantification was needed. We decided to test the visual percentage estimation method (Terry and Chilingar 1955). We investigated the other probable source of error by studying comparative materials and initiating a program of description, photographic documentation, and classification of larger phytoliths.

Naomi Miller (personal communication, 1985) suggested to me that visual estimation figures developed for improving and standardizing estimates of percentages of components in sedimentary rocks might be useful for estimating phytolith quantities (Fig. 5.27). The figures, among those published by Terry and Chilingar (1955), illustrate how scattered components of sedimentary rock of certain areas (expressed as percentage cover) appear in a circular microscope field. By comparing an actual field to these standards, the scanner can more accurately estimate the percentage. We tested the use of visual estimation to quantify phytolith occurrence with comparative samples and soil samples from the Philippines (Bodner 1986; Pearsall 1986a; Pearsall et al. 1985).

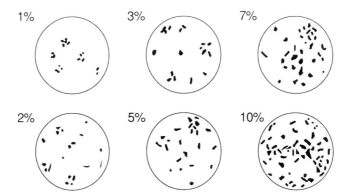

Figure 5.27 Visual estimation figures used for describing the composition of sedimentary rocks. By comparing the view in a microscope field to these standards for percentage cover (i.e., 1% of area, 2%, 3%, and so on), a more accurate estimation of area is possible (after Terry and Chilingar 1955:230–231).

The visual estimation method is a four-step process: (1) making a standard slide mount, (2) selecting fields for scanning, (3) estimating percentage cover of phytoliths in each field, and (4) averaging percentage cover data over all fields. Slide mounts are prepared as described earlier. It is important that an equal quantity of phytolith extract be mounted on each slide so that differences in abundance of types among slides can be ascribed to differences in occurrence in samples, not to quantity of material mounted. To determine abundance of phytoliths on a slide, short of scanning every field, one must choose a representative subsample. Probability sampling helps ensure representativeness of scanned fields. We choose fields to be observed by a basic probability sampling technique: coordinates of 720 fields (our slide universe at 250×) are written on slips of paper; we draw 60, the number needed for scanning.

We determined the number of fields needed to give a valid estimate of phytoliths on a slide using Monte Carlo simulation (Pearsall et al. 1985). The aim of the simulation was to find the minimum number of fields that should be read for any given phytolith type in order to produce a valid estimate of the frequency of occurrence of the type. For the simulation, 120 fields, or about a 17% sample, were scanned on a test slide. The goal of the simulation was to find, if possible, a smaller number of fields which would be nearly as reliable for future use. The following assumptions were implicit in the simulation:

1. One hundred and twenty fields are sufficient to determine the true frequency of each phytolith type in the assemblage.
2. Convergence to the asymptotic value (the point at which cumulative

average area approaches a straight line) in the total sample of 120 can be determined by eye.
3. Convergence to asymptotic value, determined from the main sample, can be determined by examination of randomly selected subsamples.

Percentage cover is graphed as follows. The first value is the area (percentage cover) of the phytolith type in the first field, the second is the average of the first and the second, the third is the average of the first three, and so on, with the last being the average of all 120. We expect to see wide fluctuations in the first fields, but, if there are no problems of bias in the sample and if 120 fields are sufficient to estimate population frequency, then we should observe an asymptote at some point before 120. Figure 5.28 shows an asymptote at approximately 60 fields.

The results of carrying out simulation of six phytolith types, chosen to represent the range of abundances in the test sample, were that approximately 60 fields seemed sufficient to approximate the frequencies observed in 120 for most types. Simulation of one very rare type showed that scanning more fields did little to change the cumulative average. In some runs, as few as 30 fields gave an average close to that obtained for 120. This suggests that, while scanning much larger numbers of fields might increase precision for very rare types, doubling or tripling the number of fields does not lead to substantial improvement. Thus 60-field scan results are weakest for very infrequent types.

To use the 60-field scan method, we examine each field as follows. First, an inventory is made of all phytoliths occurring in the field. Figure 5.29 illustrates the form used for recording these data. Each quadrant is used to record one field. Phy-

Figure 5.28 Monte Carlo simulation of phytolith scanning for the classic honeycomb type. An asymptote is reached in the running average of fields scanned at about 60 fields.

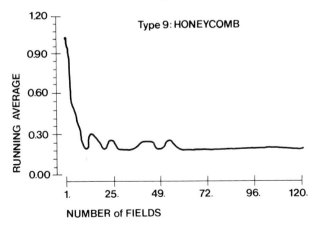

60 FIELD SCAN FORM

Project__________ Scanner__________

taxon__________ Date__________

slide #__________ Page_____of 8

Coordinates	Number	Tally	%	Coordinates	Number	Tally	%

Figure 5.29 Missouri lab 60-field scanning form.

toliths are tallied by lab number in order by major category (10, 20, and so on). Any new types are described, assigned a number, and photographed. We then make a visual estimate of percentage of the field occupied by each phytolith type by comparing abundance of the type to standard figures for visually estimating area (Terry and Chilingar 1955). Percentage cover less than 1% (0.75, 0.50, 0.25, and 0.125) is estimated with figures adapted from those of Terry and Chilingar (Fig. 5.30). We also record number of phytoliths of each type present, since it is sometimes informative to compare count and area data.

One could also determine area covered by each phytolith type with an eyepiece micrometer calibrated for area estimation. The procedure could be further streamlined with computer image analysis in which the area of each phytolith is determined by tracing its outline on a monitor display of the microscope field. Area could then be entered directly into computer memory.

After scanning is completed, we enter raw data on a computer spreadsheet. Phytoliths are ordered by major category as columns on the spreadsheet. Percentage occurrences are entered for each field (rows on the sheet). For each phytolith type, total area occupied is summed, then averaged over 60 fields; this gives a mean occurrence figure for each type, expressed as a percentage. Comparison of these figures among samples serves as a basis for interpretation, especially of relative abundance of larger phytoliths.

The scanning and counting procedures described here are the current state of the art at the Missouri lab. They are still being developed. We are still experimenting with the 60-field scan technique, for instance; it is quite time consuming, both in scanning time and data tabulation. I hope to explore digitizing area data to speed both aspects of the process. We might also incorporate diagnostic, low-redundancy

Figure 5.30 Visual estimation figures for estimating very small (0.125–0.75%) percentage cover figures (adapted from Terry and Chilingar 1955:230).

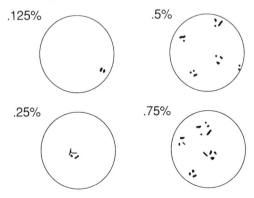

larger phytoliths into a standard count. Recent work in phytolith taxonomy, especially that of Piperno (1983, 1985d, 1988a), in conjunction with our ongoing comparative work, has resulted in new diagnostics that can be reliably counted (e.g., forms are small or strongly silicified, reducing breakage problems). The important point is this: the objective of scanning is to describe what phytoliths occur in a soil sample or comparative plant specimen and to determine the relative abundances of those phytoliths. To reach this objective, the researcher must (1) define what is to be included in the phytolith count and what is to be excluded, (2) examine a sufficient number of phytoliths falling within defined boundaries of the count to get an accurate estimate of their occurrence on a slide, and (3) provide clear descriptions of what is being counted. The last point can be achieved by using published keys such as those of Brown (1984) and Piperno (1988a) or by providing photographs or drawings and descriptions of types.

This brings us back to the concept of the phytolith assemblage, which I view as analogous to the pollen sum. I originally defined the phytolith assemblage as the total range of phytolith types occurring in a sample expressed quantitatively (counts, percentage occurrence, or ratios). I now think more in terms of multiple phytolith assemblages in a sample, since it has proven difficult to quantify all types of phytoliths within one assemblage. The case in point is the problem, described in detail above, of comparing relative abundances of short cells, larger grass phytoliths (long cells, bulliforms, and the like), and the diverse diagnostic and amorphous nongrass phytoliths found in many forbs and woody plants. As the palynologist describes what is included in different pollen sums, the phytolith analyst must define the parameters of counts, establishing what is included in one or more phytolith assemblages.

Alternative Scanning and Counting Procedures

Several mounting media other than Canada Balsam have been found to give phytolith mounts which are easy to scan and which store well. Piperno (1983) and Mulholland and Rau (1985), among others, use Permount as mounting medium. The Mulholland and Rau procedure for mounting liquid phytolith extract using Permount is as follows:

1. Select a clean microscope slide (3" × 1") and cover slip (24 × 30 mm), and label the slide with project name, sample number, fraction, vial date, and date of mounting. Place the slide on a preheated warming tray or hot plate set at low to medium heat.
2. Agitate the processed phytolith sample (phytoliths in 95% ethanol) by shaking the storage vial, then transfer four drops of solution by disposable pipette to the labeled slide. Warm the slide until the ethanol evaporates and the phytolith residue is dry.

3. Add six or seven drops of Permount to the dried phytolith material and warm for a few seconds. Remove the slide from the heat source and add a cover slip. Press air bubbles to one edge with a toothpick.
4. Allow mounted slides to air dry flat for 1 to 2 weeks. After one day, ring cover slips with Permount.

Mulholland and Rau (1985) caution that Permount shrinks with age, and that this shrinkage is accentuated by excessive heating. The mounting medium should extend to the edges of the cover slip to inhibit formation of air bubbles.

Huber (1982) reports the following procedure for preparing slide mounts using benzyl benzoate:

1. Warm a clean microscope slide on a slide warmer.
2. Transfer two drops of agitated phytolith solution to the slide and allow the liquid (ethanol) to evaporate. Remove the slide from the warmer.
3. Place two drops of benzyl benzoate on the phytolith residue and mix well using a clean toothpick.
4. Add a cover slip and press down with a toothpick to press out air bubbles and spread the mounting medium to the edge of the cover slip. Care must be taken to push the medium past the cover slip edges.
5. Using fingernail polish, seal the edges of the cover slip. After the fingernail polish is dry, add a second sealing coat to ensure minimal escape of fumes from the slide.
6. Label slides and clean with xylene.

Twiss *et al.* (1969) and Carbone (1977) report using Cadex as mounting medium for dried phytolith extract.

Kaplan and Smith (1980) describe how to wet mount spodograms from comparative plant tissue in Canada Balsam. Following tissue clearing, spodograms are floated onto slides and dehydrated and infiltrated using the solution series described above, which results in specimens in xylene. Canada Balsam diluted with xylene is then added to the specimen, a cover slip is added, and the slide is warmed on a warming table until the medium is set.

Phytolith material is also easily mounted on SEM stubs for examination. Kaplan and Smith (1980) suggest the following procedure:

1. Agitate phytolith residue in ethanol by shaking the storage bottle, then transfer several drops to a clean glass slide. This should be done in a dust-free area and the transfer should be observed through a microscope.
2. After the ethanol has evaporated, touch an SEM stub coated with the mounting medium microstick to the dried sample. Coat samples with carbon and gold-palladium and examine.

Another procedure for preparation of SEM stubs is presented by Thomson (1982):

1. Place numbered SEM stubs (numbered on the bottom) in an aluminum drying plate on a hot plate.
2. Agitate each phytolith solution sample and transfer one or two drops of extract from the vial to its corresponding stub. Use a clean eyedropper for each vial.
3. Allow phytolith extract to dry, removing each stub from the drying plate as it is completely dried. Drying may take up to one hour.
4. Coat and view samples.

Thomson (1982) notes that samples that are not pH neutral charge rapidly during SEM examination, making photography difficult. She recommends neutralizing such samples by soaking extract overnight in 25 to 30 ml of distilled water and 7 drops of $1M$ KOH, and then using a pH meter, $1M$ KOH, and 10% HCl to bring the solution in the beaker to pH 7. The residue is then centrifuged, rinsed in water, and transferred back to ethanol.

If phytolith material being mounted has been extracted from plant material, residual organic material may cause the sample to bead up on the stub and flake off when dry (Thomson 1982). This can be remedied by roughening the surface of the stub with aluminum oxide polishing powder.

Thomson (1982) also describes how to mount using Microstik. Her procedure differs from that of Kaplan and Smith (1980) in that wet phytolith extract is dropped directly onto a drop of Microstik on the stub. Mounts are placed in an aluminum drying plate on a hot plate and dried. Thomson (1982) cautions that mounting should be done on cold stubs, with heat applied only at the end, since drying Microstik gives off fumes. I recommend using a fume hood for drying mounts.

Other mounting media are available. A good medium should have a refractive index that permits phytoliths to be viewed easily and should be stable over time. If phytolith extract is stored wet, mounting medium and storage liquid (ethanol, acetone, xylene) must be soluble; otherwise, the extract should be dried prior to addition of mounting medium. All media described above dry to form a permanent mount in which phytoliths are "frozen" in position. This is necessary for counting procedures. However, since Permount takes 1 to 2 weeks to dry completely, phytoliths can be rotated somewhat by tapping the incompletely dry slide while viewing. The three-dimensional morphology of phytoliths can also be studied by preparing impermanent mounts—using glycerin, oil for oil immersion microscopy, or the like—which allow free rotation of phytoliths.

It should be possible to create standard slide mounts of phytolith material with wet as well as dry residue. This would involve simply storing a known quantity of extract (standardized by weight or volume) in a known volume of liquid and then transferring the same number of drops of solution on each slide mount.

Procedures for counting and scanning phytoliths also vary among analysts. Scans of as few as 50 short cells per slide have been used to characterize archaeological phytolith assemblages, and scans of 1000 phytoliths and more are reported for comparative plant studies. Little published information is available on scanning and counting procedures used by other researchers, however. In her recent book, Piperno (1988a) consistently uses a data presentation format patterned after pollen diagrams (e.g., Fig. 5.31, pp. 404–406), which demonstrates standardization of counting procedures. I discuss this format further in the next section; here, I simply indicate the counting technique implicit in this approach.

First, phytoliths are counted within one sum, with the exception of nondescript carbon inclusions, which are tabulated separately (Fig. 5.31). Counts of 100 to 200 silica bodies are made in silt fraction samples (fine silt, 5–20 microns; coarse silt, 20–50 microns). Counts include diagnostic arboreal forms (e.g., Palmae, spherical; Palmae, conical), herbaceous forms, including grasses (e.g., short cells, bulliforms, elongate spiney), other herbs such as Marantaceae, Compositae, and Cyperaceae, and distinctive phytoliths that are more widely produced (certain spheres, elongates, and epidermal types). All these are types that can be readily counted (i.e., discrete forms such as short cells, spheres, hairs, hair bases, and so forth).

Not all phytoliths encountered in samples are counted. Platelike and honeycomb types are not tabulated, for example. Unknown types are omitted from the diagram but not from the count.

Although a separate count of grasses is not part of Piperno's presentation format, grass indicators are subtotalled within the overall sum. It is therefore possible to compare relative abundance of grasses against other vegetation, as well as compare relative abundance of different grass forms.

Abundance categories (abundant, rare, not present) are used to indicate occurrence of Compositae in the sand fraction (>50 microns). Piperno (1985d) also uses abundance categories to characterize amounts of silica present in comparative specimens; her categories are (1) abundant; every low-power field has large numbers of phytoliths, (2) common; every field has some phytoliths, (3) not common; every field does not have phytoliths, and (4) rare; scanning of several fields is necessary to observe phytoliths. As discussed above, estimating abundance is a very efficient way to characterize silica occurrence quickly.

All counting procedures discussed above deal with relative counts—abundance of types based on a standard sum. Absolute counting procedures, such as those discussed in Chapter 4, are also being experimented with. Piperno (1988b) reports encouraging preliminary results calculating phytolith influx data in lake core sediments by means of total slide counts. Adding exotics (distinctive phytoliths, pollen,

404 • Chapter 5 Phytolith Analysis

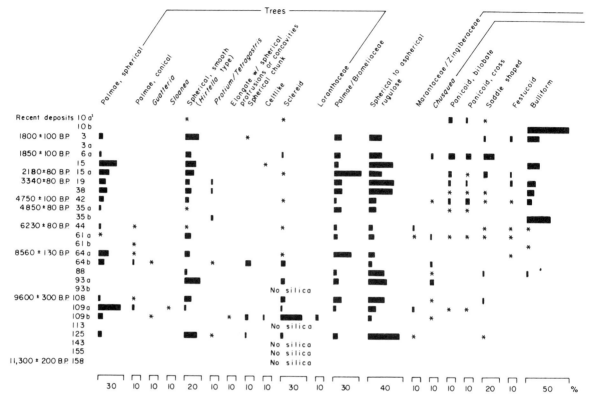

Figure 5.31 Percentage phytolith frequencies from a Gatun Lake core, Panama: (a) fine silt fraction; (b) coarse silt fraction; no notation, fine and coarse silt; * = observed. Compositae categories: A, abundant in sand fraction; R, rare in sand fraction; NP, not observed in sand fraction (from Piperno 1988a:202–203).

or standard spheres) may be a fruitful approach; what to use and when to add the exotic to samples needs to be worked out.

Presenting and Interpreting Results

In a developing field, such as archaeological phytolith analysis, analysts are faced with presenting results of research to a professional audience that is sometimes skeptical, and often poorly informed, about a new technique. This situation places a burden on practitioners not only to present research results clearly, in terms that nonspecialists can understand, but also to include details about procedures and methods of interpretation. Little can be assumed to be common knowledge or standard procedure in the field of phytolith analysis.

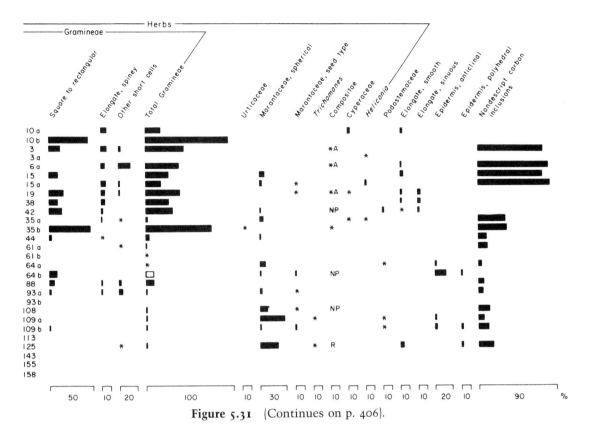

Figure 5.31 (Continues on p. 406).

In this section, I first present some general suggestions for presenting phytolith results and give examples of different presentation formats. I then illustrate how phytolith data can be used to address issues of importance in archaeology by reviewing quantitative approaches and by discussing two case studies, one that applies the phytolith technique for identifying maize, and the other on reconstructing paleoenvironment.

Presenting Results

There are clear similarities between this discussion on phytolith research and the previous discussions of data presentation for pollen and macroremain analysis. Because phytoliths are microremains, presentation formats developed for pollen analysis often lend themselves well to phytoliths. Among lessons I have learned from analyzing macroremains are several useful for phytolith analysis—for example, the importance of explaining level of precision of identification.

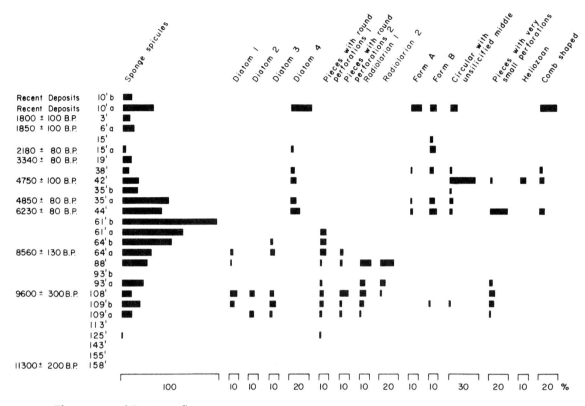

Figure 5.31 (Continued).

A good report of results of phytolith analysis should have four components, (1) statement of objectives and goals of analysis, (2) description of procedures utilized in the study, including field procedures, sampling decisions, lab procedures, identification methods, and quantitative measures, (3) presentation of results, including descriptions of types, and data tables or graphs, and (4) interpretation of results in relation to objectives and goals of the study. I have already discussed types of problems that can be approached using phytolith analysis and summarized current field and laboratory procedures. I focus here, therefore, on how to present data effectively.

Presentation of data has two aspects: discussion of types of phytoliths identified, and presentation of quantified results. Types of phytoliths recovered in samples should be described or illustrated by line drawings or photographs as part of the presentation of results of analysis. Because similar phytoliths occur in a variety of plants (e.g., short cells in the Gramineae), and identification criteria are not widely known, I advise against simply listing plant taxa identified. It is better to present types recovered and then to discuss the taxa these represent. As a general rule, use published terminology to describe phytoliths whenever possible; it is desirable to keep proliferation of names to a minimum. If it is necessary to devise a name for an undescribed phytolith, precision in description is important. For example, in our current desktop key, the old type "horned tower" (see Fig. 5.23) has now been placed in a larger category "spiked cells" in the group "two spiked" in category 30 (short cells). These are further divided by nature of the base of the phytolith and relative proportion of height to width. This system allows us to distinguish phytoliths of similar, but not identical, appearance, but at the same time to indicate that they are closely related. Care must be taken when describing phytolith shape, size, surface characteristics, or edge characteristics to use terminology in its standard, dictionary usage. Detailed descriptions of surface characteristics of pollen grains (e.g., Faegri and Iversen 1975; Kapp 1969) can be used as a guide, as can standard botanical terminology for leaf edge character and other anatomical features of plants.

I agree with Piperno's suggestion that purely descriptive names be replaced whenever possible with designators based on the cell in which the phytolith formed, with modifiers describing shape, surface reticulation, and other features. Some examples of such names are "segmented hair cell," "stellate hair base," and "anticlinal epidermal cell" (Piperno 1985d).

In summary, discussion of types of phytoliths requires clear description of bodies observed in samples, using either published names or a clear description accompanied by drawings or photographs (e.g., Fig. 5.32). Identification of the source of phytoliths, that is, what plants contributed them to a soil assemblage, is made by referring to keys or descriptions of comparative material. If reporting a new diagnostic phytolith type, be sure it is well illustrated and carefully described (including measurements). Assess its diagnostic value carefully; is it diagnostic to order, family, genus, or species level?

Presentation of quantitative data is another important aspect of reporting results of analysis, since it is often on the basis of relative abundances of phytoliths that we draw conclusions about changing patterns of vegetation or presence of cultivated plants such as maize. This necessitates presentation of numerical data in the form of counts, percentages, or ratios.

A common way to present phytolith data is in tabular form using percentages calculated from a standard count. Table 5.10 illustrates short-cell occurrence in soil samples from the El Bronce site, Puerto Rico. Total phytoliths counted are tabulated (third column); note that 200 short cells were rarely present in these samples. This table includes provenience information (in this case, feature or level numbers),

408 • Chapter 5 Phytolith Analysis

```
10IBa200Ab: Epidermal Quadrilateral,
long, slightly sinuous edges, smooth
surface, no perforations, no projections,
short edges pointed and interdigitating.
Found in Palmaceae.
```

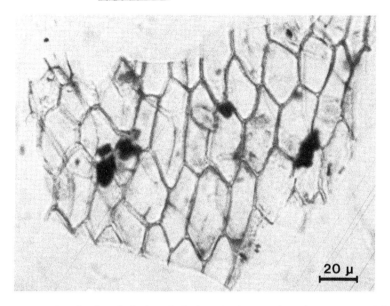

Figure 5.32 Example of a detailed phytolith description used at the Missouri lab.

presents samples in stratigraphic or feature-number order, and includes subtotals of short cells by grass subfamily (to aid interpretation of grass formation type) and information on number of slide rows scanned. In general, I hesitate to calculate relative abundance figures for short cells unless 100–200 are present. In this case, counts as low as 75 were used for calculations. For very low short-cell occurrence, presence (P) only is noted.

In Table 5.11, which presents results from Cueva de los Ladrones, phytoliths included in the assemblage are the rows and analyzed soil samples are the columns. Occurrence is given in percentages, using a sum that includes arboreal, grass, and other herbaceous phytolith types. A subtotal for occurrence of grass short cell phytoliths is included.

Table 5.12, an example of a computer spreadsheet table of 60-field scan results, presents phytolith occurrence at the Shriver site, Missouri. Phytolith types form

Table 5.10 • Short Cell Phytolith Occurrence in Archaeological Samples from the El Bronce Site, Puerto Rico (from Pearsall 1985a:B41)

	Archaeological Provenience	Short Cells	Total grass short cells #	Dumbbells %	Nodular %	Crenates %	Angular %	Cross %	Rotated %	Half-dumbbells %	Total Panicoid %	Round/Oblong %	Rectangular/Square %	Rotated %	Total short Festicoid %	Saddles %	Rotated %	Total Chloridoid %	Horned towers %	Flat towers %	Regular spools %	Irregular spools %	Angles %	Half-rotated %	Rows counted #
483	F34:12 H/I	1331	23				P																		10
534	F576	1051	72						1.4		1.4	45.8	47.2	P	94.4				1.4		2.8				10
491	F609	1050	48	P								P	P	P					P	P					11
531	F647	1057	204	0.5	0.5		0.5		1.0		2.0	50.0	43.1	0.5	94.0				3.0	1.0					10½
533	F669	1032	199	6.5					2.0	1.5	11.0	35.0	30.0	5.0	70.0	3.5		3.5	11.5	1.5				2.5	8½
536	F685	1326	100	4.0							4.0	55.0	23.0	7.0	85.0				6.0	3.0				2.0	11
492	F699	1500	144	8.3		P			9.7		18.0	43.1	37.5		80.6				1.4						11
493	F721	1623	10									P	P	P					P	P					11
494	F735	1047	31				P		P			P	P	P											11
495	F764	1334	42				1.2				8.3	46.4	29.8	9.5	85.7	3.6		3.6	1.2	1.2					11
496	F816	1058	84	7.1		0.6	1.2			1.9	4.9	48.0	34.0	3.0	85.0	1.0		1.0	3.7	3.7	0.7			1.0	11
535	F834	1041	162	1.2			1.2	0.7	1.5		2.2	54.8	37.2	2.2	94.2				2.2	0.7				0.7	10
530	F844	1025	135									P	P	P											11
484	5J	1332	12				P					P	P	P											11
479	11J	1327	13									P	P	P											11
480	11K	1328	8									P	P	P											11
481	13I	1329	14									P	P	P					P	P					11
482	13J	1330	11	P								P	P	P					P	P					10
478	73B	1052	42	P				P	P			P	P	P					1.8	P	0.9				11
477	73C	1053	110	5.5			5.5	0.9	0.9		12.8	21.8	51.8	3.6	77.2				5.5		0.9		1.8		11

Table 5.11 • Archaeological Phytolith Frequencies from Cueva de los Ladrones

	7–12		17–22		27–32		37–42		47–52		52–57		59–65		67–72		77–82		87–92		97–102		107–113		117–122		127–132		137–142		147–152		157–162	
	a	b	a	b	a	b	a	b	a	b	a	b	a	b	a	b	a	b	a	b	a	b	a	b	a	b	a	b	a	b	a	b	a	b
Palmae					1		1	1					1		3		2		2		7	1					1							
Spherical, rugulose	15		24		15	1	34		34	1	75		73		61	2	85		40		47	1	43		64		70		78	1	67	1		
Spherical, smooth	7		8		6		11		37		6		2		16		5		20		5		29		7		9		4		14			
Spherical, with cavity	1								2		2		2								1		1		4				1					
Marantaceae	6				2								1				1				2										4			
Palmae/Bromeliaceae			4		1		3				4				6				10		7		6				4		1					
Gramineae																																		
Panicoid	30		24		26		28		13		5		12		4		3		14		16		17		16		3		3		4			
Chloridoid	8		10		6		4		1		4		1		1		2				1				2		6							
Festucoid	2				2		6				1		1		1												1							
Other	4, 93		2, 95		5, 97		2, 96		4, 89			96	2, 93		1, 94		1, 99		4, 98		5		4, 98		7, 92		7, 98		6, 96		6, 96			
Total Gramineae	44		36		39		40		18		10		16		7		6		18		22 95		21		25		12		15		10			
Elongated, smooth	13	5	12		20		6		6	6	3	4	3	7	5	4	1	1	4		3	2			5		4		1	2		2		
Form stippled			2		2	2		*													1													
Irregular rod			2		3										2						1													
Sclereid					1	*	1												2															
Circular cyst-like	14		12		10	*	3	1	3				2								4		*	1		*		1						
Irregular, stalk						*		1						*								1	*	1										
Irregular, no stalk																																		
Circular chunks									4											1														
Heliconia																		*														1		
Segmented hair								*										*					*		*									
Hair base								*																										
Diatom																							*											

All levels in centimeters below surface. Samples from 7 to 32 cm are associated with Monagrillo, and possibly later ceramics. Those from 37 to 72 cm are associated with Monagrillo ceramics. The levels from 77 to 152 cm are preceramic, and 157 to 162 cm is culturally sterile. Soil samples were recovered as column samples.

Source: Piperno (1985c:256).

Table 5.12 • Shriver Site 60-field Scan Results, Category 10 Phytoliths (from Dinan 1988)

	10IAa100A	10IAa100B	10IAa101	10IAb	10IAc	10IBa1	10IBa200	10IBb
#600	0.7	0.05				0.44	0.72	0.02
#236	0.004			0.008			0.004	
#601	0.24					0.083	0.28	0.127
#598	0.006		0.017				0.058	0.013
#602	0.004							
#599	0.165		0.004			0.03	0.323	0.039
#603	0.66	0.169			0.17	0.07	0.42	0.029

	10IBc	10IBd	10ICa2	10IIAa2	10IICa	10IICb	10IIEa	10IIIBa100
#600	0.12		0.49		0.03		0.15	
#236				0.025			0.08	
#601	0.04		0.017		0.0125		0.28	0.038
#598	0.042						0.15	
#602	0.067						0.071	
#599	0.104	0.006	0.054	0.042	0.004	0.017	0.133	
#603			0.09				0.083	

	10IIIBa102	10IIIBa300	10IIIBa301	10IIIBb101	10IIIBb2	10IIIBb300	10IIICa	10IV
#600			0.46	0.096				0.002
#236								
#601			0.058					
#598								
#602								
#599	0.03	0.03	0.367			0.025	0.054	0.004
#603		0.04	0.14		0.015			

columns of the table and are grouped by major category. Rows correspond to individual samples, arranged in stratigraphic order. Data are percentages, but rather than percentage by count these figures represent area (total area occupied by each type, averaged over 60 fields).

Unlike the three examples of phytolith data tables presented above, each of which focused on broad assemblages of phytoliths, Tables 5.2, 5.13, and 5.14 illus-

Table 5.13 • Archaeological Cross-Shaped Phytoliths, Cueva de los Vampiros and Cueva de los Ladrones Sites, Panama

	Cross-shaped phytolith size			% Crosses	Variant					n	Dumbbell–cross ratio
	Small	Medium	Large		1	2	6	7	8		
Cueva de los Vampiros Preceramic deposits	5 / 25%	15 / 75%	0 / 0%	8	19	—	1	—	—	20	7.5 : 1
Cueva de los Ladrones 7–12 b.s.	9 / 45%	8 / 40%	3 / 15%	7	20	—	—	—	—	20	4 : 1
Monagrillo and possibly later ceramics 17–22 b.s.	11 / 27%	27 / 68%	2 / 5% (1 x-large)	9	22	—	17	—	1	40	1.5 : 1
27–32 b.s.	14 / 35%	20 / 50%	6 / 15% (1 x-large)	6	32	—	8	—	—	40	1.6 : 1
Monagrillo ceramics 37–42 b.s.	14 / 41%	16 / 45%	5 / 14% (1 x-large)	12	24	1	10	—	—	35	1.4 : 1
47–52 b.s.	7 / 35%	11 / 55%	2 / 10%	2	11	—	9	—	—	20	2.4 : 1
52–57 b.s.	5 / 50%	5 / 50%	0 / 0%	1	5	—	5	—	—	10	2.3 : 1
59–65 b.s.	4 / 20%	13 / 65%	3 / 15%	4	16	2	1	—	1	20	4 : 1
67–72 b.s.	6 / 60%	3 / 30%	1 / 1%	1	5	—	5	—	—	10	1.8 : 1
Preceramic 77–82 b.s.	5 / 25%	12 / 60%	3 / 15%	1	13	—	7	—	—	20	2.7 : 1
87–92 b.s.	18 / 45%	19 / 47%	3 / 8%	2	31	4	5	—	—	40	3.4 : 1

Source: adapted from Piperno (1984:376).

trate presentation of detailed information on one type of phytolith, the short cell cross. Table 5.2 (Pearsall 1982a) presents cross data by size category using my original maize identification technique (Pearsall 1979, 1982a). Data from modern comparative material, modern surface soil, and archaeological soil from Ecuador are presented in the same format so that occurrences of crosses of different sizes can be compared. In Table 5.13, Piperno (1984) presents not only size classification of crosses from sites in Panama but also data on abundance of crosses in the assemblage, their distribution in each variant category, and ratio of dumbbells to crosses. Table 5.14 illustrates another format for presenting combined size and variant data for crosses, here with data from raised fields at Peñón del Rio (Pearsall 1987b).

Tabular data presentation sometimes includes too much information for easy assimilation. Graphic presentation of results can help illustrate patterning in data and aid interpretation of results. In Figure 5.33, for example, percentages of large cell grass phytoliths are presented by histogram by excavation level for four sites and two control columns from the Waimea-Kawaihae project. Histograms were used to condense data presented in tabular form in the final report and to illustrate patterns of change in the phytolith record, in this case, abundance of larger phytoliths from grasses relative to occurrence of large cells from forbs and trees.

Figure 5.31, from Piperno (1988a), illustrates effective use of a pollen diagram format to present stratigrahic phytolith data. In this figure, occurrence through time of each phytolith type is depicted by a series of histograms. The same scale is used throughout the diagram, making it easy to compare occurrence of different phytoliths in each level. Occurrence of grass phytoliths is presented both by individual type and by total grass. A pollen diagram format does not eliminate all confusion resulting from putting many data on one figure, but patterns are easier to see than in a large table of percentages. Tables are easier to read for exact percentages; histograms are better to illustrate general trends.

An alternative to individual histograms is the cumulative frequency diagram, illustrated by Figures 5.34 and 5.35. Figure 5.34 shows cumulative frequency curves for six comparative grass specimens. The x axis is percentage, the y axis shape categories, in this case shape of the "top" (planar view) of twelve short cell phytolith types. Differences among samples in relative abundance of phytolith types are expressed by differences in shape of curves. Note, for example, the similarity between curves for *ichu* and blanco, one of the maize cobs tested. Both have high levels of circular and oval short cells. Other maize cobs are more similar to each other, having higher proportions of lobate forms. Such a format can also be used for soil phytolith assemblages by drawing a curve for each sample. Figure 5.35 shows some of the short-cell data from the El Bronce site illustrated in Table 5.10.

Table 5.14 • Cross-Shaped Phytoliths from Earliest Raised Fields, Peñón del Rio Site, Ecuador (from Pearsall 1987b:291)

	STAGE I													
Location*	D19(L)	D19(R)	D12(L)	D12(L)	D12(L)	D12(L)	D12(L)	D12(L)	D12(R)	D12(R)	D12(R)	D12(R)
Meters	21	18	12	11	10	9	8	7	5	2	1	0
Bolsa No.	283	269	255	249	241	229	209	187	95	73	49	33
Lab No.	497	498	500	501	502	503	504	505	506	507	508	509
SMALL												
Total	0	1	0	0	0	4	1	0	0	1	0	2
variant 1						1	1					2
variant 2		1				2						
variant 5/6						1				1		
variant 7												
MEDIUM												
Total	0	10	0	1	1	0	2	2	2	1	0	4
variant 1		6			1			2	1	1		2
variant 2												
variant 5/6		4					2					
variant 7				1					1			1
LARGE												
Total	0	1	1	0	0	1	3	0	1	0	1	2
variant 1		1					3		1		1	1
variant 2												
variant 5/6			1									
variant 7						1						1
EXTRA LARGE												
Total	0	0	0	0	1	0	0	0	0	1	0	0
variant 1												
variant 2					1					1		
variant 5/6												
variant 7												
TOTAL CROSSES	0	12	1	1	2	5	6	2	3	3	1	8
% Large	0	8.3	100	0	0	20	50	0	33	0	100	25
% Extra Large	0	0	0	0	50	0	0	0	0	33	0	0

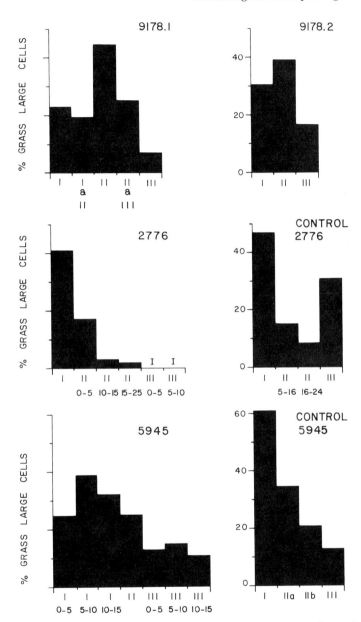

Figure 5.33 Histograms presenting large-cell phytolith percentages from the Waimea-Kawaihae project. Site numbers and control column numbers are in upper right corners. Deposits are numbered I–III in descending depth. Depths are measured in inches from the top of each deposit (from Pearsall and Trimble 1984:127).

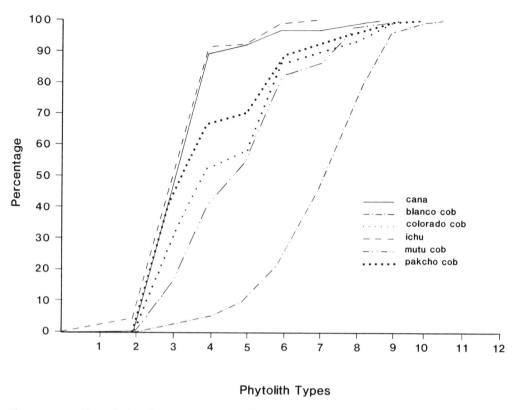

Figure 5.34 Cumulative frequency curves illustrating percentage occurrences of short-cell phytoliths: 1, rectangular; 2, quadrilateral; 3, circular; 4, oval; 5, crenellated; 6, two-lobed; 7, dumbbell; 8, three-lobed; 9, four-lobed; 10, cross-shaped; 11, five-lobed; 12 saddle-shaped. Each line represents a comparative sample (maize races and wild grasses) (from Chiswell 1984).

Presenting quantitative data to illustrate how conclusions have been reached is an important aspect of presenting results of phytolith analysis. It is often necessary to use data tables, especially when whole phytolith assemblages must be discussed. Graphic presentation, on the other hand, can often help the reader seeing patterning in percentage occurrence data. Whatever format is used to present phytolith counts, ratios, or percentages, it is important that sufficient information on sample provenience and date and identification of phytoliths be presented, so that the reader can clearly see patterning in data and follow the reasoning behind the conclusions.

Interpreting Phytolith Data

There have been few attempts to apply quantitative methods to phytolith data. In most studies, changes in occurrence of phytolith types over time, or among site

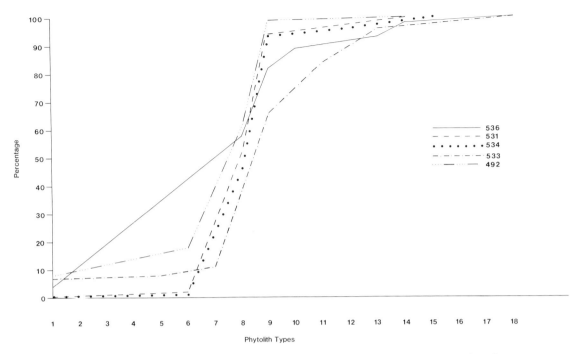

Figure 5.35 Cumulative frequency curve of soil phytolith assemblages from the El Bronce project, Puerto Rico: 1, dumbbells; 2, nodular; 3, crenate; 4, angular; 5, cross; 6, rotated panicoid; 7, half-dumbbells; 8, round/oblong; 9, rectangular/square; 10, rotated festucoid; 11, saddle; 12, rotated chloridoid; 13, horned towers; 14, flat towers; 15, regular spools; 16, irregular spools; 17, angles; 18, half-rotated. Each line represents an archaeological sample (data from Pearsall 1985a).

contexts, are demonstrated by shifts in percentage occurrences. Significance tests are rarely used to demonstrate statistically significant change (but see the discussion of Chiswell 1984, below); rather, studies base interpretation of archaeological or paleoenvironmental phytolith assemblages on modern vegetation analogues. For example, a shift of a certain magnitude in occurrence of grass phytoliths is interpreted as a change in relative abundance of forest versus grass cover. As is the case in pollen analysis, demonstrating that a change in relative abundance values is statistically significant may be useful, but establishing corresponding relationships between vegetation and phytolith assemblages has more interpretive power. Such correspondences have been made "by eye" in phytolith analysis to this point rather than by using the quantitative approaches discussed in Chapter 4.

Chiswell (1984) used the Kolmogorov-Smirnov two-sample procedure to test statistical similarity of pairs of phytolith cumulative frequency curves. This test involves comparison of maximum absolute difference of cumulative frequency curves of two samples (calculated D value) to a critical value of D that depends on size of samples. The null hypothesis, that two curves are the same, can be rejected or upheld at a predetermined level of significance. Chiswell applied the Kolmogorov-Smirnov test to all pairs of curves for comparative grasses in her study from Peru and demonstrated which grasses could be distinguished from one another at the .01 level of significance (Fig. 5.34).

Piperno and Starczak (1985) applied discriminant function analysis to the problem of determining whether maize cross-shaped phytoliths could be discriminated statistically from those of wild grasses. Using a population of 23 races of maize, 6 varieties of teosinte, and 39 wild panicoid grasses, data on five attributes known intuitively to be significant were compiled: size of Variant 1 crosses (length of shortest axis; mean for each species), percentage of Variant 1 crosses in relation to all crosses, size of Variant 6 crosses (mean for each species), percentage of Variant 1 crosses in relation to Variant 6 crosses, and percentage of crosses in relation to total crosses and dumbbells (formerly the cross–dumbbell ratio). The result of calculating Fisher's linear discriminant analysis on these data was that maize and wild grasses could be separated into two statistically distinctive groups based on three of these attributes, Variant 1 size, Variant 6 size, and percentage of Variant 1 crosses (Fig. 5.36).

Subsequently, Pearsall and Piperno (1986) applied this technique to classify crosses in archaeological samples from Ecuador as a test of the identification of maize at Las Vegas (Piperno 1986) and Real Alto (Pearsall 1979) using cross size only. Figure 5.37 illustrates the results of reanalyzing samples from seven levels at the Vegas type site, OGSE-80. In early Las Vegas samples (10,000–8000 B.P.), G10-11, Feature 62, F8-9, and Feature 74, cross-shaped three-dimensional structures are predominantly Variant 6, indicating presence of wild grasses. All samples fall either into the 95% confidence interval for wild grasses or between the 95% confidence intervals about the means of wild grasses and maize. One sample from late Las Vegas (8000–6600 B.P.), Feature 52, also falls between the confidence intervals. The other two late Las Vegas samples, Feature 1 and GH8-9, are classified as clearly maize, supporting presence of maize by at least 6600 B.P.

Results of discriminant function analysis of seven samples from Real Alto (Valdivia) and one sample from the Machalilla component of a nearby site, OGCh-20, are shown in Figure 5.38. Three samples test clearly as maize: scrapings from an inverted, in situ Valdivia 2 vessel, a Valdivia 3 pit, and a Valdivia 5 pit. These data confirm the presence of maize at the site beginning in at least Valdivia 2 times (4500 B.P.). Two samples fall into the area between the 95% confidence intervals for maize and wild grasses; three samples test as wild grass. The Valdivia 1 sample that fell into the area between confidence intervals contained one extra-large Variant 1 cross,

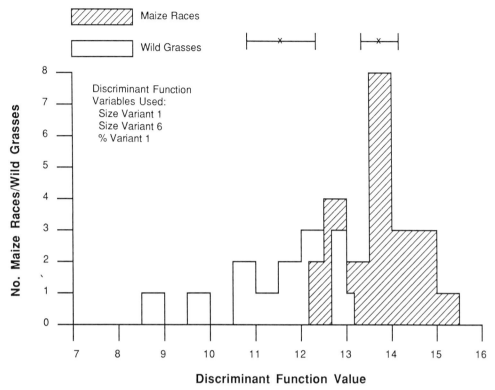

Figure 5.36 Separation of maize and wild panicoid grasses by discriminant function analysis (from Pearsall and Piperno 1986).

observed only in maize, indicating that maize was present but masked in the discriminant function analysis by abundant wild grasses, perhaps from decayed roof thatch.

Case Study: Identifying Maize at Peñón del Rio

I have already reviewed the development of a technique for identifying maize using phytolith analysis, discussed in general terms how to apply the method, and presented results of a reanalysis by Piperno and me of our earlier studies which illustrated that identification using a combination of cross size and three-dimensional morphology allows maize to be identified with confidence in the New World tropics. Because this technique is of considerable interest among archaeologists, I here summarize the steps in applying it, using as an example a study of raised fields at Peñón del Rio. I also discuss how the maize identification technique can be applied outside the tropics.

One of the principal objectives of the Peñón del Rio Archaeological-Agri-

420 • Chapter 5 Phytolith Analysis

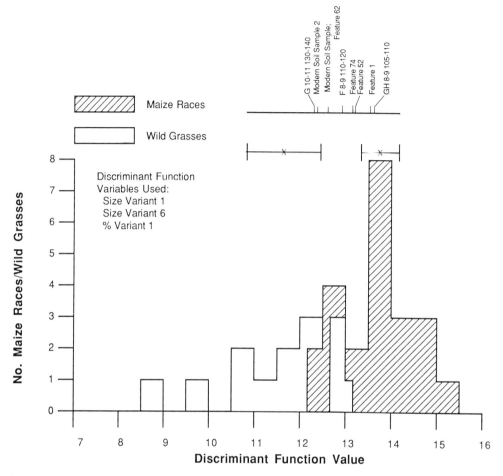

Figure 5.37 Discriminant function values for archaeological samples from Las Vegas, Ecuador. Discriminant function values are superimposed on the maize–wild panicoid grass graph (Fig. 5.36) to illustrate which archaeological samples are maizelike and which segregate as wild grass (from Pearsall and Piperno 1986).

cultural project was the study of *camellones* associated with the Peñón del Rio site, located on the left bank of the lower Babahoya River (Buys and Muse 1987; Marcos 1987; Martinez 1987; Muse and Quintero 1987; Pearsall 1987b). Ethnobotanical research had as a primary goal recovery of botanical remains from the site and its associated field complex. Consequently, soil samples for flotation and phytolith analysis were taken systematically during excavations in habitation and field areas.

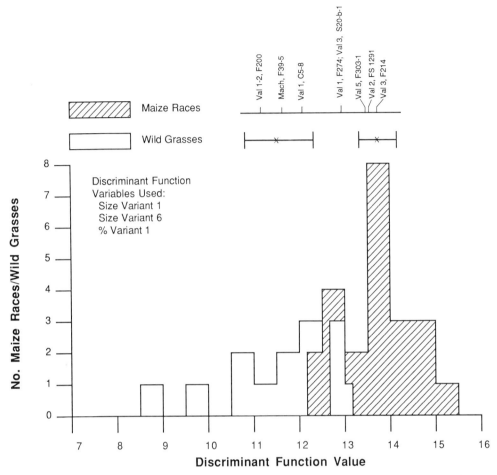

Figure 5.38 Discriminant function values for archaeological samples from Real Alto, Ecuador. Discriminant function values are superimposed on the maize–wild panicoid grass graph (Fig. 5.36) (from Pearsall and Piperno 1986).

Analysis to date has focused on the question of whether maize was cultivated in the raised field complex. Evidence for maize, both charred remains and phytoliths, occurs in habitation areas. Final identification of numerous distinctive phytoliths of probable tree and forb origin in field and site samples awaits completion of ongoing comparative plant studies.

Soil samples for phytolith analysis were taken from raised field contexts by Martinez (1987), who utilized a long profile cut through two extant fields. As Figure 5.39 shows, excavation revealed four small buried fields, deposits 19 and 12, beneath

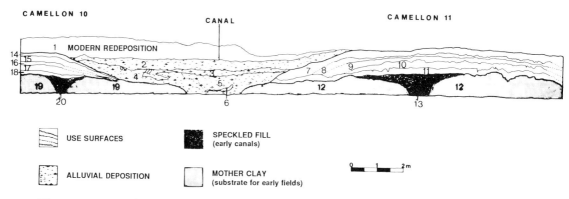

Figure 5.39 South profile of *camellones* 10 and 11, Peñón del Rio site area. The deposits are numbered. Deposits 13 and 20 are buried swales; deposits 19 and 12 are early, buried fields (illustration by V. Martinez and M. Muse; from Pearsall 1987b:283).

the later extant fields. These have been dated to the Regional Developmental period (500 B.C.–A.D. 500) by associated pottery. Soil samples were taken from each deposit at 1-m intervals along the 21-m profile. It proved impossible to open horizontal surfaces on fields because of heavy rains of the 1982–1983 El Niño.

Following standard processing of a subset of soil samples from the field profile, we carried out short-cell scans of samples using the procedures described above. Table 5.15 illustrates results of counts for the earlier fields, deposits 19 and 12. We used the modified Twiss *et al.* (1969) classification illustrated in Figure 5.23 and attempted to count 200 short cells per slide. Percentage occurrence figures were calculated for all samples with 75 or more short cells.

Central to short-cell scanning for maize identification is correct classification of crosses. I define a cross as a panicoid phytolith that is indented on at least three of four sides and is no more than 9.16 microns longer than wide (see Fig. 5.14). We measured every cross encountered during the short-cell scan, including those in the short-cell assemblage (i.e., Table 5.15); after completing 200-counts, we scanned the entire slide and counted and measured any additional crosses.

The final steps in classifying crosses are to determine the variant, following Piperno (1984), and to place crosses into size categories on the basis of length of shortest side, following Pearsall (1978, 1979, 1982a). To determine variant, we used descriptions and photographs in Piperno (1983, 1984). These are illustrated in Figure 5.16. Magnification of 400× or higher should be used to determine cross variant, since distinctions among some forms are subtle. Once crosses are classified by variant, a table showing distribution of crosses by size and variant is made (e.g., Table 5.14).

Table 5.15 • Short Cell Phytoliths from Earliest Raised Fields, Peñón del Rio Site, Ecuador (from Pearsall 1987b:288)

							STAGE 1					
Location*	D19(L)	D19(R)	D12(L)	D12(L)	D12(L)	D12(L)	D12(L)	D12(L)	D12(R)	D12(R)	D12(R)	D12(R)
Meters	21	18	12	11	10	9	8	7	5	2	1	0.0
Bolsa No.	283	269	255	249	241	229	209	187	95	73	49	33
Lab No.	497	498	500	501	502	503	504	505	506	507	508	509
# Total Grass Short Cells	200	200	200	76	41	75	200	146	50	30	55	203
% Dumbbells	6.0	13.0	12.0	5.0	P**	4.0	10.0	5.0	P	P	P	14.0
% Nodular			1.0	3.0			0.5		P			
% Crenates	5.0		5.0				0.5				P	
% Angular		4.0	0.5			1.0	2.0	1.0	P	P	P	3.0
% Cross		4.0	0.5	1.0	P	7.0	3.0	1.0	P	P	P	1.0
% Rotated	3.5	8.0	5.0	16.0	P	15.0	7.0	8.0	P			12.0
% Half-Dumbbells	0.5	1.0	2.0				1.0					1.0
% Total Panicoid Cells	15.0	28.0	26.0	25.0		27.0	25.0	15.0				31.0
% Round/Oblong	23.0	13.0	18.0	13.0	P	20.0	15.0	15.0	P	P	P	18.0
% Rectangular/Square	19.0	15.0	13.0	17.0	P	31.0	33.0	29.0	P	P	P	19.0
% Rotated	21.0	8.0	18.0	12.0	P	8.0	8.0	6.0	P	P	P	7.0
% Total Festucoid Short Cells	63.0	36.0	48.0	43.0		59.0	54.0	50.0				44.0
% Saddles	10.0	12.0	10.0	11.0	P	10.0	8.0	15.0	P	P	P	5.0
% Rotated		1.0	3.0	5.0		1.0	1.0	1.0	P			1.0
% Total Chloridoid Cells	10.0	13.0	13.0	16.0		11.0	9.0	16.0				6.0
% Horned Towers	4.0	10.0	9.0	5.0	P	1.0	6.0	11.0	P	P	P	10.0
% Flat Towers		7.0		3.0	P	1.0	2.0	3.0	P		P	8.0
% Regular Spools	2.0	5.0	1.0	4.0	P		3.0	3.0		P	P	0.5
% Irregular Spools	2.0		3.0	3.0			1.0		P			0.5
% Angles	4.0											
% Half-Rotated	1.0	1.0		1.0		1.0		1.0	P			
# Rows Counted	1.5	4.0	.75	11	11	11	10.5	11	11	11	11	8

Because the Peñón del Rio site is located in coastal Ecuador, where wild panicoid grasses have been studied and a method developed to distinguish them from maize, determining whether short-cell data in Tables 5.14 and 5.15 may be interpreted to indicate presence of maize is straightforward. I compared the archaeological assemblages to the following phytolith assemblage traits, which characterize maize rather than wild panicoid grasses (Pearsall 1979, 1982a; Piperno 1983, 1984):

1. a short-cell assemblage dominated by panicoid phytoliths
2. panicoid assemblage with low ratios of dumbbells to crosses or a high percentage of crosses compared to all crosses and dumbbells.
3. a cross assemblage with high percentages of large-size cross-shaped phytoliths
4. occurrence of extra-large cross-shaped phytoliths
5. cross assemblage with a high percentage of Variant 1 crosses
6. cross assemblage with a low percentages of Variant 6 crosses
7. occurrence of large and extra-large Variant 1 crosses

Using the earlier raised fields at Peñón as an example, I now consider each assemblage trait in turn (refer to Tables 5.14 and 5.15):

1. Panicoid short-cell occurrence is variable, ranging from 15 to 31%; this is second in abundance behind festucoid types (36–63%). This pattern is repeated in all field samples. Clearly, wild festucoid species are present and contribute significantly to the phytolith assemblages. The very moist conditions in the field area favor festucoid grasses as field weeds. Panicoid species would not dominate in this situation.

2. The seven samples for which dumbbell–cross ratios can be calculated give the following results: #498, 3.3:1; #500, 24:1; #501, 5:1; #503, 0.6:1; #504, 3.3:1; #505, 5:1; #509, 14:1. Comparing these to the dumbbell–cross ratios in Table 5.13, which illustrates archaeological assemblages interpreted by Piperno (1984) as representing maize, three Peñón samples have low ratios (#498, #503, #504), two (#500, #509) have high ratios, and two (#501, #505) have low to moderate ratios.

3. Taking all samples together ($N = 44$ crosses), 22.7% of all crosses recovered from the oldest fields fall into the large-size category; this is a high level of occurrence.

4. Taking all samples together, 4.5% of all crosses are extra-large.

5. Variant 1 crosses dominate the cross assemblage; 54.5% of all crosses are Variant 1.

6. Variant 5/6 crosses are less abundant: 34.1% of all crosses are Variant 5/6.

7. Large Variant 1 crosses occur in five samples. There are no extra-large Variant 1 crosses.

Taking all seven traits into account, there is clear indication of maize on the earliest raised fields. Because panicoid phytoliths do not dominate the assemblage and cross occurrence is somewhat low, it is probable that maize was not the only plant growing in the fields. Invasion of local grasses is clearly indicated; multicropping with other annuals or tree crops is another factor we are investigating.

A logical question at this point is whether this method of identifying maize can be applied outside the region for which it was developed, the New World tropics. How can one determine whether maize is present at a site in North America, for example? If phytolith occurrence in the archaeological samples under study fits the maize pattern outlined above, it may be possible to conclude that maize was present if no wild panicoid grass in the study area produces a similar assemblage. For grasses whose cross characteristics have not been studied, the general nature of phytolith production can be determined by consulting Metcalfe (1960). If grasses producing panicoid phytoliths are present in local vegetation, especially if these are common in the area under study, specimens should be obtained and silica extracted. Short-cell counts of these grasses give an assessment of the relative abundance of crosses as well as of their size and variant characteristics. It should then be clear whether any local wild panicoid grasses overlap in phytolith character with maize.

Another point relevant to applying phytolith maize identification outside the lowland New World tropics is variability in phytolith assemblages of maize races. Both Piperno (1983) and I (Pearsall 1979) noted differences in phytolith assemblages of local maize varieties included in our comparative studies. Mulholland (1986b), studying Mandan yellow flour, a maize race native to the plains of the United States, discovered an overall lower percentage of crosses as well as lower quantities of large crosses in that race in comparison to Latin American varieties. Low levels of large crosses in soils tested from the Big Hidatsa site, a village site where maize use is known from macrobotanical remains, may be due to lower quantities of large crosses in the race thought to have been utilized. This example illustrates the importance of conducting region-specific studies not only of wild grasses but also of local maize varieties as part of archaeological phytolith analyses aimed at testing for maize.

Case Study: Paleoenvironmental Reconstruction in Panama

As I discussed earlier, paleoenvironmental reconstruction was one of the earliest applications of phytolith analysis. Much of the earliest work, done by soil scientists, utilized bulk density of silica in soil to indicate presence or absence of grass cover. Interesting work that included quantification by count has also been done in con-

426 • Chapter 5 Phytolith Analysis

junction with analyses of silicoflagellates in deep sea cores (Bukry 1979, 1980). Phytolith analysts working on archaeological soils or sedimentary deposits studied in conjunction with archaeological projects use the phytolith assemblage approach to distinguish among major vegetation types and to draw inferences about climatic conditions.

The use of phytolith data recovered from archaeological sites to reconstruct paleoenvironment raises the issue of contextual bias in sampling. Phytoliths recovered from cultural contexts (cooking hearths, butchering areas, storage pits, living floors) are rarely an unbiased source of data on naturally occurring vegetation. The same can be said of pollen or macroremains. Occurrence of phytoliths in habitation deposits is skewed in favor of utilized plants and weedy or early succession plants growing on the upper stratum of a site immediately after abandonment. Phytoliths from many components of the floral community may never be introduced into deposits of a site.

As an example of paleoenvironmental reconstruction based on phytolith assemblages, I now discuss Piperno's (1983, 1985a, 1988a) analysis of sediments from Gatun Lake in eastern Panama. Figure 5.31 illustrates the phytolith data from this study. Raw counts and percentage calculations are available in an appendix in Piperno (1983). I choose this study for discussion in part because it deals with clearly noncultural deposits, which are the best contexts for paleoenvironmental reconstruction.

An analysis of twenty samples from deep core sediments from Gatun Lake was undertaken in order to compare phytolith data with pollen data from the same core. The pollen study was conducted by Bartlett and Barghoorn (1973). Samples for phytolith analysis were selected from four different cores and arranged in the stratigraphic sequence shown in Figure 5.31 on the basis of ten radiocarbon dates. Placement of samples not directly associated with a radiocarbon date is explained clearly in Piperno's (1983, 1985a, 1988a) discussions.

Bartlett and Barghoorn's pollen study led them to propose four periods of distinctive vegetation in the coring region during the 11,300 years covered by the cores:

11,300–9000 B.P.: mature tropical rainforest
9000–7300 B.P.: mangrove swamp, with nearby tropical rainforest
7300–4200 B.P.: transition from mangrove vegetation to freshwater swamp
4200 B.P.–present: freshwater swamp; establishment of slash and burn agriculture

To compare phytolith and pollen data, Piperno utilizes the same temporal divisions in her discussion.

All silica remains present in Piperno's diagrams with the exception of the category "Nondescript carbon inclusions" are included in one phytolith sum of 100 to 200+ silica bodies. Silica bodies making up the sum are derived from a diverse group of plant and animal sources. Plant silica selected for inclusion in the sum come from dicotyledon and monocotyledon families where Piperno has identified diagnostic phytoliths. She states that, among dicotyledons producing diagnostic phytoliths, silica often occurred in quantities similar to those observed in monocotyledons. It is still the case, however, that silica from monocots is overrepresented, and that from dicots underrepresented, in core samples (Piperno 1985a).

The interpretive problem caused by differential production of phytoliths is analogous in many ways to that caused by differential pollen production (see Chapter 4). Piperno addresses this problem by referring to work done in modern vegetation formations in her study region, where vegetation cover can be correlated with soil phytolith occurrence. In the case of tropical forests, for example, palms, Marantaceae herbs, and bromeliads among other forest taxa are well represented in phytolith assemblages. Occurrence of phytoliths from silica-accumulating forest taxa in combination with low occurrence of grasses and other indicators of open areas indicates presence of mature forest.

For the first of Bartlett and Barghoorn's periods, 11,300–9000 B.P. (samples from 158′ through 109′), there is good agreement between pollen and phytolith data. Only two of the six samples analyzed produced appreciable quantities of silica (125′, 109′). It can be clearly seen from the phytolith diagram that forest taxa (grouped in the left portion of Fig. 5.31) make up the majority of phytoliths present, with low quantities of a variety of silica bodies of aquatic origin in the uppermost sample. The virtual absence of grass phytoliths in this interval reinforces the interpretation that a mature forest was present. Although Piperno comments that the low levels of silica present in four of the six samples is difficult to explain, I have observed a similar pattern in other regions for samples deposited under forest cover. Sometimes silica in forested samples is mostly platelike or honeycomb forms whose occurrence cannot be easily quantified by count and which are not diagnostic to family. Quantities of such material, with some forest diagnostics and an absence of grass phytoliths, may also represent forest cover. Of course, this may not be the situation in the Panamanian case.

In the second period, 9000–7300 B.P., five samples were analyzed (108′ through 61′). In this interval *Rhizophora* pollen are abundant. In the biogenic silica section of the diagram there is an increase in sponge spicule occurrence and a continuation of the other aquatic forms observed late in the previous period. These are thought to represent marine forms (Diatoms 1, 2, and 3; radiolarians), since they do not occur when pollen indicates a shift to freshwater conditions (see below) and have been observed in modern samples from intertidal environments (Piperno 1985a: 16). The slight increase in grass phytolith occurrence reinforces the interpretation that a more open vegetation formation was present.

Clark, J. T., and P. V. Kirch (editors)
- 1983 *Archaeological investigations of the Mudland–Waimea–Kawaihae Road Corridor, Island of Hawaii: An interdisciplinary study of an environmental transect.* Report 83-1, *Departmental Report Series.* Honolulu, Hawaii: Department of Anthropology, Bernice P. Bishop Museum.

Denevan, William M., and Kent Mathewson
- 1983 Preliminary results of the Samborondon raised field project, Guayas Basin, Ecuador. In *Drained field agriculture in Central and South America*, edited by J. P. Darch, pp. 167–181. BAR International Series 189.

Dinan, Elizabeth
- 1988 The changing prairie at the Shriver site, N.W. Missouri: Phytolith analysis as an aid in environmental reconstruction. Ms. on file, American Archaeology Division, University of Missouri–Columbia.

Dinan, Elizabeth, and Deborah M. Pearsall
- 1988 The University of Missouri phytolith classification system. Paper presented at the 3rd Phytolith Research Workshop, University of Missouri–Columbia.

Dunn, Mary Eubanks
- 1983 Phytolith analysis in archaeology. *Mid-Continental Journal of Archaeology* 8:287–297.

Edman, G., and E. Soderberg
- 1929 Auffindung von Reis in einer Tonscherbe aus einer etwas fünftausendjährigen Chinesischen Siedlung. *Bulletin of the Geological Society of China* 8:363–365.

Faegri, Knut, and Johs. Iversen
- 1975 *Textbook of pollen analysis* (3rd ed., revised by Knut Faegri). New York: Hafner.

Folger, D. W., L. H. Burckle, and B. C. Heezen
- 1967 Opal phytoliths in a North Atlantic dust fall. *Science 155:1243–1244.*

Geis, J. W.
- 1973 Biogenic silica in selected species of deciduous angiosperms. *Soil Science* 116:113–130.

Helbaek, H.
- 1959 How farming began in the Old World. *Archaeology* 12:183–189.
- 1961 Studying the diet of ancient man. *Archaeology* 14:95–101.

Huber, J. K.
- 1982 Phytolith procedures. Laboratory procedures, Archaeometry Laboratory, University of Minnesota–Duluth.

Izumi, Seiichi, and Kazuo Terada
- 1972 *Andes 4: Excavations at Kotosh, Peru, 1963 and 1966.* Tokyo: University of Tokyo Press.

Jackson, M. L.
- 1956 *Soil chemical analysis—advanced course.* Published by the author, Department of Soils, University of Wisconsin, Madison. Cited in Twiss *et al.* 1969.

Jones, L. H. P., and K. A. Handreck
- 1967 Silica in soils, plants, and animals. In *Advances in Agronomy*, Vol. 19, edited by A. G. Norman, pp. 107–149. New York: Academic Press.

Jones, L. H. P., and A. A. Milne
- 1963 Studies of silica in the oat plant: I. Chemical and physical properties of silica. *Plant and Soil* 18:207–220.

Jones, L. H. P., A. A. Milne, and S. M. Wadham
    1963   Studies of silica in the oat plant: II. Distribution of silica in the plant. *Plant and Soil* 18:358–371.

Kaplan, Lawrence, and Mary B. Smith
    1980   Procedures for phytolith reference materials. Paper distributed at the 45th Annual Meeting of the Society for American Archaeology, Philadelphia.

Kapp, Ronald O.
    1969   *How to know pollen and spores.* Dubuque, Iowa: William C. Brown.

Karlstrom, P.
    1978   Epidermal leaf structures in species of Strobilantheae and Petalidieae (Acanthaceae). *Botaniska Notiser* 131:423–433.
    1980   Epidermal leaf structures in species of Asystasieae, Pseuderanthemeae, Graptophylleae and Odontonemeae (Acanthaceae). *Botaniska Notiser* 133:1–16.

Kaufman, P. B., W. C. Bigelow, R. Schmid, and N. S. Ghosheh
    1971   Electron microprobe analysis of silica in epidermal cells of *Equisetum*. *American Journal of Botany* 58:309–316.

Kilmer, V. J., and L. T. Alexander
    1949   Methods of making mechanical analyses of soils. *Soil Science* 68:15–24.

Klein, R. L., and J. W. Geis
    1978   Biogenic silica in the Pinaceae. *Soil Science* 126:145–156.

Kurman, Marie H.
    1983   *An opal phytolith and palynomorph study of extant and fossil soils in Kansas.* M.A. thesis, Division of Biology, Kansas State University, Manhattan, Kansas. Cited in Rovner 1983.

Labouriau, L.
    1983   Phytolith work in Brazil: A minireview. *Phytolitharien Newsletter* 2:6–10.

Lewis, Rhoda Owen
    1978   Use of opal phytoliths in paleo-environmental reconstruction. *Wyoming Contributions to Anthropology* 1:127–132.
    1979   *Use of opal phytoliths in paleo-environmental reconstruction.* M.A. thesis, University of Wyoming, Laramie.
    1981   Use of opal phytoliths in paleo-environmental reconstruction. *Journal of Ethnobiology* 1:175–181.

Lippi, Ronald D., Robert M. Bird, and David M. Stemper
    1984   Maize recovered at La Ponga, an early Ecuadorian site. *American Antiquity* 49:118–124.

MacDonald, Linda L.
    1974   *Opal phytoliths as indicators of plant succession in north central Wyoming.* M.A. thesis, University of Wyoming, Laramie. Cited in Rovner 1983.

Marcos, Jorge G.
    1987   Los campos elevados de la Cuenca del Guayas, Ecuador: El proyecto Peñón del Rio. In *Pre-Hispanic agricultural fields in the Andean region* (Part 2), edited by William M. Denevan, Kent Mathewson, and Gregory Knapp, pp. 217–224. *BAR International Series* 359.

Martin, Larry D., and B. Miles Gilbert
    1978   Excavations at Natural Trap Cave. *Transactions of the Nebraska Academy of Science* 6:107–116.

Martinez, Valentina L.
    1987   Campos elevados al norte del sitio arqueológico Peñón del Rio, Guayas, Ecuador. In *Pre-Hispanic agricultural fields in the Andean region* (Part 2), edited by William Denevan, Kent Mathewson, and Gregory Knapp, pp. 267–278. *BAR International Series* 359.

Matsutani, Akiko
    1972   Spodographic analysis of ash from the Kotosh site. In *Andes 4: Excavations at Kotosh, Peru, 1963 and 1966,* edited by Seiichi Izumi and Kazuo Terada, pp. 319–326. Tokyo: University of Tokyo Press.

Maynard, N. G.
    1976   Relationship between diatoms in surface sediment of the Atlantic Ocean and the biologic and physical oceanography of overlying waters. *Paleobiology* 2:99–121.

Mehra, P. N., and O. P. Sharma
    1965   Epidermal silica cells in the Cyperaceae. *Botanical Gazette* 126:53–58.

Metcalfe, C. R.
    1960   *Anatomy of the monocotyledons: I. Gramineae.* London: Oxford University Press.
    1971   *Anatomy of the monocotyledons: V. Cyperaceae.* London: Oxford University Press.

Miller, Arlene V.
    1980   *Phytoliths as indicators of farming techniques.* Paper presented at the 45th Annual Meeting of the Society for American Archaeology, Philadelphia.

Miller-Rosen, Arlene V.
    1987   Phytolith studies at Shiqmim. In *Shiqmim I: Studies concerning Chalcolithic societies in the northern Negeu Desert, Israel (1982–84),* edited by T. E. Levy, Part (i) text: 243–249, Part (ii) Figures adn Plates:547–548. *BAR International Series* 356.

Mulholland, Susan C.
    1986a   Classification of grass silica phytoliths. In *Plant opal phytolith analysis in archaeology and paleoecology,* edited by Irwin Rovner, pp. 41–52. Occasional Papers of the Phytolitharien 1. Raleigh: North Carolina State University.
    1986b   Phytolith studies at Big Hidatsa, North Dakota: Preliminary results. In *The prairie: Past, present, and future: I. Pre-European people on the prairies,* edited by Gary K. Clambey and Richard H. Pemble, pp. 21–24. Fargo, North Dakota: Tri-College University Center for Environmental Studies.

Mulholland, Susan C., and George Rapp, Jr.
    1988   *A general phytolith classification emphasizing grass silica-bodies.* Paper presented at the 3rd Phytolith Research Workshop, University of Missouri–Columbia.

Mulholland, Susan C., G. Rapp, Jr., and J. A. Gifford
    1982   Phytoliths. In *Troy: The archaeological geology, Supplementary Mongraph 4,* edited by G. Rapp, Jr., and J. A. Gifford. pp. 117–137. Princeton, New Jersey, Princeton University Press.

Mulholland, Susan C., and Daniel Rau
　1985　Extraction of opal phytoliths from sediment samples: VII. Laboratory procedure. Ms. on file, Archaeometry Laboratory, University of Minnesota, Duluth.

Muse, Michael, and Fausto Quintero
　1987　Experimentos de reactivación de campos elevados, Peñón del Rio, Guayas, Ecuador. In *Pre-Hispanic agricultural fields in the Andean region* (Part 2), edited by William M. Denevan, Kent Mathewson, and Gregory Knapp, pp. 249–266. *BAR International Series* 359.

Netolitzky, F.
　1900　Mikroskopische Untersuchung Gänzlich verkohlter vorgeschichtlicher Nahrungsmittel aus Tirol. *Zeitschrift für Untersuchung der Nahrungs- und Genussmittel* 3:401–407.
　1914　Die Hirse aus antiken Funden. *Sitzbuch der Keiserliche Akadamie für Wissenschaft der Mathematisch-Naturwissenschaften* 123:725–759.

Parmenter, Carol, and D. W. Folger
　1974　Eolian biogenic detritus in deep sea sediments: A possible index of equatorial Ice Age aridity. *Science* 185:695–698.

Parsons, James J.
　1969　Ridged fields in the Rio Guayas Valley, Ecuador. *American Antiquity* 34:76–80.

Pearsall, Deborah M.
　1976　Evidence of maize from Real Alto, Ecuador: Preliminary results of opal phytolith analysis. Paper presented at the 41st Annual Meeting of the Society for American Archaeology, St. Louis.
　1977　Maize and beans in the Formative period of Ecuador: Preliminary report of New Evidence. Paper presented at the 42nd Annual Meeting of the Society for American Archaeology, New Orleans.
　1978　Phytolith analysis of archeological soils: Evidence for maize cultivation in Formative Ecuador. *Science* 199:177–178.
　1979　*The application of ethnobotanical techniques to the problem of subsistence in the Ecuadorian Formative.* Ph.D. dissertation, University of Illinois–Urbana. Ann Arbor: University Microfilms.
　1980　Phytolith evidence of achira and maize cultivation in Formative Ecuador. Paper presented at the 45th Annual Meeting of the Society for American Archaeology, Philadelphia.
　1981　Phytolith analysis at Natural Trap Cave. Ms. on file, American Archaeology Division, University of Missouri–Columbia.
　1982a　Phytolith analysis: Applications of a new paleoethnobotanical technique in archeology. *American Anthropologist* 84:862–871.
　1982b　Phytolith analysis of soil samples from the Range site (11-S-47), Madison County, Illinois. Report submitted to the FAI-270 Project, University of Illinois, Urbana.
　1982c　Maize phytoliths: A clarification. *Phytolitharien Newsletter* 1(2):3–4.
　1984a　Phytolith content in selected soil samples. In *Subsistence and conflict in Kona, Hawaii: An archaeological study of the Kuakini Highway realignment corridor*, edited by Rose Schilt, pp. 397–407. Department of Anthropology, Bernice P. Bishop Museum, Departmental Report Series 84-1.

1984b   Progress report on ethnobotanical research for the ancient agricultural technologies project (A.I.D./ESPOL No. 936-5542). Ms. on file, American Archaeology Division, University of Missouri–Columbia.

1984c   Informe del análisis de fitolitos y semillas carbonizadas del sitio Cotocollao, Provincia de Quito, Ecuador. Ms. on file, American Archaeology Division, University of Missouri–Columbia.

1985a   Analysis of soil phytoliths and botanical macroremains from El Bronce archaeological site, Ponce, Puerto Rico. In *Archaeological data recovery at El Bronce, Puerto Rico: Final report, phase 2*, by Linda S. Robinson, Emily R. Lundberg, and Jeffery B. Walker. pp. B1–46. Prepared for U.S. Army Corps of Engineers, Jacksonville District.

1985b   Ethnobotanical research for the ancient agricultural technologies project: Phytolith analysis of raised field soils at Peñón del Rio, Ecuador. Paper presented at the 45th International Congress of Americanists, Bogota, Colombia.

1986a   Phytolith analysis of soil from the Lubok (I-1982-WS) and Bekes (I-1982-X5) sites, Mountain Province, Philippines. In *Prehistoric agricultural intensification in Central Bontoc, Mountain Province, Philippines*, by Connie C. Bodner, pp. 526–586. Ph.D. dissertation, Department of Anthropology, University of Missouri–Columbia.

1986b   Final report on analysis of plant macroremains and phytoliths from the Nueva Era site, Pichincha Province, Ecuador. Ms. on file, American Archaeology Division, University of Missouri–Columbia.

1987a   Phytolith analysis of soil samples from raised fields in the Yumes Area, Daule River floodplain, Ecuador. Ms. on file, American Archaeology Division, University of Missouri–Columbia.

1987b   Evidence for prehistoric maize cultivation on raised fields at Peñón del Rio, Guayas, Ecuador. In *Pre-Hispanic agricultural fields in the Andean region* (Part 2), edited by William Denevan, Kent Mathewson, and Gregory Knapp, pp. 279–296. BAR International Series 359.

1987c   Phytolith analysis of vessels 2, 4, and 9. In *Putney Landing: Archaeological investigations at a Havana-Hopewell settlement on the Mississippi River, west-central Illinois*, edited by C. W. Markman, pp. 416–420. Northern Illinois University Department of Anthropology Report of Investigations 15.

Pearsall, Deborah M., and Dolores R. Piperno
1986   Antiquity of maize cultivation in Ecuador: Summary and reevaluation of the evidence. Paper presented at the 51st Annual Meeting of the Society for American Archaeologist, New Orleans.

Pearsall, Deborah M., and Michael K. Trimble
1983   Phytolith analysis of soil samples: Report 18. In *Archaeological investigations of the Mudlane–Waimea-Kawaihae Road Corridor, Island of Hawaii: An interdisciplinary study of an environmental transect*, edited by J. T. Clark and P. V. Kirch, pp. 472–497. Department of Anthropology, Bernice P. Bishop Museum Departmental Report Series, 83–1.

1984   Identifying past agricultural activity through soil phytolith analysis: A case study from the Hawaiian Islands. *Journal of Archaeological Science* 11:119–133.

Pearsall, Deborah M., Michael K. Trimble, and Robert A. Benfer
- 1985 A methodology for quantifying phytolith assemblages in comparative and archaeological contexts. Paper presented at the 2nd Phytolith Research Workshop, Duluth, Minnesota.

Piperno, Dolores R.
- 1979 *Phytolith analysis of archaeological soils from central Panama.* M.A. thesis, Department of Anthropology, Temple University, Philadelphia.
- 1980 Phytolith evidence for maize cultivation in central Panama during the Early Ceramic (Monagrillo) period. Paper presented at the 45th Annual Meeting of the Society for American Archaeology, Philadelphia.
- 1983 *The application of Phytolith analysis to the reconstruction of plant subsistence and environments in prehistoric Panama.* Ph.D. dissertation, Department of Anthropology, Temple University, Philadelphia.
- 1984 A comparison and differentiation of phytoliths from maize and wild grasses: Use of morphological criteria. *American Antiquity* 49:361–383.
- 1985a Phytolithic analysis of geological sediments from Panama. *Antiquity* 59:13–19.
- 1985b Phytolith records from prehistoric agricultural fields in the Calima region, Colombia. *Pro Calima* 4:37–40.
- 1985c Phytolith taphonomy and distributions in archaeological sediments from Panama. *Journal of Archaeological Science* 12:247–267.
- 1985d Phytolith analysis and tropical paleo-ecology: Production and taxonomic significance of siliceous forms in New World plant domesticates and wild species. *Review of Palaeobotany and Palynology* 45:185–228.
- 1986 The analysis of phytoliths from the Vegas site, OGSE-80, Ecuador. In *The Vegas culture: Early prehistory of southwestern Ecuador*, edited by Karen E. Stothert. Guayaquil: Museo Antropológico del Banco Central del Ecuador.
- 1988a *Phytolith analysis: An archaeological and geological perspective.* San Diego: Academic Press.
- 1988b Preliminary phytolith records of a 13,000 year-old lake sequence from central Pacific Panama. Paper presented at the 3rd Phytolith Research Workshop, University of Missouri–Columbia.

Piperno, D. R., and K. Husum-Clary
- 1984 Early plant use and cultivation in the Santa Maria Basin, Panama: Data from phytoliths and pollen. In *Recent advances in Isthmian archaeology*, edited by F. Lange, pp. 85–121. Proceedings: 44th International Congress of Americanists; British Archeological Report, International Series 212.

Piperno, Dolores R., and V. Starczak
- 1985 Numerical analysis of maize and wild grass phytoliths using multivariate techniques. Paper presented at the 2nd Phytolith Research Workshop, Duluth, Minnesota.

Piperno, Dolores R., Karen H. Clary, Richard G. Cooke, Anthony J. Ranere, and Doris Weiland
- 1985 Preceramic maize in central Panama: Phytolith and pollen evidence. *American Anthropologist* 87:871–878.

Posteck, M. T.
- 1981 The occurrence of silica in the leaves of *Magnolia grandiflora* L. *Botanical Gazette* 142:124–134.

Powers, A. H., and D. D. Gilbertson
    1987   A simple preparation technique for the study of opal phytoliths from archaeological and Quaternary sediments. *Journal of Archaeological Science* 14:529–535.

Rattel, Cheryl
    1983   Extraction of opal phytoliths from plant samples. *Phytolitharian Newsletter* 2(3):7–8.

Robinson, Ralph L.
    1980   Environmental chronology for central and south Texas: External correlations to the Gulf Coastal Plain and the Southern High Plains. Paper presented at the 45th Annual Meeting of the Society for American Archaeology, Philadelphia.

Roosevelt, Anna
    1984   Problems interpreting the diffusion of cultivated plants. In *Pre-Colombian plant migration*, edited by Doris Stone, pp. 1–18. *Papers of the Peabody Museum of Archaeology and Ethnology, Harvard University*, Vol. 76.

Rovner, Irwin
    1971   Potential of opal phytoliths for use in paleoecological reconstruction. *Quaternary Research* 1:343–359.
    1972   Note on a safer procedure for opal phytolith extraction. *Quaternary Research* 2:591.
    1975   Plant opal phytolith analysis in midwestern archaeology. *Michigan Academician* 7:129–137.
    1980   The history and development of plant opal phytolith analysis. Paper presented at the 45th Annual Meeting of the Society for American Archaeology, Philadelphia.
    1983a   Plant opal phytolith analysis: Major advances in archaeobotanical research. In *Advances in archaeological method and theory* (Vol. 6), edited by Michael B. Schiffer, pp. 225–266. New York: Academic Press.
    1983b   Final report of preliminary study of phytolith assemblages from three Kentucky sites (15CK89, 15CK146, 15CK147). Ms. on file, Department of Sociology and Anthropology, North Carolina State University, Raleigh.
    1983c   Phytolith assessment of Alta Toquima Village, Nevada: First preliminary report. Ms. on file, Department of Sociology and Anthropology, North Carolina State University, Raleigh.
    1984   Vertical movement of phytoliths in stable soils: A non-issue. Paper presented at the 1st Phytolith Research Workshop, North Carolina State University, Raleigh.
    1986   Downward percolation of phytoliths in stable soils: A non-issue. In *Plant opal phytolith analysis in archaeology and paleoecology*, edited by Irwin Rovner, pp. 23–30. Occasional paper of the Phytolitharien 1. Raleigh: North Carolina State University.

Rovner, Irwin, and Judith Zurita Noguera
    1986   Miscellaneous modifications of phytolith laboratory procedures. *Phytolitharien Newsletter* 4(1):5–6.

Rowlett, Ralph M., and Deborah M. Pearsall
    1980   Thermoluminescence beyond ceramics. Paper presented at the American Nuclear Society Meetings, Las Vegas.

    1988   Phytolith dating by thermoluminescence. Paper presented at the 3rd Phytolith Research Workshop, University of Missouri–Columbia.

Russ, J., and I. Rovner
    1987   Stereological verification of *Zea* phytolith taxonomy. *Phytolitharien Newsletter* 4:10–18.

Schellenberg, H. C.
    1908   Wheat and barley from the North Kurgan, Anau. In *Explorations in Turkestan* (Vol. 3), edited by R. Pumpelly, pp. 471–473. Washington, D.C.: Carnegie Institution.

Schreve-Brinkman, E. J.
    1978   A palynological study of the upper Quaternary sequence in the El Abra corridor and rock shelters (Colombia). *Palaeogeography, Palaeoclimatology, Palaeoecology* 25:1–109.

Scurfield, G., C. A. Anderson, and E. R. Segnit
    1974   Silica in woody stems. *Australian Journal of Botany* 22:211–229.

Shafer, H. J., and R. G. Holloway
    1979   Organic residue analysis in determining stone tool function. In *Lithic use-wear analysis*, edited by Brian Hayden, pp. 385–399. New York: Academic Press.

Smith, Richard T., and Kenneth Atkinson
    1975   *Techniques in pedology*. London: Paul Elek.

Starna, William A., and Donald A. Kane, Jr.
    1983   Phytoliths, archaeology, and caveats: A case study from New York State. *Man in the Northeast* 26:21–31.

Stothert, Karen E.
    1985   The preceramic Las Vegas culture of coastal Ecuador. *American Antiquity* 50:613–637.
    1986   (editor) *The Vegas culture: Early prehistory of southwestern Ecuador*. Guayaquil: Museo Antropológico del Banco Central del Ecuador.

Terry, Richard D., and George V. Chilingar
    1955   Summary of "Concerning some additional aids in studying sedimentary formations," by M. S. Shretsov. *Journal of Sedimentary Petrology* 25:229–234.

Ter Welle, B. J. H.
    1976   Silica grains in woody plants of the Neotropics, especially Surinam. *Leiden Botanical Series* No. 3: pp.107–142. Cited in Piperno 1988.

Thompson, Robert
    1988   The recovery of opal phytoliths from food residues in utilized ceramics. Paper presented at the 3rd Phytolith Research Workshop, University of Missouri–Columbia.

Thomson, Margaret C.
    1982   Preparation of opal phytolith SEM stubs. Archaeometry Laboratory document, University of Minnesota–Duluth.

Tomlinson, P. B.
    1961   *Anatomy of the monocotyledons: II. Palmae*. London: Oxford University Press.
    1969   *Anatomy of the monocotyledons: III. Commelinales–Zingiberales*. London: Oxford University Press.

Twiss, Page C.
    1983   Dust deposition and opal phytoliths in the Great Plains. *Transactions of the Nebraska Academy of Sciences* 11 (Special Issue):73–82.

Twiss, Page C., Erwin Suess, and R. M. Smith
    1969   Morphological classification of grass phytoliths. *Proceedings: Soil Science Society of America* 33:109–115.

Veintimilla, Cesar, Marcelle Umlauf, and Deborah M. Pearsall
    1985   Resultados preliminares de flotación y análisis de fitolítos por el proyecto arqueológico–etnobotánico San Isidro. Paper presented at the 45th International Congress of Americanists, Bogota, Colombia.

Verma, S. D., and R. H. Rust
    1969   Observations on opal phytoliths in a soil biosequence in southeastern Minnesota. *Proceedings: Soil Science Society of America* 33:749–751.

Walker, A., H. Hoeck, and L. Perez
    1978   Microwear of mammalian teeth as an indicator of diet. *Science* 201:908–910.

Watanabe, N.
    1968   Spodographic evidences of rice from prehistoric Japan. *Journal of the Faculty of Science of the University of Tokyo (Section V)* 3(3):217–235.
    1970   A spodographic analysis of millet from prehistoric Japan. *Journal of the Faculty of Science of the University of Tokyo (Section V)* 3(5):357–379.

Wilding, L. P.
    1967   Radiocarbon dating of biogenetic opal. *Science* 156:66–67.

Wilding, L. P., and L. R. Drees
    1968   Biogenic opal in soils as an index of vegetative history in the prairie peninsula. In *The Quaternary of Illinois*, edited by R. E. Bergstrom, pp. 96–103. University of Illinois College of Agriculture, Special Publication 14.
    1971   Biogenic opal in Ohio soils. *Proceedings: Soil Science Society of America* 35:1004–1010.

Wilding, L. P., N. E. Smeck, and L. R. Drees
    1977   Silica in soils: Quartz, cristobalite, tridymite, and opal. In *Minerals in soil environments*, Edited by J. B. Dixon, S. B. Weed, *et al.* pp. 471–552. Madison: Soil Science Society of America.

Wilson, Samuel M.
    1985   Phytolith analysis at Kuk, an early agricultural site in Papua, New Guinea. *Archaeology in Oceania* 20(3):90–97.

Witty, J. E., and E. G. Knox
    1964   Grass opal in some chestnut and forest soils in north central Oregon. *Proceedings: Soil Science Society of America* 28:685–688.

Yeck, R. D., and F. Gray
    1972   Phytolith size characteristics between udolls and ustolls. *Proceedings: Soil Science Society of America* 36:639–641.

Chapter 6 Integrating Paleoethnobotanical Data

Introduction

In Chapters 2 through 5 I presented detailed information on how to recover, identify, and interpret archaeological macroremains, pollen, and phytoliths. In this final chapter I return to a broader view to discuss how the results of analyzing diverse types of botanical data can be integrated to address questions of interest in archaeology.

To do this, I first review strengths and weaknesses of each botanical data base and discuss how each complements the others. I then present examples of research which integrate different types of paleoethnobotanical data. Finally, I consider two topics of particular interest in archaeology, reconstruction of diet and reconstruction of paleoenvironments.

Strengths and Weaknesses of Paleoethnobotanical Data

Macroremains

Charred, dried, or waterlogged macroremains offer the greatest potential of all archaeological botanical materials for detailed taxonomic study and precise identification to species or varietal level. In a situation of waterlogged or dry preservation, for example, delicate attachments and surface reticulation may be preserved on seeds, allowing wild taxa to be identified to species level. Seeds of cultivated plants may be identifiable to known varieties or land races if there are modern collections for detailed comparisons of size and shape, or new forms now extinct may be discovered.

Charred remains are usually more difficult to identify to the species level, since charring often destroys delicate structures and distorts specimen size and shape. In seeds, for example, the endosperm may expand during charring, causing seed coats to crack. Shape may change dramatically, and if damaged seeds coats are lost, specimens may be unidentifiable. However, not all seeds or fruits are badly distorted by charring; it is not uncommon to identify such materials to the genus level, and to species in certain circumstances (e.g., if only a limited number of distinctive species exist). Wood is also altered by charring, but study of charred comparative specimens, with focus on characters not affected by heat distortion, permits species-level identification in some taxa.

In general, precision of identification of macroremains is limited by two factors, state of preservation of material and precision of correspondence between taxonomic divisions and differences in seed, fruit, or wood characteristics. In the former case, material may be unidentifiable because important diagnostic characteristics have been lost; in the latter, taxa assigned to different species on the basis of flower characteristics may produce identical seeds or wood and therefore be indistinguishable archaeologically.

Macroremain data are subject to deposition and recovery biases. In the case of most open-air sites, archaeological botanical materials are deposited through human activity. Plants used away from the site may leave no record, except perhaps in coprolites; this means that part of the foodstuffs used may never be deposited. Introduction may be deliberate (as firewood, foodstuffs, construction materials) or accidental (seeds adhering to clothing, in animal dung fuel). Although "seed rain," deposition of seeds by natural forces such as wind, may contribute to archaeological seed assemblages, it is usually a minor component. Most such seeds are from weedy taxa, deposited in soil by plants living on the site during occupation or after abandonment. Unless charred in a conflagration, most decay. Seed rain does become a significant contributor to prehistoric seed assemblages in open-air sites with excellent preservation, such as an inundated lakeside village or a desert site.

Recovery bias in macroremain assemblages is dependent on field and laboratory technique. Without water flotation or fine sieving, small macroremains, especially seeds, are underrepresented in collections. Bad flotation technique, faulty equipment, or rough handling of sieving can reduce assemblages by loss or destruction of remains. Although recovery bias can have a substantial impact on interpretation, it is also easily controlled.

If potential for detailed taxonomic study is a strength of macroremain data, a weakness is quantification of results. Preservation and depositional biases do not affect macroremain data randomly; this makes it very difficult to compare quantitites of remains directly. In other words, some plants are consistently underrepresented in the record while others are overrepresented. This is especially true for charred preservation. Root and tuber remains, for example, tend to be quite fragile and therefore underrepresented in relation to nut hull or corn cob fragments. Food plants that are not processed by heat may be absent from the record or very underrepresented in comparison to foods that are cooked or parched. Foods with a robust inedible portion are more likely to appear in the record than foods that are completely consumed. These biases exist even in excellent preservation conditions. Although it may be easy to count or weigh each seed or fruit type recovered from a site, it may be difficult to interpret the meaning of the quantitative data.

One way around this difficulty is to compare relative quantitites of remains in situations of like preservation. If preservation bias can be assumed to be reasonably constant, then change in relative amounts of remains may reflect different use patterns, including a changing importance of foodstuffs in the diet. Determination of the absolute importance of various taxa will remain illusive, however, since we can never be sure what is missing from the record. Another approach is to compare entire macroremain assemblages to assemblages produced by known activities, such as stages in crop processing. The comparative approach enhances interpretation of activity areas or site function.

Pollen

Archaeological pollen data offer, in general, less potential for detailed taxonomic study than do macroremain data. Very similar pollen grains may be produced by related species or genera, limiting precision of identification. In some groups, such as grasses, identification can be made only at the family level. However, detailed study utilizing size and minute exine features can sometimes result in more precise identifications, for example, in distinguishing *Zea* pollen from that of other grasses. Identification may also be limited by loss of diagnostic exine features through soil abrasion.

Pollen grains are subject to preservation bias. Being organic, like macroremains, pollen is vulnerable to destruction by soil microorganisms and by the natural environment. However, since the outer wall of the grain, the exine, is composed of sporopollenin, one of the toughest natural organic substances known, pollen survives in soil longer than uncharred macroremains. Nonetheless, not all pollen survives equally well; some taxa are known to produce pollen more resistant to decay and mechanical breakage than others. In general, there seems to be less bias introduced into the record by differential preservation than is the case for macroremains. Under waterlogged conditions (especially acidic pH), or in very dry conditions, pollen preservation may be excellent.

There are two major avenues for deposition of pollen in archaeological sites, human activities and the natural process of pollen rain. Humans contribute to pollen deposition by bringing pollinating plant material into a site (e.g., flowers for foods or medicine, seeds with adhering pollen, boughs to line pits) or by carrying pollen on feet or clothing. Although activities involving pollinating plant material usually do not represent use of pollen per se, finding pollen from edible or other useful plants, especially taxa that are not wind pollinated, may be a good indication of use or presence of plants at a site.

Unlike seed rain, pollen rain is often a major contributor to archaeological pollen assemblages. Because many plant taxa are wind pollinated, pollen is carried and mixed in the air before being deposited over the landscape. Pollen falling on lakes is eventually incorporated into bottom sediments. Analysis of lake, ocean, and bog sedimentary deposits is a traditional focus of palynology both because preservation is good in these settings and because steadily accumulating deposits provide a stratigraphic record of past vegetation. Prehistoric settlements, like other ancient land surfaces, also received pollen rain. If conditions were good for preservation (e.g., soil was accumulating, microbe activity was low), much of the pollen later extracted from archaeological soil samples may represent ancient pollen rain, not human-introduced plants. This is especially true of open areas in a site; samples taken from inside structures, pits, or ceramic vessels may be less impacted by pollen rain. Alternatively, open areas should be sampled if pollen rain data are the objective, for example, for vegetation reconstruction. Recovered pollen from plants that are not wind pollinated (e.g., zoophilous taxa) is unlikely the result of pollen rain.

Pollen data, like macroremain data, are subject to many nonrandom factors that affect the quantity of grains of any given taxon present in soil. These factors have less to do with preservation bias than with bias from differential production and deposition of grains. Plants vary in quantity of pollen produced, mechanism of dispersal, and distance grains are likely to travel. Rather than try to correct for all factors affecting quantity of grains deposited, pollen analysts use the comparative, or analogue, approach for interpreting assemblages. Assemblages produced by known vegetation formations or under known ecological conditions are compared to fossil assemblages, either by eye or through multivariate statistical techniques. A close correspondence permits modeling of past vegetation. A similar approach could be developed with ethnographic information. Alternatively, presence of indicator species, plants known to occur in a narrow ecological range, are used to reconstruct vegetation.

Phytoliths

Phytolith data are characterized by a highly variable level of taxonomic precision; some taxa produce phytoliths diagnostic to genus or species, others are identifiable only to the family level or above, and some are highly redundant in the plant kingdom. Many plants are not silica accumulators or produce only nondiagnostic phytoliths. In some silica-accumulating groups, the Gramineae, for example, phytoliths permit more precise identification than is possible with pollen. It is difficult to generalize about precision of phytolith identification; many plant groups remain to be studied.

Phytoliths are distinct from both macroremains and pollen in preservation bias. Since phytoliths are inorganic cell inclusions, they are unaffected by soil microorganisms and do not decay. Preservation may be poor in soils with very high pH (approaching 9 and above), however. Some types of phytoliths, such as large epidermal forms, break up under soil pressure and can be difficult to identify. If phytoliths are only lightly silicified, they are also more prone to mechanical damage.

As is the case for both macroremains and pollen, phytoliths are deposited in archaeological sites both through the agency of humans and by natural forces. Phytolith deposition is most similar to macroremain deposition, however. The majority of phytoliths and certainly the range of types recovered from archaeological soil result from in situ decay or burning of plant material brought to the site by its inhabitants. This material may include foodstuffs, crop residues, fuels, and materials for construction; decay of such materials releases many more phytoliths to soil than what may be present in wind-blown dust. Decay of vegetation growing on a site after its abandonment contributes phytoliths to accumulating surface soil. Although phytolith rain is less well understood than pollen rain, phytoliths do form part of wind-borne dust and are deposited with pollen. Such phytoliths are presumably picked up from open soil and redeposited by wind, or redeposited by water from sediment carried by streams.

The major source of bias in the phytolith record is from differential production of silica bodies. As mentioned above, many plants do not accumulate silica in their tissues. These taxa are absent from the phytolith record. Other taxa produce only a few diagnostic phytoliths and are thus underrepresented in counts. Finally, silica-accumulating plants are overrepresented in assemblages. As in the case of pollen data, phytolith analysts can circumvent the problem of differential production by using the comparative or indicator species approaches to interpret phytolith assemblages, whether in natural or human-influenced settings.

Examples of Paleoethnobotanical Interpretation

Paleoethnobotanical data can be used to address a variety of questions relevant to archaeology. For example, archaeological botanical data provide information on what plant foods were deposited at a site. This information can in turn be used to determine vegetation zones used for plant collecting, to postulate cultivation of certain taxa, or to model a yearly schedule of activities involving plant procurement. Under conditions of similar preservation and deposition, change though time in relative abundances of remains may indicate changing patterns of use of resources or of the nature of vegetation. These are all realistic goals, of considerable value in archaeological interpretation. Combining information from more than one data source strengthens interpretations in these areas. I give three examples—from my work at Real Alto, from Flannery's research at Guilá Naquitz, and from Piperno's work in Panama—to illustrate this point.

In the most general terms, plant utilization reveals a great deal about how human populations interact with, and adapt to, their environments—in other words, what role they play in the ecosystem. Understanding the nature of human adaptation is a basic goal of archaeology, and systematic recovery and analysis of botanical remains are integral steps toward that goal.

Real Alto: Integrating Macroremain and Phytolith Data

In my analysis of plant remains from the Formative period Real Alto village, Ecuador (Pearsall 1979), phytolith analysis of soil samples from each phase of occupation revealed that maize was present at the site from at least the Valdivia II period (2500 B.C.) and probably from Valdivia I (3000 B.C.). Maize was commonly seen in some phases. In the large Valdivia III sample, for instance, 63 of 71 soil samples (89%) contained cross-shaped phytoliths falling into the maize size range. By contrast, only 9 Valdivia III samples (13% presence) contained *achira* phytoliths. Only four fragments of charred *Canavalia* bean were recovered by flotation from 24 Valdivia III contexts (16% presence). How important was maize in subsistence relative to the root crop *achira*? What were the relative contributions of maize, *achira*, and *Canavalia* beans to the Valdivia III diet?

These questions were unanswerable directly from botanical data. I expected that maize, a species producing abundant short cell phytoliths, would leave more indication of its presence in the phytolith record than *achira*. Differences in percentage presence of these crops in the Valdivia III village might simply reflect differential phytolith production. Preservation of remains of *Canavalia* beans was dependent on accidental charring. This difference in preservation bias made it impossible to compare occurrences of charred beans directly to occurrences of maize or *achira* phytoliths.

I concluded from my work at Real Alto that Valdivia villagers pursued a mixed subsistence strategy which included cultivated plants—maize, *Canavalia* beans, *achira* (botanical remains), cotton (fiber impressions), coca, manioc (artifacts associated with use), and bottle gourd (effigy vessels)—and a variety of wild plants, fish, shellfish, and terrestrial animals. Integrating the results of macroremain and phytolith analyses, even though information was limited to presence or absence, resulted in a much fuller picture of subsistence than either data source would have given alone. My arguments for relative importance of food resources were based only on a site catchment study and modeling of potential crop yields. Real Alto's location gave easy access to both sea and riverine habitats, including alluvial land suitable for cultivation. Since easily cultivated land was limited near the village, a mixture of the higher yielding root crops with the lower yielding maize would give the most productive crop yields. I proposed a subsistence system with a strong agricultural component of mixed grains and root crops and an important hunting and fishing component.

Guilá Naquitz: Integrating Macroremain and Pollen Data

Research by Flannery and associates at Guilá Naquitz cave, in Oaxaca, Mexico (Flannery 1986), included detailed study of macroremains, pollen, and faunal remains from the site. Although I have some reservations about converting quantities of faunal remains and macroremains into 100-g food portions and using such data to reconstruct diet (Flannery 1986a) (see Chapter 3 discussion), this study is a good example of how diverse biological data can complement each other in an archaeological analysis.

Analyses by Smith (1986), Whitaker and Cutler (1986), and Kaplan (1986) of macroremains from the Archaic levels of the site, spanning the period from 8750 to 6670 B.C. (Flannery 1986b), give a picture of fairly stable use of a limited number of predominantly wild plant resources. Among the most abundant remains, by count, were acorns, seeds of mesquite, hackberry, *Opuntia* cactus, *susi* (*Jatropha neodioica*), and *nanche* (*Malpighia*), and pods from the legumes *Dalea*, *Phaseolus*, and *Leucaena*. All but four specimens of *Phaseolus* were of a wild species; the four *P. vulgaris* finds, as well as the two occurrences of cultivated paper, *Capsicum annuum*, were from possibly intrusive contexts. Cultivated squash, *Cucurbita pepo*, and gourd, *Lagenaria siceraria*, were present in good Archaic contexts. Furthermore, all wild plants recovered from the cave are available today in abundance in the area. Smith (1986) suggested that use of acorns, hackberry, and *susi* may have declined over time (these taxa are less common by count in the upper strata), but he did not see any indication of a major shift in basic resource use or in vegetation.

Pollen data (Schoenwetter and Smith 1986) give a more detailed look at vegetation and add another possible cultivated plant to the assemblage. Two pollen grains identified as cf. *Zea mays* were recovered in samples from Zone B, the uppermost Archaic zone at Guilá Naquitz. Other possible maize pollen grains were recovered from preceramic contexts at Martinez rockshelter and the Gheo-Shih site, giving a total of five possible maize pollen grains. Schoenwetter and Smith use the descriptor "cf." (similar to) on all identifications in the Maydeae (corn, teosinte, *Tripsacum*), since it is possible that another grass producing large pollen grains contributed these to the archaeological record, or that all were produced by one species in the Maydeae.

Although possible presence of maize before 6670 B.C., more than a millennium earlier than the earliest Tehuacan remains (MacNeish 1967), is an intriguing aspect of the Guilá Naquitz pollen study, detailed phase-by-phase paleoenvironmental reconstructions worked out by Schoenwetter and Smith are the major contribution of this work (see Chapter 4 discussion). By establishing correspondences between modern pollen spectra and ecological parameters of annual rainfall, annual temperature, and effective moisture, fossil pollen records could be used to reconstruct paleoecological conditions in the valley. From these data, Schoenwetter and Smith proposed a series of vegetation changes in the valley during the preceramic occupation. If both palynological and macroremain data are correct, vegetation change over the period of occupation of Guilá Naquitz did not alter patterns of basic resource utilization, however. Flannery's (1986a) reconstruction of food-procurement areas suggests why: there were many times the plant resources needed to support the small cave population within a few kilometers of the site.

Panama: Integrating Pollen and Phytolith Data

Piperno's (1985) analysis of phytoliths in the Lake Gatun cores from Panama and her work with Clary in archaeological sediments (Piperno *et al.* 1985) illustrate how pollen and phytolith data complement each other in reconstructing paleovegetation and subsistence.

I discussed Piperno's (1985) Lake Gatun analysis in some detail in Chapter 5. The point to reemphasize here is the complementarity of phytolith and pollen records for the cores. Grasses, for example, are often important environmental indicators. Whereas pollen is quite uniform in the Gramineae, its phytolith characteristics are very diverse. Data on grass phytoliths in the cores augmented limited interpretations based on presence of grass pollen. Carbon-occluded silica bodies proved to be a more sensitive indicator of early forest clearing than were shifts in pollen assemblages. Both pollen and phytolith data contributed indicator species that aided interpretation of assemblage changes.

Piperno *et al.* (1985) reported combined phytolith and pollen evidence for preceramic maize in Panama. Although maize was clearly present in lower Central America and important in subsistence by 500 B.C., when its use is associated with sedentary villages, charred maize remains had not been recovered from early ceramic or preceramic contexts—and this in spite of fine screening and water flotation on a few projects. If, however, maize was only a minor component of early mixed subsistence systems in lower Central America, lack of charred remains in a humid tropical setting is not difficult to understand. Unless maize was introduced into South America by sea, bypassing the obvious overland route, it had to be present in lower Central America by 2500 B.C. or earlier.

Soil samples from preceramic strata at two sites, Aguadulce and Cueva de los Ladrones, were analyzed. At Cueva del los Ladrones, analysis revealed the presence of cross-shaped phytoliths identifiable as maize on the basis of shape, size, and three-dimensional morphology in all levels tested. The pattern of cross occurrence did not resemble that produced by wild grasses, including teosinte, which has a distinctive cross size and three-dimensional characteristics. Maize pollen was also found in preceramic levels. Preceramic maize pollen was noticeably more fragmented and weathered than that recovered from ceramic strata. The diameters of fourteen complete preceramic grains averaged 83 microns, in the size range of maize. Minute characteristics of the exine also matched the pattern in maize. The identification of this pollen as *Zea mays* is further strenghtened by the demonstration through phytolith data that teosinte was not present (a remote possibility, since the sites are outside its natural distribution).

At the Aguadulce site, maize phytoliths were found only in ceramic period levels (2500–1000 B.C.). Pollen preservation was very poor in the preceramic strata; no maize grains were found there or in the upper levels. Piperno *et al.* (1985) hypothesize that the larger size of the Ladrones shelter and its location in an area of higher rainfall may have made it a more desirable location for a horticultural settlement than Aguadulce.

Special Topic: Reconstructing Prehistoric Diet

The studies just discussed illustrate how applying more than one paleoethnobotanical technique can strengthen data interpretation. But can the various kinds of paleoethnobotanical data, alone or in combination, be used to evaluate nutrition of prehistoric populations, to indicate importance of different plant foods in the diet, or by comparison to faunal remains, to uncover the distinct contributions of plant and animal resources? These are issues of particular concern for understanding human adaptations. Although researchers have attempted this level of analysis, these are often unrealistic goals. It should be clear from my review of the strengths and weaknesses of paleoethnobotanical data bases that phytolith and pollen data are of little utility in determining contributions of plant foods in the diet beyond what information simple presence or absence or percentage presence data can yield. The Real Alto example illustrates this point for phytolith analysis. But even macroremain data can be suspect in this regard; quantification is one of the weaker aspects of this data base. How important, for instance, was maize in Panama or Ecuador in the millennia before an abundance of charred remains makes it appear dominant in diet? Is it more important earlier than the abundance of remains seems to indicate, or is ubiquity of charred material a good indicator of importance?

Chemical analysis of human bone can help answer such questions—it can clarify the relationship of foods in diet and the conditions under which macroremain data give accurate dietary reconstruction. The three studies discussed next—Roosevelt's (1980) research at Parmana, investigations at Tehuacan, and research in Wisconsin (Bender et al. 1981; Price and Kavanagh 1982)—illustrate how chemical analyses of human bone can help refine interpretations based on ethnobiological data.

The old adage "you are what you eat" holds true for the human skeleton: the nature of the diet over the course of an individual's life is reflected in many ways in bone. Of interest here are two dietary indicators, the ratio of different isotopes of elements such as carbon and nitrogen, and the incorporation of trace elements such as strontium.

Different isotopes of the same element (i.e., atoms with the same number of protons but different numbers of neutrons) exist in nature. In the case of carbon, the two stable isotopes, ^{12}C and ^{13}C, are not absorbed by plants in the ratio of their occurrence on earth (ca. 99% ^{12}C, 1% ^{13}C). Rather, plants discriminate unequally between these isotopes, depending on the nature of their photosynthetic pathway (C_3, C_4, or CAM pathways). Animals, including humans, that eat plants which utilize a C_4 pathway absorb different ratios of ^{12}C and ^{13}C than animals that eat predominantly C_3 plants. Discrimination also occurs in absorption of stable nitrogen isotopes.

Bone also incorporates many trace elements into its mineral structure, which is mostly hydroxyapitite crystal (calcium phosphate). Of these elements, strontium has received the most attention. Plants absorb strontium roughly in the concentration in which it occurs in the environment. Animals discriminate against strontium, however; as one moves up the food chain, levels of strontium decrease. Consumed strontium that is not excreted by an animal is deposited in its bone. Thus carnivores have lower strontium levels than herbivores, and omnivores fall between. For a review of nutritional assessment from human bone, refer to Klepinger (1984).

Parmana

As a result of excavations at a number of sites in the Parmana area of the Orinoco River in Venezuela, Roosevelt (1980) defined a ceramic chronology, with associated floral and faunal remains, for early inhabitants of the region. One goal of her research was to test a hypothesis that introduction of maize into the Parmana region led to a large increase in prehistoric population density and ultimately to the rise of chiefdom-level political organization. Maize-based subsistence systems, oriented to annually renewed floodplain soils, were thought to have replaced mixed tropical forest systems based on plant gathering and cultivation of manioc and other tubers.

Leaving aside reconstruction of population densities for each time period, let us consider the botanical data from Roosevelt's excavations. Ideally, to support her hypothesis one would like to see a replacement of tropical forest taxa by maize. But tropical forest agricultural systems are dominated by roots and tubers such as manioc and sweet potato. These are just the type of plants that one would expect to be underrepresented archaeologically. This was in fact the case at Parmana.

For sites of the earliest phase, La Gruta (2100–800 B.C.), no remains of cultivated plants were recovered, although charred macroremains thought to represent wild foods were present. The artifact assemblage, which included ceramic griddles and stone grater chips, suggested that manioc was present. During the second phase, Corozal (800 B.C.–A.D. 400), charred maize remains were recovered in each of three subphases. A few maize remains were recovered in two of four excavations in Corozal I levels and in three of four Corozal II levels. By Corozal III times (A.D. 100–400), a few maize remains were seen in most excavation levels. Charred legume remains, possibly cultivated, also occurred, as did wild fruits. By the final phase, Camoruco (A.D. 400–1500), maize macroremains were recovered in most excavation units, sometimes in concentration. Domesticated *Canavalia ensiformis* (jackbean) also occurred. The only possible manioc occurrence, a charred cake on a griddle fragment, also came from the Camoruco phase.

The problem is this: one can see an increase in ubiquity and frequency of maize over time, at least in these preliminary data, which do not include flotation results; but is root consumption dropping? There are no botanical data to address this issue.

It was possible to examine this question though isotopic analysis of human bone from the sites (van der Merwe *et al.* 1981). Maize, like many tropical grasses, is a C_4 pathway plant. Manioc and most other tropical forest plants are C_3 pathway plants. As discussed above, C_3 and C_4 plants differ markedly in the abundance of ^{13}C in their tissues. Average ^{13}C level for C_4 foliage is −12.5 per mil. This value was used by van der Merwe *et al.* to represent maize. The authors tested manioc roots and obtained a ^{13}C level of −25.5 per mil. A population eating mostly maize (or grazing animals feeding on C_4 pasture—unlikely in this case) would have ^{13}C levels approaching those of C_4 plants, and a population eating root crops or wild fruits and foliage (or animals feeding on these, including fish) would have ^{13}C levels approaching that of C_3 plants.

Van der Merwe *et al.* tested five skeletons from Roosevelt's excavation: two from early Corozal (ca. 800 B.C.) and three from late Corozal (ca. A.D. 400). As discussed above, maize first appeared in the botanical record, in the form of charred macroremains, during the Corozal phase. The results of the isotopic analysis help refine interpretation of maize importance based on botanical data. For the early

Corozal skeletons, readings were close to ^{13}C levels for manioc, with an average of −31.1 per mil (this value includes a collagen-enrichment value subtracted from the raw value). It is clear that, although no remains of manioc were found, it or other roots and wild forest products dominated the diet of these individuals. This was the period when maize was already present at low levels—and so we have a good illustration that recovery of a few remains of a plant that is likely to be preserved archaeologically (i.e., maize) does not indicate its intensive use.

The late Corozal specimens gave an average ^{13}C value of −15.4 per mil. This is the likely result of a diet dominated by C_4 plants, such as maize, since consumption of fauna grazing on C_4 vegetation is an unlikely contributing factor (most bones recovered were fish). Levels of maize of 80% or higher are suggested in the diet. Again, note that this very high reading is for material that preceded the obvious increase in charred maize remains at the site, which dated to the Camoruco phase. Dietary change was, then, more rapid than suggested by the botanical data. If the few skeletons are representative of the population as a whole—not of high-status individuals who may have consumed more maize than the general population, then dietary change can be proposed.

Tehuacan

Excavation at caves in the Tehuacan region of Mexico by MacNeish and colleagues (Byers 1967) resulted in recovery of some of the first evidence for domesticated maize and a detailed record of diet and subsistence for central Mexico. I discussed in Chapter 3 the method used by MacNeish (1967) to analyze subsistence data. Briefly, floral and faunal debris were converted into liters of edible food. Total liters were summed for each archaeological phase, and percentages of wild animals, wild plants, agricultural plants, and domesticated animal in the diet were calculated (see Fig. 3.45). These data give a picture of slow, steady change from dependence on wild plants and animals to dependence on agricultural plants. Agricultural plants do not constitute a majority of the diet (>50%) until the Ajalpan phase, 1500–900 B.C. Maize constitutes 51% of the diet in that phase, although levels drop thereafter, to increase again in the final phases. I criticized this approach, being concerned about the use of "correction factors," especially in the early phases, and the wisdom of directly comparing faunal and floral data.

The scenario of dietary change revealed in coprolites (Callen 1967) is at odds with MacNeish's (1967) summary based on food volume reconstructions. Wild plants predominate in most coprolites, whether early or late, leading Callen to propose a "cave" diet for users of the shelters. Callen suggests that the cave users tended to eat wild food resources gathered nearby, as well as a few foodstuffs carried in. Maize, for example, occurs in only 4 of 94 human coprolites analyzed (4.3%). By contrast, *Setaria* grass seeds occur in 51 coprolites (54.3%). *Setaria* occurs in all phases except Abejas and Ajalpan.

Isotopic analysis was used to clarify dietary reconstruction at Tehuacan (DeNiro and Epstein 1981; Farnsworth *et al.* 1985). A series of skeletons were analyzed for ^{13}C and ^{15}N values. Twelve individuals could be analyzed. Analysis showed that the earliest skeleton, from the El Riego phase (6500–5000 B.C.), had ^{13}C values indicating a mixed diet of C_3 plants (and/or animals grazing on C_3 plants) and C_4 or CAM plants (and/or animals grazing on C_4 plants), but with C_3 intake predominating. These conclusions agree fairly well with MacNeish's reconstruction of an early diet based on wild food resources.

Isotopic data from a Coxcatlan phase (5000–3500 B.C.) skeleton showed a marked change in diet, however—one not reflected in the faunal and floral data. ^{13}C values changed dramatically, indicating an almost complete dependence on C_4 or CAM plants (and/or animals eating such plants). To give the observed values, 90% of the diet would have to have come from these sources (Farnsworth *et al.* 1985). Two interpretations of this change are proposed. First, maize is a C_4 plant; the data may indicate a marked increase in maize in the diet, that is, a switch to almost full dependence. But *Setaria*, the wild grass prevalent in the coprolites, is also C_4, and maguay and various cacti are CAM pathway plants, so an alternative interpretation is a marked increase in the use of these plants, or of animals grazing on them. Marked increase in CAM plants is unlikely, since these are assumed mostly to be starvation foods. This pattern (i.e., dependence on C_4 plants) was observed in all the remaining skeletons, from the Santa Maria (900–200 B.C.) and Venta Salada (A.D. 700–1540) phases, when maize had become a common constituent of diet in Mesoamerica. Isotopic analysis of animal remains in the cave might help clarify the contribution of meat to the ^{13}C levels observed in the human bone.

Thus the isotope data from the Coxcatlan skeleton do not parallel the floral and faunal data from this phase, which indicate a gradual change in diet, nor the coprolite data, which indicate little change between the El Riego and Coxcatlan phases. Rather, they suggest a rapid change in subsistence earlier in the sequence. Diet appears relatively stable thereafter. Lack of analyzable skeletal material from the middle phases weakens the isotopic dietary reconstruction, however.

It is rather frustrating that even with an isotopic analysis, coprolite analysis, macroremain analysis, and faunal analysis, it is still difficult to draw any conclusions about dietary changes at Tehuacan. The major problem is *Setaria*, a C_4 plant, more common in the coprolites than maize. Are observed ^{13}C values due to high *Setaria* in the diet, or is maize giving the reading? Since the caves were occupied only seasonally, local consumption of wild foods including *Setaria* ("cave diet") may have had little influence on ^{13}C values. Perhaps change did not occur until maize dominated the overall diet, especially foods consumed away from the cave. Another problem is the apparent disagreement between coprolite and macrobotanical data; much more maize is present in the cave than in the coprolites. It is possible, however, that the methods of analysis used were too different to compare occurrence directly (percentage presence in the coprolites; liters of food in the cave).

Maize Use in Wisconsin

Analysis of trace elements incorporated into human bone is another way of assessing the relative importance of foods in diet. A number of elements have been tried; most investigators have used strontium, in the form of ratios of strontium to calcium. As described earlier, measuring the level of strontium in human bone can reveal the relative contribution of plants and animals in the diet, helping resolve issues difficult to assess using botanical and faunal data alone.

Price and Kavanagh (1982) report on trace element analysis of skeletal materials from Wisconsin spanning the Late Archaic through Mississippian periods. These samples were also analyzed by Bender et al. (1981) for ^{13}C. In the latter study, the aim was to investigate the importance of maize, a C_4 plant, in the diet, especially to assess its importance during the Middle Woodland (Hopewell) period. These two studies provide an opportunity to see how strontium measurements add to the interpretation of diet gained from ^{13}C studies, and how both compare to interpretations from botanical data for these periods.

In the Bender et al. work, ^{13}C analysis of skeletal material from sites in the Middle Woodland period clustered with material from the earlier Archaic site, giving ^{13}C readings that fell into the range of high C_3 plant utilization. There was clear separation of this group from a group of later Late Woodland and Mississippian sites. Skeletal material from the later sites clustered at a level indicating high consumption of C_4 plants, presumably maize, Bender et al. also analyzed animal bone and vegetable material from one site; the ^{13}C levels from the animals, presumably eating C_3 vegetation, were very similar to the levels from the Archaic human population and quite dissimilar to levels observed in maize. It appears that the later cultures were consuming both C_4 plants and C_3 animals, but that plant consumption dominated.

The strontium analysis (Price and Kavanagh 1982) included human skeletal material from sites from Archaic, Middle Woodland, and Mississippian periods. Deer (herbivore) and dog (omnivore) remains were also analyzed. Strontium–calcium ratios from the Archaic site averaged 0.319 for four samples. Comparing this to herbivore (deer, average 0.858) and omnivore (dog, average 0.683) seems to clearly indicate a strongly meat-oriented diet (i.e., carnivore readings lower than omnivore). Combined with the ^{13}C data, which indicates high C_3 intake, it suggests reliance on animals grazing in C_3 vegetation.

Data from the two Middle Woodland sites gave average strontium–calcium ratios of 0.647 and 0.754. These values contrast sharply with the earlier Archaic data, suggesting an increase in the vegetable component of the diet (i.e., a herbivore diet). ^{13}C data, which are virtually identical for the Middle Woodland and Archaic skeletons, suggest that this change was due to an increased use of C_3 plants, in combination with C_3 animals, not due to an increase in C_4 plant consumption. In other words, increasing reliance on plant foods preceded increasing reliance on maize. Paleoethnobotanical data, such as that of Johannessen (1984) and Asch and Asch (1985), suggest that small seeds such as maygrass, knotweed, and chenopod were a likely source for the increasing vegetable component of the diet.

Finally, strontium–calcium data from Mississippian site skeletons are very similar to the Middle Woodland data. An average of four samples gave 0.747, in the same range as the Middle Woodland sites. Omitting the value for a possibly intrusive burial raised the average to 0.882, above the value of average herbivores from the site. This supports the suggestion of a diet almost entirely of plant food. The ^{13}C data, which show a marked change with increased input from C_4 plants, strongly suggest that maize was the plant that increased in the diet. This change may have occurred before the Mississippian period—in the Late Woodland according to data from the Ledder's site (Bender et al. 1981).

From these data we can propose a model of dietary change in the Midwest from animal-oriented subsistence in combination with nut use in the Archaic, to a mixed diet of animals and C_3 plants, with small seeds predominating over nuts, and finally to decreased use of animal foods and an increase in plant food consumption, but this time with a rapid shift to maize. This model incorporates the full range of subsistence activities, integrating use of faunal and floral resources. Such a model could be tested at other sites by an integrated approach, incorporating floral and faunal analysis and nutritional analysis of bone. Unlike the Tehuacan and Parmana examples discussed earlier, patterning of change in assemblages of charred botanical remains from sites in the Midwest parallel the timing and magnitude of change suggested by nutritional analyses of bone.

Routine application of flotation and uniform methods of analysis have given archaeologists in the Midwest replicate botanical data for each prehistoric period. We do not have to rely on a few scattered macroremains to model subsistence; data are abundant and patterns of plant occurrence are repeated at numerous sites. By combining data from floral, faunal, and human skeletal analyses, archaeologists can model dietary mix precisely and test those models. These are vital steps in understanding the nature and success of human adaptation. Only by combining all available data does a complete picture emerge. Chemical analyses cannot tell us what people ate (specific plants and animals); floral and faunal data cannot tell us how much they ate (dietary mix).

Special Topic: Reconstructing Paleoenvironment

One of the concerns of archaeology is to model the functioning of human populations in their ecosystems and to understand the nature of their adaptations. I have discussed one important aspect of adaptation, the nature of diet; I now consider another, the environments or landscapes in which human populations lived. My focus is on the plant world; clearly there are other important components of ecosystems to be considered.

I have already discussed how botanical data are used to reconstruct past vegetation. The discussions of Hansen et al.'s (1984), Schoenwetter and Smith's (1986), and Piperno's (1985) work served to illustrate application of pollen and phytolith data to this goal. Here I review a number of basic considerations for reconstructing paleoenvironment and discuss a topic of special interest for research on adaptations of human populations in the Holocene—the impact of humans on vegetation.

At the outset, let me reemphasize that the only direct sources of information on the nature of past vegetation are archaeological and geological records of the remains of that vegetation—that is, pollen, phytoliths, and macroremains. Recovery and analysis of botanical data are essential for any investigation of paleoenvironments.

Analysis of botanical materials from archaeological sites is the focus of most work in paleoethnobotany. Archaeological sites are by definition "cultural contexts," the loci of human activities. We focus on cultural contexts because we hope to elucidate human–plant relationships; remains of activities involving humans and plants are mostly to be found in habitation sites and other human-modified landscapes such as gardens.

Archaeological sites are also sources of data on the environment in which human–plant interactions were taking place. In other words, we can also learn about plants that were not "involved" with humans, those making up the "background" vegetation. As Schoenwetter and Smith's (1986) work illustrates, useful vegetation data can be derived from even severely human-altered landscapes. Often elements of vegetation remain to serve as indicators of the nature of past vegetation or as markers of environmental parameters such as moisture.

The key to using archaeological data to reconstruct paleoenvironment is distinguishing between the background vegetation and those plants whose presence or abundance are the result of human activities. Humans select plants for various purposes and bring them into habitation sites. This can increase the frequency with which remains of those plants occur in deposits. Plants brought in from considerable distances may appear as exotics (i.e., plants not expected in the local flora) in the record. The very nature of human habitation sites leads to both increases and decreases in deposition of plant taxa. For example, the environmental disturbance created by a village favors weedy plant taxa, whose occurrences increase, while construction of houses and other structures inhibits deposition of wind-borne pollen in those contexts. Comparison of archaeological pollen, phytolith, or macroremain assemblages to assemblages from various known vegetation formations and loci of human activities allows background components to be separated from those influenced by cultural selection or human disturbance.

Analysis of botanical materials from geological deposits is a major source of data for paleoecological reconstruction. Sampling locations are chosen to avoid biases introduced by human activity at the sampling locus. For example, it is unlikely that macroremains, pollen, or phytoliths recovered in a lake core were carried there by humans. Cultural selection has no impact on such deposits. Vegetation changes recorded in geological deposits may be the result of human activities taking place away from the sampling locus, however. As is discussed further below, land clearing for agriculture is a major cause of vegetation change. Determining whether vegetation change is due to natural causes (e.g., changes in rainfall, destruction from volcanic eruption) or is human-induced can be difficult, especially in Holocene deposits. Use of vegetation and activity analogues, as discussed above, will be essential in some instances.

Before turning to a discussion of human impact on vegetation, I want to re-emphasize the importance of utilizing all types of botanical data for paleoenvironmental reconstructions. Use of pollen, phytoliths, or macroremains alone does not permit the precision modeling achievable from the full range of available botanical data. By the same token, it may prove impossible to test models adequately without multiple indicators of vegetation patterning. This holds true for both archaeological and geological sediments. Admittedly, constraints of time or funding may require the researcher to choose among techniques, or to emphasize one approach at the expense of others. In this situation, samples can be stored for future analysis. The vital step is to collect appropriate samples in the field and document stratigraphy; once a site is destroyed, one cannot return to take samples for another analysis. Quality of preservation of remains may also limit the utility of pursuing an approach.

Little Tennessee River Valley

When archaeologists and paleoethnobotanists work to reconstruct the vegetation setting of an archaeological site and its environs, it is important that they realize that humans are not passive components of the landscape; rather, they play an active role in shaping that landscape, especially the vegetation growing there. In turn, modification of vegetation has a direct, and sometimes dramatic, impact on the nature and success of human adaptation. So when we try to model the nature of past vegetation, it is important that we consider humans as interactive components of the landscape.

Human impact on the plant world takes many forms. Humans have camped, constructed villages, and built cities. A camp used by an extended family for a few days has a very different effect on local vegetation than a village occupied for generations. People have fed themselves by hunting and gathering, casual gardening, and full-scale agriculture. Again, different food-procurement activities have very different effects on local vegetation. How does one evaluate the nature of human impact on vegetation using the archaeological and geological record?

A study by Delcourt *et al.* (1986) of the impact of humans on the environment of the Little Tennessee River valley of Tennessee can serve as an example of an integrated botanical approach to this problem. Unfortunately, this study does not utilize phytolith data, but both macroremains and pollen were analyzed. Data spanning 10,000 years were gathered from twenty-five archaeological sites in the Holocene alluvial terrace of the river. Botanical remains were in the form of charred macroremains (wood, nuts, seeds). In addition to archaeological sampling, two small ponds, chosen to provide data on local upland and lower stream terrace vegetation, were cored. Macroremains and pollen were analyzed from these deposits.

Analysis of charred macroremains from the archaeological sites permitted detailed reconstruction of plant use by the prehistoric and historic period inhabitants of the valley. Among findings relevant to an evaluation of the nature of human impact on environment was an increase in occurrence of disturbance-favored wood taxa over time. Although percentages of various wood species recovered from sites could not be used to indicate the proportion of those trees in the environment (given the operation of cultural selection), a long-term trend away from the use of bottomland and upland taxa toward the use of taxa preferring disturbed habitats is clearly shown. In the late Late Archaic (ca. 4500–3000 B.P.), 25% of wood charcoal recovered was of disturbance-favored taxa, and cultivated plants such as maygrass, sunflower, squash, and gourd were encountered for the first time. Maize was not observed until the Middle Woodland and did not become common until the Mississippian period. Use of disturbance-favored wood taxa continued to rise through the Woodland and Mississippian periods.

Paleoecological data from the two ponds were used to reconstruct local vegetation. Pollen data from the pond in the upland area show high arboreal–nonarboreal pollen ratios until nearly the time of Euroamerican settlement. This is interpreted as indicating minimal forest clearing in the upland zone until the Historic period. By contrast, data from the pond in the third terrace (bottomland forest zone) indicate that forest clearing on the lower terraces dates to at least the Late Woodland period (the oldest period represented in the deposits). This interpretation is based on high percentages of ragweed pollen. Maize pollen was also observed. By the Mississippian period, a rich assemblage of pollen and macroremains of weedy plants are documented in the deposits of this pond.

The authors propose that significant human impact on the environment dates to the Late Archaic, when forest clearing and limited cultivation began on the floodplain and lower river terraces. This is documented first in the archaeological macroremain record, by a significant increase in occurrence of disturbance-favored wood taxa and the appearance of cultivated plants in the late Late Archaic. No pollen record exists for this period, but independent pollen core data from the Late Woodland period indicate that forest clearing had occurred in the lower terraces. There is no indication of clearing in the uplands, however, until the period of Euroamerican settlement. This is very suggestive; sufficient agricultural land must have been available in the bottomlands to support the prehistoric population.

References

Asch, David L., and Nancy E. Asch
    1985   Prehistoric plant cultivation in west-central Illinois. In *Prehistoric food production in North America*, edited by Richard I. Ford, pp. 149–203. Museum of Anthropology, University of Michigan, Anthropological Papers 75.

Bender, M. M., D. A. Baerreis, and R. L. Steventon
    1981   Further light on carbon isotopes and Hopewell agriculture. *American Antiquity* 46:346–353.

Byers, D. S. (editor)
    1967   *The prehistory of the Tehuacan Valley: I. Environment and subsistence.* Austin: University of Texas Press.

Callen, E. O.
    1967   Analysis of Tehuacan coprolites. In *The prehistory of the Tehuacan Valley: I. Environment and subsistence*, edited by D. S. Byers, pp. 261–289. Austin: University of Texas Press.

Delcourt, Paul A., Hazel R. Delcourt, Patricia A. Cridlebaugh, and Jefferson Chapman
    1986   Holocene ethnobotanical and paleoecological record of human impact on vegetation in the Little Tennessee River valley, Tennessee. *Quarternary Research* 25:330–349.

DeNiro, Michael J., and S. Epstein
    1981   Influence of diet on the distribution of nitrogen isotopes in animals. *Geochimica et Cosmochimica Acta* 45:341–351.

Farnsworth, Paul, James E. Brady, Michael J. DeNiro, and Richard S. MacNeish
    1985   A re-evaluation of the isotopic and archaeological reconstructions of diet in the Tehuacan Valley. *American Antiquity* 50(1):102–116.

Flannery, Kent V.
    1986a   Food procurement area and preceramic diet at Guilá Naquitz. In *Guilá Naquitz: Archaic foraging and early agriculture in Oaxaca, Mexico*, edited by K. V. Flannery, pp. 303–317. Orlando, Florida: Academic Press.
    1986b   Radiocarbon dates. In *Guilá Naquitz: Archaic foraging and early agriculture in Oaxaca, Mexico*, edited by K. V. Flannery, pp. 175–176. Orlando, Florida: Academic Press.

Flannery, Kent V. (editor)
 1986 *Guilá Naquitz: Archaic foraging and early agriculture in Oaxaca, Mexico.* Orlando, Florida: Academic Press.

Hansen, Barbara C. S., H. E. Wright, Jr., and J. P. Bradbury
 1984 Pollen studies in the Junín area, central Peruvian Andes. *Geological Society of America Bulletin* 95:1454–1465.

Johannessen, Sissel
 1984 Paleoethnobotany. In *American Bottom archaeology,* edited by Charles J. Bareis and James W. Porter, pp. 197–214. Urbana: University of Illinois Press.

Kaplan, Lawrence
 1986 Preceramic *Phaseolus* from Guilá Naquitz. In *Guilá Naquitz: Archaic foraging and early agriculture in Oaxaca, Mexico,* edited by K. V. Flannery, pp. 281–284. Orlando, Florida: Academic Press.

Klepinger, Linda L.
 1984 Nutritional assessment from bone. *Annual Review of Anthropology* 13:75–96.

Lambert, J. B., C. B. Szpunar, and J. E. Buikstra
 1979 Chemical analysis of excavated human bone from Middle and Late Woodland sites. *Archaeometry* 21(2):115–129.

MacNeish, Richard S.
 1967 A summary of the subsistence. In *The prehistory of the Tehuacan Valley: I. Environment and subsistence,* edited by D. S. Byers, pp. 290–309. Austin: University of Texas Press.

Pearsall, Deborah M.
 1979 *The application of ethnobotanical techniques to the problem of subsistence in the Ecuadorian Formative.* Ph.D. dissertation, Department of Anthropology, University of Illinois. Ann Arbor: University Microfilms.

Piperno, Dolores R.
 1985 Phytolithic analysis of geological sediments from Panama. *Antiquity* 59:13–19.

Piperno, Dolores R., K. Clary, R. Cooke, A. J. Ranere, and D. Weiland
 1985 Preceramic maize in central Panama: Phytolith and pollen evidence. *American Anthropologist* 87:871–878.

Price, T. Douglas, and Maureen Kavanagh
 1982 Bone composition and the reconstruction of diet: Examples from the midwestern United States. *Mid-Continental Journal of Archaeology* 7(1):61–79.

Roosevelt, Anna C.
 1980 *Parmana: Prehistoric maize and manioc subsistence along the Amazon and Orinoco.* New York: Academic Press.

Schoenwetter, James, and Douglas Landon Smith
 1986 Pollen analysis of the Oaxaca Archaic. In *Guilá Naquitz: Archaic foraging and early agriculture in Oaxaca, Mexico,* edited by K.V. Flannery, pp. 179–237. Orlando, Florida: Academic Press.

Smith, C. Earle, Jr.
 1986 Preceramic plant remains from Guilá Naquitz. In *Guilá Naquitz: Archaic foraging and early agriculture in Oaxaca, Mexico,* edited by K. V. Flannery, pp. 265–274. Orlando, Florida: Academic Press.

van der Merwe, Nikolaas, J., Anna C. Roosevelt, and J. C. Vogel
    1981   Isotopic evidence for prehistoric subsistence change at Parmana, Venezuela. *Nature (London)* 292:536–538.

Whitaker, Thomas W., and Hugh C. Cutler
    1986   Cucurbits from preceramic levels at Guilá Naquitz. In *Guilá Naquitz: Archaic foraging and early agriculture in Oaxaca, Mexico,* edited by K. V. Flannery, pp. 275–279. Orlando, Florida: Academic Press.

Index

Absolute count, *see* Count
Abundance measure, *see also* Comparison ratio; Count; Density ratio; Food value measure; Multivariate analysis; Percentage; Presence analysis; Species diversity; Standard score; Weight
and macroremain interpretation, 226–227
Acetolysis, 271–272, 277
Achira, 444–445
Agriculture, *see also* Dietary reconstruction; Raised field agriculture; Subsistence reconstruction
and vegetation change, 428, 454–457
crop identification, 181–183, *see also* Achira; Banana; Beans (*Phaseolus*); *Canavalia* bean; Corn; Cotton; *Lepidium*; Manioc; Rice; Squash; Sunflower; Sweet potato; Wheat
field identification, by phytolith data, 336–338, 352–353

Aguadulce site, Panama, 331, 336, 447
Air sampling, pollen, 267–268, *see also* Sampling strategies
Ali Kosh site, Iran, 24
Alto Salaverry site, Peru, 205–209
American Bottom, *see* FAI-270 project
Analogue approach, *see* Comparative approach
Analogy, *see* Ethnographic modeling
Animal pollination, 252–253, 300, *see also* Water pollination; Wind pollination
Ankara flotation system, *see* Water sieve/separator
Aperture, pollen, 249
Apple Creek flotation system, *see* Manual flotation
Archaeological palynology, *see* Palynology
Artifact wash
phytolith, 349
pollen, 264
Ashing, *see* Comparative material, dry ashing

Banana, phytolith, 322–323
Barrel flotation, *see* Manual flotation
Beans (*Phaseolus*), 80, 153–154, *see also* Canavalia bean
Bias, depositional
macroremain, 228, 440–441
phytolith, 425, 443
pollen, 254–255, 299–300, 442
Bias, preservation
macroremain, 202–205, 208–212, 228–229, 440–441
phytolith, 340, 344, 443
pollen, 254–256, 262–263, 266, 302–303, 441
Bias, production
phytolith, 344, 352–354, 427, 442–443
pollen, 441–442
Bias, recovery, macroremain, 16–19, 22–23, 228, 440
Biozone, *see* Pollen diagram
Blanket sampling, macroremains, 95–96, *see also* Column sampling; Pinch sampling; Point sampling
Bohrer and Adams sampling procedure, *see* Profile sampling

461

462 • Index

Bryant and Holloway sampling procedure, *see* Profile sampling
Bulk soil processing, *see* Fine sieving; Screening; Water sieve/separator

C_3/C_4/CAM photosynthesis pathways, and isotope analysis, 193, 448–453
Cambridge froth flotation system, *see* Froth flotation
Canavalia bean, 444–445, 449, *see also* Beans (*Phaseolus*)
Carbon sample form, *see* Flotation form
Carbone phytolith processing procedure, *see* Soil extraction
Charcoal, *see* Wood charcoal
Chemical analysis, human bone, 448, *see also* Isotope analysis; Trace element analysis
Chemical flotation, macroremain, *see also* Soil extraction, phytolith
  as an alternative to hand sorting, 125–126
  recovery procedure, 24–25, 32, 47, 86–91
Chemical (wet) oxidation, *see* Comparative material
Chenopod, 5, 16, 119, 146–149, 200–201, 215–217
Clark, J.G.D., 5
Classification, *see* Multivariate analysis; Phytolith, classification
Clay, *see* Deflocculation of clay
Climatic reconstruction, *see also* Vegetation reconstruction
  utilizing phytoliths, 339
  utilizing pollen, 298–299
Cluster analysis, *see* Multivariate analysis
Collecting, *see* Plant collecting
Column sampling, macroremains, 98–99, *see also* Blanket sampling; Pinch sampling; Point sampling; Profile sampling
Comparative approach, 455, *see also* Ethnographic modeling; Indicator species approach; Representation factor (*R*-factor) approach
  and phytolith interpretation, 352–354, 427
  and pollen interpretation, 295–296, 301–302, 442
Comparative material, *see also* Plant collecting
  chemical (wet) oxidation, phytolith
    comparison to dry ashing, 375–377
    procedure, 380–384
  dry ashing procedure, phytolith, 377–379
  epidermal peel and spodogram, phytolith, 379–380
  for macroremain analysis, 128–129, 138–144
  for palynology, 277–278
  for phytolith analysis, 375–384
  Kaplan and Smith procedures, phytolith, 379–380, 383–384
Comparison ratio, 201–205, *see also* Count; Density ratio; Food value measure; Multivariate analysis; Percentage; Presence analysis; Species diversity; Standard score; Weight
Coprolite, 230–231, 253, 259–260, *see also* Dung
  from Mammoth Cave, 217, 220–221
  from Newt Kash Hollow, 5
  from Salts Cave, 5, 289, 292, 300
  from Tehuacan, 450–452
  processing, 276–277
Coring, *see also* Profile sampling; Sampling strategies
  phytolith, 350–351
  pollen, 264–266
Corn, 79–80, 110, 116, 150, 192, 212–214, 217–218, 315, 444–453
  identification by phytolith, 328–336, 418–425, 428
Cost, of flotation, 63, 74–79
Cotton, 175–177
Count, counting, *see also* Comparison ratio; Density ratio; Food value measure; Multivariate analysis; Percentage; Presence analysis; Species diversity; Standard score; Weight
  absolute
    phytolith, 403–404
    pollen, 281–283
  macroremain, 196–197
  phytolith, 387–400, 403–404
  pollen, 254–255, 257, 279–285, 287
  quick-scanning, 388–392
  relative, pollen, 280–281
  short-cell scanning, 387, 392–394, 403, 408–409
Crawford froth flotation system, *see* Froth flotation
Cross-shaped phytolith, *see also* Short cell phytolith
  corn identification, 328–336, 418–425
  data presentation, 412–414
  definition, 329
  variant, 333–336
Cueva de los Ladrones site, Panama, 336, 408, 410–412, 447
Cultivated plant, *see also* Achira; Banana; Beans (*Phaseolus*); *Canavalia* bean; Corn; Cotton; *Lepidium*; Rice; Squash; Sunflower; Sweet potato; Wheat
  identification, macroremains, 181–183
  seed electrophoresis, 191–192
Cultural resource management (CRM) project, access to botanical data, 6

Index • 463

Cumulative frequency diagram, 413, 416–417, see also Histogram; Phytolith diagram; Pollen diagram; Tabular presentation

Dating
  by pollen assemblage, 257
  of phytoliths, 312
Dayton Museum flotation system, see Manual flotation, IDOT system
Dean pollen procedure, see Soil extraction
Deflocculation of clay, see also Soil extraction
  before flotation, 50, 85–86
  during phytolith extraction, 343, 369–370
  during pollen extraction, 271
Dehydration procedure, 185–186
Density ratio, 197–198, 200, also Comparison ratio; Count; Food value measure; Multivariate analysis; Percentage; Presence analysis; Species diversity, Standard score; Weight
Depositional bias, see Bias, depositional
Desiccated macroremain, see also Desiccated wood
  mold control, 127–128
  recovery, 79–82
  sorting, 128
Desiccated wood, 162–163
Diagram, see Cumulative frequency diagram; Phytolith diagram; Pollen diagram
Dietary reconstruction, see also Agriculture; Subsistence reconstruction
  by chemical analysis of bone
    maize use, Wisconsin, 452–453
    Parmana, 448–450
    Tehuacan, 450–452
    limitation, 447–448
  using macroremain data, 205–210

  using pollen data, 259–260, 300
Dimbleby, G. W., 3
  sampling procedure, see Profile sampling
Discriminant analysis, see Multivariate analysis
Dried macroremain, see Desiccated macroremain
Dry ashing, see Comparative material
Dry sieving, see Fine sieving
Drying, see Plant collecting
Dung, as source of seeds, 197, 225–226, 228, see also Coprolite

Ecological reconstruction, see Vegetation reconstruction
Efficiency, flotation, 76–79, see also Recovery test
El Bronce site, Puerto Rico, 87–88, 408–409
Electron microscopy, see also Scanning electron microscopy; Transmission electron microscopy
  macroremain application, 161, 188–189
Electrophoresis, 191–192
Elutriator, 28–29, 67
Embedding, sectioning, and grinding
  application
    root identification, 172
    wood identification, 162–164
  procedure, 183–189
Environmental reconstruction, see Vegetation reconstruction
Epidermal peel, see Comparative material
Ethnobiology, see also Ethnobotany; Paleoethnobotany
  definition, 2
  society, 2–3
Ethnobotanical cover sheet, see Flotation form
Ethnobotany, see also Ethnobiology;

Paleoethnobotany
  definition, 1
  status of field, 6–9
Ethnographic modeling, see also Comparative approach
  for macroremain interpretation, 7, 218–220, 225, 230–231
  for pollen interpretation, 260–261
Exine, 247–251
Exotics, as markers
  in phytolith counting, 403–404
  in pollen counting, 278, 281–282
Experimentation, and macroremain interpretation, 230

FAI-270 archaeological project, Illinois, 26, 49, 85, 87–90, 197–198, 202–204, 212–215
Factor analysis, see Multivariate analysis
Fiber
  identification, 175–177
  structure, 175–176
60-Field phytolith scanning, 394–399, 410, see also Count form, 398
Fiber optic light, 115
Fine sieving, 17–19, 79–85, see also Froth flotation; In situ recovery; Manual flotation; Screening; Water sieve/separator
Flotation, see also Cost, flotation; Efficiency, flotation; Fine sieving; In situ recovery; Screening
  for macroremain recovery, see Chemical flotation; Froth flotation; Manual flotation; Paraffin flotation; Water sieve/separator
  identification, see Cultivated plant; Fiber; Leaf and stem; Root and tuber; Seed; Wood

Flotation (continued)
  in phytolith processing, see Soil extraction, phytolith sorting, see Sorting
Flotation data record, see Flotation form
Flotation form, 108–114
Flower structure, 145
Food value measure, 205–211, see also Comparison ratio; Count; Density ratio; Multivariate analysis; Percentage; Presence analysis; Species diversity; Standard score; Weight
Form, see Flotation form; Quick-scan form; 60-field scan form; Short-cell form
Fractionation, 364–365, 370, 372
Franchthi Cave flotation system, see Water sieve/separator
Freeze fracturing, 171–172, 186
  procedure, 190–191
Froth flotation, see also Manual flotation; Water sieve/separator
  bubbler unit, 70
  Cambridge system, 20, 27, 32, 68–79, 92
  Crawford system, 33, 68–74
  cell, 68–70
  cost, 76–79
  development, 32–33, 69
  procedure, 70–74
  recovery, efficiency, 92–94
  recovery, improvement, 74–75
  settling tank, 70
Fruit
  identification, 151–155
  sorting procedure, 110
  structure, 150–152
Furrow, see Aperture

Gatun Lake, Panama
  integrating pollen and phytolith data, 425–428, 446
  phytolith core, 403–406

Gilmore, Melvin, 5
Grass phytolith, see Short cell phytolith; Long cell phytolith
Grinding, see Embedding, sectioning, and grinding
Guilá Naquitz site, Mexico, 205–207, 210, 287, 289–291, 295–296, 301–302, 445–446

Hair, as distinguished from fiber, 177
Harshberger, J. W., 1
Heavy fraction sorting, see Sorting
Helbaek, Hans, 1, 4, 24
Hickory, identification, 154–155
Histogram, see also Cumulative frequency diagram; Phytolith diagram; Pollen diagram; Tabular presentation
  phytolith data, 404–406, 413, 415
Historical overview
  flotation, 23–35
  paleoethnobotany, 3–6
  phytolith analysis, 328–340
  pollen analysis, 256–258

IDOT (Illinois Department of Transportation) flotation system, see Manual flotation
Identification, see also Cultivated plant; Electron microscopy; Electrophoresis; Embedding, sectioning, and grinding; Fiber; Isotope analysis; Keying; Leaf and stem; Morphometric analysis; Pollen; Phytolith; Root and tuber; Seed; Starch grain analysis; Taxonomic guide; Wood
  embedding and sectioning, 185

  fiber, 175
  fruit and nut, 155
  leaf and stem, 178–179
  limit, macroremain, 148–149, 154–155, 164–165, 173
  phytolith, 322–324, 406
  pollen, 249, 268–269
  root, 173
  seed, 119, 150
  wood, 156, 165
Image analysis, see Morphometric analysis
In situ recovery, macroremain, 16, see also Fine sieving; Froth flotation; Manual flotation; Screening; Water sieve/separator
Incinerator flotation system, see Manual flotation, IDOT system
Indiana flotation machine, see Water sieve/separator, Franchthi Cave system
Indicator species approach, pollen, 297, 442, see also Comparative approach; Representation factor (R-factor) approach
Insect pollination, see Animal pollination
Integrating paleoethnobotanical data
  for dietary reconstruction, 447–453
  for reconstructing paleoenvironment, 454–457
  macroremain and phytolith data, 444–445
  macroremain and pollen data, 445–446
  pollen and phytolith data, 424–428, 446–447
  strength and weakness of data, 340–345, 439–443
Interpretation, see also Macroremain; Phytolith; Pollen
  issues, macroremain, 224–231
Intine, 247–248

Isotope analysis, *see also* Chemical analysis, human bone; Trace element analysis
  of cooking residue, 192–193
  of human bone, 448–453
Izum flotation system, *see* Water sieve/separator

Jarman, Legge, and Charles froth flotation system, *see* Froth flotation, Cambridge system
Jones, Volney, 1, 5

Kaplan and Smith phytolith processing procedures, *see* Comparative material
Kenward, Hall, and Jones wet sieving system, 83–85
Keying, *see also* Identification, manuals; Taxonomy guide
  fruit, desktop, 151–154
  phytolith, 322–324, 407
  pollen, 249, 268–269
  wood, desktop, 143–144
Kolmogorov-Smirnov procedure, 417–418
Krum Bay site, U.S. Virgin Islands, 36–37, 39, 94, 155

Laboratory, basic equipment
  macroremain, 108–120
  phytolith, 356–363
  pollen, 269–270
Lake Junín, pollen core, 285–287, 293–294, 298–299
Lake sampling, *see* Coring
Landscape modification, 455–457
Las Vegas site, Ecuador, 331–333, 418–420
Leaf and stem identification, 177–181
Lepidium, identification, 214–215
Light fraction sorting, *see* Sorting
Light microscope, *see also* Metallurgical microscope; Scanning electron microscopy; Transmission electron microscopy
  for flotation, sieve sample sorting, 115
  for phytolith scanning, 360–363
  for viewing macroremain thin section, 184
  for wood charcoal identification, 162–163
Lipid analysis, 183
Little Tennessee River valley sites, 455–457
Long cell phytolith, 312–316, *see also* Short cell phytolith
Loss, macroremain, *see also* Bias
  by wetting and drying, 51–52, 79–81
  prior to flotation, 101

Machine-assisted flotation, *see* Froth flotation; Water sieve/separator
Macroremain, *see* Fiber; Leaf and stem; Macroremain analysis; Root and tuber; Seed; Wood
Macroremain analysis, *see also* Agriculture; Dietary reconstruction; Subsistence reconstruction; Vegetation reconstruction
  depositional bias, 228, 440–441
  identification, 107–194
  interpretation, 194–231
  preservation bias, 202–205, 208–212, 228–229, 440–441
  recovery, 15–102
    bias, 16–19, 22–23, 228, 440
Maize, *see* Corn
Mammoth Cave site, Kentucky, 217, 220–221
Manioc, 449–450
Manual flotation, *see also* Froth flotation; Water sieve/separator
  Apple Creek system, 20, 24, 26, 35–36, 38, 47
  bucket/tub, 35–37
  cost, 76–79
  definition, 20–21
  development, 24–27
  hand sieve, 38–39
  IDOT (Illinois Department of Transportation) system, 26–27, 37, 39–40, 43–45, 48–50, 76–79, 86, 92
  Minnis and LeBlanc system, 25, 39, 48
  procedure, 39–50
  recovery, efficiency, 92–94
  recovery, improvement, 50–52
  Schock system, 27
  Stewart and Robertson system, 25, 49
  tank/barrel, 39
Mechanized flotation, *see* Froth flotation; Water sieve/separator
Metallurgical microscope, wood identification, 161, *see also* Light microscope; Scanning electron microscopy; Transmission electron microscopy
Microscope, use, *see* Electron microscopy; Light microscope; Metallurgical microscope; Scanning electron microscopy; Transmission electron microscopy
Microtome, use, 184–187
Minnis and LeBlanc flotation system, *see* Manual flotation
Missouri phytolith scanning and counting procedures, 384–400, *see also* Soil extraction, Pearsall (Missouri) phytolith processing procedure
Monagrillo site, Panama, 331, 412
Monte Carlo simulation, *see also* Sorting test
  for subsampling flotation, 122–123

Monte Carlo simulation (*continued*)
  for testing 60-field scanning, 396–397
Morphometric analysis (image analysis), 193–194, 362
Muffle furnace
  for ashing, 337–338
  for charring, 138–140
Mulholland and Rau phytolith processing procedure, *see* Soil extraction
Multivariate analysis, *see also* Comparison ratio; Count; Density ratio; Food value measure; Percentage; Presence analysis; Species diversity; Standard score; Weight
  classification, 217–218
  cluster analysis, 218
  discriminant function analysis, 218–219, 296, 301–302, 418–421
  factor analysis, 220–221
  of macroremains, 217–221
  of phytoliths, 418–421
  of pollen, 294, 296, 301–302

Naked eye recovery, *see* In situ recovery
Nueva Era site, Ecuador, 214
Numerical approach, *see* Multivariate analysis
Nut, *see* Fruit

Old Monroe site, Missouri, 121–123
Opaline silica, *see* Phytolith; Phytolith analysis
Organic material, removal
  phytolith processing, 343, 370
  pollen processing, 271–272, 277
Oven, use, 140, *see also* Muffle furnace

Pachamachay site, Peru, 196, 211–212, 230

Paleoenvironmental reconstruction, *see* Vegetation reconstruction
Paleoethnobotany, *see also* Ethnobiology; Ethnobotany; Macroremain analysis; Palynology; Phytolith analysis
  definition, 1–2
  historical overview, 3–7
  status, 6–9
Palynology, *see also* Agriculture; Dietary reconstruction; Subsistence reconstruction; Vegetation reconstruction
  archaeological (soil), 245–246, 255–261, 286–292, 299–304
  field sampling, 258–269
  history, 256–258
  laboratory analysis, 269–284
  nature and production of pollen, 246–256
  presenting and interpreting results, 284–304
  stratigraphic (sediment), 246, 256–258, 264–266, 284–287, 292–299
Panaulauca Cave site, Peru, 198–200, 214–217, 225–228
Paraffin flotation, 84–85, *see also* Chemical flotation; Froth flotation; Manual flotation; Water sieve/separator
Parenchyma
  root and tuber, 173
  wood, 159–161
Parmana sites, Venezuela, 448–450
Payne, S., 22, 27–28
Pearsall (Missouri) phytolith processing procedure, *see* Soil extraction
Peel, *see* Comparative material, epidermal peel; Root and tuber, identification

Peñón del Rio site, Ecuador, 337, 351–353, 413–414, 419–425
Percentage, macroremain, 197–200, *see also* Comparison ratio; Count; Density ratio; Food value measure; Multivariate analysis; Presence analysis; Species diversity; Standard score; Weight
Pest control, 136, 141
Phloem fiber
  root, 173
  stem, 175–176
Phytolith, *see also* Phytolith analysis; Phytolith assemblage; Phytolith rain
  bias, 340, 344, 352–354, 425–427, 442–443
  classification, 387–389
  composition, 312
  deposition, 339–342, 350
  description, 406–408
  extraction
    plant, 375–384
    soil, 344, 356–360, 363–375
  identification, corn, 328–336
  morphology
    dicotyledon, 326–328, 341
    grass, 312–324
    other monocotyledon (not grass), 322–326, 341
  mounting, 384–387, 400–403
  preservation, 340, 344, 443
  production, 344–345
  recovery, 342–343
  redundancy, 343
  sampling, 337–338, 345–356
  scanning and counting, 387–400, 403–404
Phytolith analysis, *see also* Agriculture; Dietary reconstruction; Subsistence reconstruction; Vegetation reconstruction
  complementarity with pollen analysis, 338, 340–341
  field sampling, 345–356
  historical overview, 328–340

Index • 467

laboratory technique, 356–404
nature and occurrence of phytoliths, 312–328
presenting and interpreting results, 404–428
status, 340–345
Phytolith assemblage, definition, 400
Phytolith diagram, 404–407, 413, *see also* Cumulative frequency diagram; Histogram; Pollen diagram; Tabular presentation
Phytolith rain, 339, *see also* Pollen rain; Seed rain
Pinch (composite) sampling, *see also* Column sampling; Point sampling; Profile sampling
macroremain, 96–98
phytolith, 347–353
pollen, 266–267
Piperno phytolith procedure, *see* Soil extraction
Plant collecting, *see also* Comparative material
equipment, 129–130
phytolith specimens, 355–356
pollen specimens, 268
pressing and drying, 133–137
procedures, 130–133
voucher identification, 136–138
Plant opal, *see* Phytolith; Phytolith analysis
Plant specimen, *see* Comparative material
Point sampling, macroremain, 98, 100, *see also* Column sampling; Pinch sampling
Pollen, *see also* Palynology; Pollen accumulation value; Pollen concentration value; Pollen diagram; Pollen rain; Pollen sum
bias, 254–256, 262–263, 266, 299–303, 441–442
counting, 254–257, 279–283

data presentation, 284–292
destruction, 253–255
dispersal, 251–253
distribution, soil, 254–255, 299–300
extraction, 269–278
identification, 268–269, 283–284
interpretation, 292–302
mounting, 270–279
preservation, 254–256, 262–263, 266, 302–303
sampling, 256, 258–269
shape, 247
structure, 247–251
Pollen accumulation value, 281–283
Pollen analysis, *see* Palynology
Pollen concentration value, 281–283
Pollen diagram, *see also* Cumulative frequency diagram; Histogram; Phytolith diagram; Tabular presentation
form, 284–292
interpretation
climate, 298–299
pollen zones (biozones), 292–294
vegetation, 294–297
Pollen rain, 245, 251–252, 258–259, 442, *see also* Phytolith rain; Seed rain
Pollen sum, 254–255, 257, 279–285, 287
Pollen zone (biozone), *see* Pollen diagram
Pollination, *see* Animal pollination; Water pollination; Wind pollination
Poppy seed test, *see* Recovery test
Pore, *see* Aperture, pollen; Vascular tissue
Potato, identification, 173
Presence analysis, *see also* Comparison ratio; Count; Density ratio; Food value measure; Multivariate anal-

ysis; Percentage; Species diversity; Standard score; Weight
as a sorting shortcut, 120
in macroremain interpretation, 195, 212–217, 229
Preservation bias, *see* Bias, preservation
Pressing, *see* Plant collecting
Principal components analysis, *see* Multivariate analysis, factor analysis
Production bias, *see* Bias, production
Profile sampling, *see also* Pinch sampling
Bohrer and Adams procedure, pollen, 263–264
Bryant and Holloway procedure, pollen, 262
Dimbleby procedure, pollen, 262–263
phytolith, 346–347, 350
pollen, 262–264
Proof and falsification, macroremain, 229–231

Qualitative approach, macroremain, 194–195
Quantitative approach, *see* Comparison ratio; Count; Density ratio; Food value measure; Multivariate analysis; Percentage; Presence analysis; Species diversity; Standard score; Weight
macroremain analysis, 195–221, 226–227
phytolith analysis, 344, 416–419
pollen analysis, 279–284, 293–294, 301–302
Quick-scan form, 391
Quick-scanning, *see* Count

Raised field agriculture, identification by phytolith data, 419–425
Ratio, *see* Comparison ratio; Density ratio; Percentage

Ray
  root and tuber, 173
  wood, 159
Real Alto site, Ecuador, 329–332, 355, 367, 418–419, 421, 444–445
Recovery bias, *see* Bias, recovery
Recovery techniques, macroremains, *see also* Chemical flotation; Fine sieving; Froth flotation; *In situ* collection; Manual flotation; Paraffin flotation; Screening; Water sieve/separator
  choosing among techniques, 75–85
Recovery test, *see also* Efficiency, flotation
  seed (poppy seed recovery test), 51, 91–94
  saltwater flotation, 94–95
  sieving/screening, 17–18, 27–28
Recycling, flotation, 56–57, 61–62, 70
Relative count, *see* Count
Report preparation
  macroremain, 221–224
  phytolith, 405–416
  pollen, 284–292
Representation factor (R-factor) approach, 296–297, *see also* Comparative approach; Indicator species approach
Rescue archaeology, access to botanical data, 6
Rhizome, *see* Root and tuber
Rice, identification, 315, 317, 328
Riffle box, use, 120–121
Root and tuber
  identification, 170–174
  preservation, 169–170
  sorting, 110
  structure, 165–169
Rovner phytolith processing procedure, *see* Soil extraction

60-field phytolith scanning, 394–399, 411, *see also* Count form, 398
SMAP (Shell Mound Archaeological Project) flotation system, *see* Water sieve/separator
Salmon Ruin Pueblo, New Mexico, 110, 260–261
Salts Cave site, Kentucky, 5, 289, 292, 300
Saltwater flotation, 94–95
Sample size, *see also* Monte Carlo simulation
  flotation, 98–101
  phytolith count, 392
  pollen count, 280–281
  seed count, 120–125
Sampling strategies, 258–261, 346–351, *see also* Air sampling; Artifact wash; Blanket sampling; Column sampling; Coring; Pinch sampling; Point sampling; Profile sampling; Sample size; Subsampling flotation
Sawing, *see* Embedding, sectioning, and grinding
Scanning, *see* Count; 60-field phytolith scanning
Scanning electron microscopy, *see also* Electron microscopy; Transmission electron microscopy
  contrasted to transmission electron microscopy, 184
  macroremain
    identification, 117, 149, 171–172, 189
    sample preparation, 189–190
  phytolith
    identification, 362
    sample preparation, 401–402
Schock flotation system, *see* Manual flotation
Sclerenchyma, 175
Screening, *see also* Fine sieving; Froth flotation; *In situ* collection; Manual flotation; Water sieve/separator
  contrasted to flotation, 22–23
  macroremain, 17–19
Sculpturing, 250–251
Sectioning, *see* Embedding, sectioning, and grinding
Sedimentation, *see* Fractionation
Seed
  comparative collection, 140–143
  destruction, 51–52, 79–81
  distortion, 141, 149–150
  identification, 145–150
  recognition, 118–119
  recovery test, 51, 91–94
  sorting, 110–116
  source, 224–226
  structure, 144–145
Seed blower, 81–82, 90–91, 126
Seed–nut ratio, *see* Comparison ratio
Seed rain, 224–226, *see also* Phytolith rain; Pollen rain
Self-pollination, 253
Setaria, 451
Shackley pollen procedure, *see* Soil extraction
Shackley slide-mounting procedure, *see* Slide mounting
Shannon-Weaver information index, 211–212
Short-cell form, 393
Short cell phytolith, 317–324, *see also* Count; Cross-shaped phytolith; Long cell phytolith
Short-cell scanning, *see* Count
Sieving, *see* Fine sieving; Screening; Water sieve/separator
Silica body, *see* Phytolith; Phytolith analysis
Simpson index, 211
Siphon, use in flotation, 46, 64, 66
Siraf flotation system, *see* Water sieve/separator
Slide mounting, *see also* Scanning electron microscopy

macroremain thin section, 187
phytolith, 384–387, 400–401
pollen, 278–279
Shackley pollen procedure, 278–279
Smith, C. Earle, Jr., 195
Smoky Creek Cave site, 17–18
Soil extraction, *see also* Chemical flotation; Froth flotation; Manual flotation; Water sieve/separator
  Carbone phytolith procedure, 367–368
  Dean pollen procedure, 273–274
  Mulholland and Rau phytolith procedure, 372–375
  Pearsall (Missouri) phytolith procedure, 357–360, 368–371
  phytolith, 356–360, 363–375
  Piperno phytolith procedure, 371–372
  pollen, 270–275
  Rovner phytolith procedure, 330, 365–367
  Shackley pollen procedure, 274–275
  Twiss, Suess, and Smith phytolith procedure, 364–365
Soil moisture, effect on flotation, 50, 71, 86, 101
Sorting, macroremain, *see also* Count; Sorting test
  alternative to hand sorting, 125–126
  by hand, 108–120
  desiccated sample, 126–128
  waterlogged sample, 126–128
Sorting test, *see also* Monte Carlo simulation; Sorting
  to determine minimum wood charcoal size, 116–117
  to determine seed sample size, 121–123
Species diversity, 211–212, *see also* Comparison ratio; Count; Density ratio; Food value measure; Multivariate analysis; Percentage; Presence analysis; Standard score; Weight
Splitting flotation/sieve samples, 110, 115
Spodogram, 315, 379–380
Sporopollenin, 247–248, 254
Squash, phytolith, 326
Staining
  pollen, 273–275, 278–279
  starch grain, 191
  thin-section, 187
Standard score, 200–201, *see also* Comparison ratio; Count; Density ratio; Food value measure; Multivariate analysis; Percentage; Presence analysis; Species diversity; Weight
Standardization
  in macroremain procedure, 107
  in phytolith slide mounting, 385–387
Starch grain
  preparation, 191
  root and tuber, 173–174
Statistical analysis, *see* Multivariate analysis; Proof and falsification
Stem, *see* Leaf and stem
Stewart and Robertson flotation system, *see* Manual flotation
Storage
  comparative collection, 141–143
  macroremain, 118
Stratigraphic palynology, *see* Palynology
Strontium analysis, *see* Trace element analysis
Struever, Stuart, 20, 24
Subsampling flotation, 120–125, *see also* Sampling strategies
Subsistence reconstruction, *see also* Agriculture; Climatic reconstruction; Dietary reconstruction; Vegetation reconstruction
  integrating macroremain and phytolith data, 444–445
  integrating macroremain and pollen data, 445–446
  integrating pollen and phytolith data, 446–447
  macroremain, 194–217
  phytolith, 328–336, 419–425
  pollen, 300
Sunflower, 5, 181
Sweet potato, 173
Swiss lake village sites, 4, 16

Tabular presentation, *see also* Cumulative frequency diagram; Histogram; Phytolith diagram; Pollen diagram
  macroremain data, 196–197, 222–223
  phytolith data, 407–414
  pollen data, 289, 292
Tagging
  flotation soil, 101
  sorted sample, 118
Taxonomy guide, 129, *see also* Identification, manuals; Keying
Tectum, 250–251
Tehuacan sites, Mexico, 195, 205–210, 450–452
Teosinte, 335
Trace element analysis, 448, 452–453, *see also* Chemical analysis, human bone; Isotope analysis
Transmission electron microscopy, *see also* Electron microscopy; Scanning election microscopy
  sample preparation, 184–189
Tub flotation system, *see* Manual flotation, Apple Creek flotation system
Tuber, *see* Root and tuber
Twiss, Suess, and Smith phytolith classification system, 317–322, 386–387
Twiss, Suess, and Smith phytolith processing procedure, *see* Soil extraction

Ubiquity, *see* Presence analysis
Underrepresentation, *see* Bias, production, preservation

Vascular tissue
  root and tuber, 173
  wood, 157–159
Variant maize identification technique, *see* Cross-shaped phytolith
Vegetation reconstruction, *see also* Climatic reconstruction; Dietary reconstruction; Subsistence reconstruction
  by phytolith analysis, 338–340, 350–351, 425–428
  by pollen analysis, 256–259, 265–266, 294–297, 301–302, 426–428, 456–457
  integrating macroremain and pollen data, 454–457
Vegetation sampling, *see* Plant collecting
Vessel, *see* Vascular tissue
Visual percentage estimation, *see* 60-field phytolith scanning
von Post, L., 256–257

Waimea-Kawaihae project, Hawaii, 320–322, 337, 352–354, 370, 386–387, 395, 413, 415
Walnut, identification, 151–153
Water flotation, *see* Froth flotation; Manual flotation; Water sieve/separator

Water pollination, 253, *see also* Animal pollination; Wind pollination
Water sieve/separator, *see also* Froth flotation; Manual flotation
  Ankara flotation system, 20–22, 28–29, 52–63
  compared to water screening, 20–22
  cost, 76–79
  flot box, 29, 56–58
  Franchthi Cave (Indiana) flotation system, 29–30, 53, 58–59, 94
  history, 27–35
  Izum flotation system, 31–32
  main box/barrel, 29, 52–55
  procedure, 61–66
  pump/plumbing, 58–62
  recovery, efficiency, 92–94
  recovery, improvement, 67–68
  SMAP (Shell Mound Archaeological Project) flotation system, 20, 22, 32–35, 53–68, 76–79, 92
  screen insert, 29, 54–56
  Siraf flotation machine, 29–31, 53, 55–59, 69
Water screening, *see* Screening
Waterlogged macroremain
  distortion, 126
  identification, 164
  recovery, 82–85
  sorting, 127
Weight, macroremain, 196–197, *see also* Comparison ratio;

Count; Density ratio; Food value measure; Multivariate analysis; Percentage; Presence analysis; Species diversity; Standard score
Wet oxidation, *see* Comparative material, chemical (wet) oxidation
Wet sieving, *see* Fine sieving; Water sieve/separator
Wheat, identification, 219, 336
Wind pollination, 251–252, *see also* Animal pollination; Pollen rain; Water pollination
Wood charcoal, *see also* Desiccated wood; Waterlogged macroremain; Wood Structure
  comparative, 138–144
  destruction, 51
  distortion, 161–162
  identification, 161–163
  loss in froth flotation, 74–75
  removal from pollen samples, 272
  sorting, 110, 116–117
Wood structure, 155–161

Zea mays, *see* Corn
Zinc chloride flotation, *see* Chemical flotation